# Electromagnetic Scintillation
## I. Geometrical Optics

Electromagnetic scintillation describes the phase and amplitude fluctuations imposed on signals that travel through the atmosphere. These volumes provide a modern reference and comprehensive tutorial for this subject, treating both optical and microwave propagation. Measurements and predictions are integrated at each step of the development.

This first volume deals with phase and angle-of-arrival measurement errors. These are accurately described by geometrical optics. Measured properties of tropospheric and ionospheric irregularities are reviewed first. Electromagnetic fluctuations induced by these irregularities are estimated for a wide range of applications and compared with experimental results in each case. These fluctuations limit the resolution of astronomical interferometers and large single-aperture telescopes. Synthetic-aperture radars and laser pointing/tracking systems are also limited by such effects. Phase errors ultimately limit the accuracy of laser metrology and GPS location. Similar considerations will become important as terrestrial and satellite communications move to higher frequencies. Scintillation measurements now provide an accurate and economical way to establish atmospheric properties. Amplitude and intensity fluctuations are addressed in the second volume.

This volume will be of particular interest to astronomers, applied physicists and engineers developing instruments and systems at the frontier of technology. It also provides a unique reference for atmospheric scientists and scintillation specialists. It can be used as a graduate textbook and is designed for self-study. Extensive references to original work in English and Russian are provided.

DR ALBERT D. WHEELON has been a visiting scientist at the Environmental Technology Laboratory of NOAA in Boulder, Colorado, for the past decade. He holds a BSc degree in engineering science from Stanford University and a PhD in physics from MIT, where he was a teaching fellow and a research associate in the Research Laboratory for Electronics. He has published thirty papers on radio physics and space technology in learned journals.

He has spent his entire career at the frontier of technology. He made important early contributions to ballistic-missile and satellite technology at TRW, where he was director of the Radio Physics Laboratory. While in government service, he was responsible for the development and operation of satellite and aircraft reconnaissance systems. He later led the development of communication and scientific satellites at Hughes Aircraft. This firm was a world leader in high technology and he became its CEO in 1986.

He has been a visiting professor at MIT and UCLA. He is a Fellow of the American Physical Society, the IEEE and the AIAA. He is also a member of the National Academy of Engineering and has received several awards for his contributions to technology and national security including the R. V. Jones medal. He is currently a trustee of Cal Tech and the RAND Corporation. He was a member of the Defense Science Board and the Presidential Commission on the Space Shuttle Challenger Accident. He has been an advisor to five national scientific laboratories in the USA.

# Electromagnetic Scintillation

## I. Geometrical Optics

Albert D. Wheelon

Environmental Technology Laboratory
National Oceanic and Atmospheric Administration
Boulder, Colorado, USA

CAMBRIDGE
UNIVERSITY PRESS

PUBLISHED BY THE PRESS SYNDICATE
OF THE UNIVERSITY OF CAMBRIDGE
The Pitt Building, Trumpington Street, Cambridge, United Kingdom

CAMBRIDGE UNIVERSITY PRESS
The Edinburgh Building, Cambridge CB2 2RU, UK
40 West 20th Street, New York, NY 10011-4211, USA
10 Stamford Road, Oakleigh, VIC 3166, Australia
Ruiz de Alarcón 13, 28014 Madrid, Spain
Dock House, The Waterfront, Cape Town 8001, South Africa

http://www.cambridge.org

First published 2001

Printed in the United Kingdom at the University Press, Cambridge

*Typeface* CMR 10/12     *System* LaTeX [HBA]

*A catalogue record for this book is available from the British Library*

*Library of Congress Cataloguing in Publication data*
Wheelon, Albert D. (Albert Dewell), 1929–
    Electromagnetic scintillation/Albert D. Wheelon.
        p. cm.
    Includes bibliographical references and index.
    Contents: Vol. 1. Geometrical optics.
    ISBN 0 521 80198 2
        1. Electromagnetic waves–Transmission. I. Title.

QC665.T7.W48 2001
539.2–dc21   00-065167

ISBN 0 521 80198 2 hardback

*These volumes are
dedicated to Valerian Tatarskii who taught us all*

# Contents

# Preface

## History

Quivering of stellar images can be observed with the naked eye and was noted by ancient peoples. Aristotle tried but failed to explain it. A related phenomenon noted by early civilizations was the appearance of shadow bands on white walls just before solar eclipses. When telescopes were introduced, scintillation was observed for stars but not for large planets. Newton correctly identified these effects with atmospheric phenomena and recommended that observatories be located on the highest mountains practicable. Despite these occasional observations, the problem did not receive serious attention until modern times.

## How it began

Electromagnetic scintillation emerged as an important branch of applied physics after the Second World War. This interest developed in response to the needs of astronomy, communication systems, military applications and atmospheric forecasting. The last fifty years have witnessed a growing and widespread interest in this field, with considerable resources being made available for measurement programs and theoretical research.

Radio signals coming from distant galaxies were detected as this era began, thereby creating the new field of radio astronomy. Microwave receivers developed by the military radar program were used with large apertures to detect these faint signals. Their amplitude varied randomly with time and it was initially suggested that the galactic sources themselves might be changing. Comparison of signals measured at widely separated receivers showed that the scintillation was uncorrelated, indicating that the random modulation was imposed by ionized layers high in the earth's atmosphere. Careful study of this scintillation now provides an important new tool for examining ionospheric structures that influence reflected short-wave signals and transionospheric propagation.

Vast networks of microwave relay links were established soon after the Second World War to provide wideband communications over long distances. The effect of scintillation on the quality of such signals was investigated and found not to be important for the initial systems. The same question arose later in connection with the development of communication satellites and gave rise to careful research. These questions are now being revisited as terrestrial and satellite links move to higher frequencies and more complicated modulation schemes.

Large optical telescopes were being designed after the war in order to refine astronomical images. It became clear that the terrestrial atmosphere places an unwelcome limit on the accuracy of position and velocity measurements. The same medium limits the collecting area for coherent signals to a small portion of the apertures for large telescopes. A concerted effort to understand the source of this optical noise was begun in the early 1950s. Temperature fluctuations in the lower atmosphere were identified as the source. When high-resolution earth-orbiting reconnaissance satellites were introduced in 1960 it was feared that the same mechanism might limit their resolution.

Development of long-range missiles began in 1953 and early versions relied on radio guidance. Astronomical experience suggested that microwave quivering would limit their accuracy. This concern encouraged the performance of terrestrial experiments to measure phase and amplitude fluctuations induced by the lower atmosphere. The availability of controlled transmitters on earth-orbiting satellites after 1957 allowed a wider range of propagation experiments designed to investigate the structure of the atmosphere.

The presence of refractive irregularities in the atmosphere suggested the possibility of scattering microwave signals to distances well beyond the optical horizon. This was confirmed experimentally in 1955 and became the basis for scatter-propagation communications links using turbulent eddies both in the troposphere and in the ionosphere. Because of its military importance, considerable research performed in order to understand the interaction of microwave signals with atmospheric turbulence was sponsored.

## Understanding the phenomenon

The measurement programs that investigated these applications generated a large body of experimental data that bore directly on the scattering of electromagnetic waves in random media. There was a need to develop theoretical understanding in order to explain these results. The first attempts used geometrical optics combined with space-correlation models for the turbulent atmosphere to describe the electromagnetic propagation. Temporal variations of the field were included by assuming a frozen random medium carried past the propagation path on prevailing winds. These models were successful at describing phase and angle-of-arrival measurements, but failed to explain amplitude and intensity variations. The next step was to exploit the Rytov approximation to describe the influence of random media on electromagnetic waves. This technique includes diffraction effects and provides a reliable description of weak scattering conditions.

Our understanding of refractive irregularities in the lower atmosphere benefited greatly from basic research on turbulent flow fields. Kolmogorov was able to explain the most important features of turbulence using dimensional arguments. His approach was later used to describe the turbulent behavior of temperature and humidity, which influence electromagnetic waves. These models now provide a physical basis for describing many of the features observed at optical and microwave frequencies.

## The second wave of applications

The development of coherent light sources took scintillation research into a new and challenging regime. With laser sources it is possible to form confined beams of optical radiation. These beam waves find important applications in military target-location systems. Their ability to deliver concentrated forms of optical energy onto targets at some distance soon led to laser weapons. Wave-front tilt monitors and corrective mirror systems were combined to correct the angle-of-arrival errors experienced by such signals and were later applied to large optical telescopes. Rapidly deformable mirrors were developed later to correct higher-order errors in the arriving wave-front. These applications stimulated research on many aspects of atmospheric structure and wave propagation.

Radio astronomy has moved from 100 MHz to over 100 GHz in the past forty years. Microwave interferometry has become a powerful technique for refining astronomical observations using phase comparison of signals received at separated antennas. The lower atmosphere defines the inherent limit of angular accuracy that can be achieved with earth-based arrays. Considerable effort has gone into programs devoted to measuring the phase correlation as a function of the separation between receivers and these results have been used to guide the design of large interferometric arrays.

The development of precision navigation and location techniques using constellations of earth satellites focused attention on phase fluctuations at 1550 MHz. Ionospheric errors are removed by scaling and taking the difference of the phases of two signals at nearby frequencies. The ultimate limit on position determination is thus set by phase fluctuations induced by the troposphere, which are the same errors as those that limit the resolution of interferometric arrays.

Coherent signals radiated by spacecraft sent to explore other planets have been used to examine the plasma distribution in our own solar system. Transmission of spacecraft signals through the atmospheres of planets and their moons provides a unique way to explore the atmospheres of neighboring bodies. The discovery of microwave sources far out in the universe led to exploration of the interstellar plasma with scintillation techniques. Comparing different frequency components of pulsed signals that travel along the same path provides a unique tool with which to study this silent medium. One can also use scintillation measurements to estimate the size of distant quasars with surprising accuracy.

An extension of scatter propagation occurred when radars become sensitive enough to measure backscattering by turbulent irregularities. The structure of atmospheric layers from the troposphere to the ionosphere has been established using high-power transmitters and large antennas. It was later found that scanning radars can detect turbulent conditions over a considerable area, thereby providing a valuable warning service to aircraft. The same phenomenon is now making an important contribution to meterology. A network of phased-array radars was installed in the USA in order to measure the vertical profiles of wind speeds and temperature by sensing the signals returned from irregularities and their Doppler shifts.

Acoustic propagation is a complementary field to the one we will examine. Problems with long-range underwater acoustic detection have provoked important experimental and theoretical research. Controlled experiments can be done in the ocean that are not often possible with electromagnetic signals in the atmosphere. Theoretical descriptions of acoustic propagation have helped us to understand the strong scintillation observed at optical frequencies. Studies on the acoustic and electromagnetic problems are mutually supporting endeavors.

These recent applications have stimulated further theoretical research. Laser systems often operate in the strong-scattering regime, for which the Rytov approximation is no longer valid. This encouraged a sustained effort to develop techniques that can describe saturation effects. This effort has taken three directions. The first is based on the Markov approximation which provides differential equations for moments of the electric field strength. The second approach is an adaptation of the path-integral method developed for quantum mechanics. The third approach relies on Monte Carlo simulation techniques in which the random medium is replaced by a succession of phase screens.

## Resources for learning about scintillation

After fifty years of extensive experimental measurements and theoretical development this subject has become both deep and diverse. Many are asking *"How can one learn about this expanding field?"* Where does one go to find results that can be applied in the practical world? Despite its growing importance, one finds it difficult to establish a satisfactory understanding of the field without an enormous investment of time. That luxury is not available to most engineers, applied physicists and astronomers. They must find reliable results quickly and apply them.

There are few reference books in this field. A handful of early books were written in the USSR, where much of the basic work was done. These were influential in shaping research programs but subsequent developments now limit their utility. A later series published in the USSR summarized the theory of strong scattering but made little contact with experimental data. Several books on special topics in propagation in random media have appeared recently. Even with these references, it takes a great deal of time to establish a confident understanding of what is known and not known.

## The origin of the present volumes

These volumes on electromagnetic scintillation came about in an unusual way. They resulted from my return to a field in which I had worked as a young physicist. My life changed dramatically in 1962 and the demands of developing large radar, reconnaissance and communication systems at the frontier of technology took all of my energy for several decades. That experience convinced me how important research in this field has become. When I returned to scientific work in 1988, I resolved to explore the considerable progress that had been made during my absence.

I was immediately confronted with an enormous literature scattered in many journals. Fortunately, the Russian-language journals had been translated into English and were available at MIT, where I was then teaching. As an aid to my exploration, I began to develop a set of notes with which I could navigate the literature. My journal grew steadily as I added detail, made corrections and included new insights. It soon became several large notebooks. I was invited to work with the Environmental Technology Laboratories of NOAA in 1990. This coincided with the arrival from Moscow of several leaders in this field, which has made Boulder the premier center of research. I shared my notebooks with colleagues there who encouraged me to bring them to book form.

In reviewing the progress made over the past thirty years, I found a number of loose ends and apparent conflicts. To resolve these issues, I spent a good deal of time examining the field. Several areas needed clarification and this resulted in some original research, which is reported here for the first time.

## Approach and intended users

The purpose of these volumes is to provide an understanding of the underlying principles of electromagnetic propagation through random media. We shall focus on transmission experiments in which small-angle forward scattering is the dominant mechanism. The elements common to different applications are emphasized by focusing on fundamental descriptions that transcend their boundaries. I hope that this approach will serve the needs of a diversified community of technologists who need such information. Measurements and theoretical descriptions are presented together in an effort to build confidence in the final results. In each application, I have tried to identify critical measurements that confirm the basic expressions. These experiments are often summarized in the form of tables that readily lead one to the original sources. Actual data is occasionally reproduced so that the reader can judge the agreement for himself. In some cases, I give priority to the early experiments to recognize pioneering work and to provide a sense of historical development. In other cases, I have used the most recent and accurate data for comparison.

It is important to understand how the series is organized. The goal is always to present the *simplest* description for a measured quantity. We only advance to more sophisticated explanations when the simpler models prove inadequate. The first volume explores this subject with the most elementary description of electromagnetic radiation, which is provided by geometrical optics. We find that it gives a valid description for phase and angle-of-arrival fluctuations for almost any situation. In the second volume we introduce the Rytov approximation which includes diffraction effects and provides a significant improvement on geometrical optics. With this approximation one can describe weak fluctuations of signal amplitude and intensity over a wide range of applications. The third volume is devoted to strong scattering, which is encountered at optical and millimeter-wave frequencies. That regime presents a greater analytical challenge and one must lean more heavily on experimental results to understand it.

This presentation emphasizes scaling laws that show how the measured quantities vary with the independent variables – frequency, distance, aperture size, separation between receivers, time delay, zenith angle and frequency separation. It is often possible to rely on these scaling laws without knowing the absolute value for a measured quantity. That is important because the level of turbulent activity changes diurnally and seasonally. These scaling laws are expressed in closed form wherever possible; numerical computations are presented when it is not. Brief descriptions of the special functions needed for these analyses are provided in appendices and referenced in the text. This should allow those who have studied mathematical physics to proceed rapidly, while providing a convenient reference for those less familiar with such techniques. Problems are included at the end of each chapter. They are designed to develop additional insights and to explore related topics.

The turbulent medium itself is a vast subject about which much has been written. Each new work on propagation attempts to summarize the available information in order to lay a foundation for describing the electromagnetic response to turbulent media. I looked for ways to avoid that obligatory preamble. Alas, I could find no way out and my summary is included in the first volume. There is good reason to use the Kolmogorov model to characterize the inertial range. All too often, however, we find that large eddies or the dissipation range plays an important role. I have tried to avoid promoting particular models for the large-eddy regime since we have no universal model with which to describe it. My position is that it should be investigated by performing experiments.

Any attempt to describe the real atmosphere must address the reality of anisotropy. We know that plasma irregularities in the ionosphere are elongated in the direction of the magnetic field. In the troposphere, irregularities near the surface are correlated over greater distances in the horizontal direction than they are in the vertical direction. That disparity increases rapidly with altitude. One cannot ignore the influence of anisotropy on signals that travel through the atmosphere and much of the new material included here is the result of recent attempts to include this effect.

## Acknowledgments

This series is the result of many conversations with people who have contributed mightily to this field. Foremost among these is Valerian Tatarskii, who came to Boulder just as my exploration was taking on a life of its own. He has become my teacher and my friend. Valery Zavorotny also moved to Boulder and has been a generous advisor. Hal Yura reviewed the various drafts and suggested several ingenious derivations. Rich Lataitis has been a steadfast supporter, carefully reviewing my approach and suggesting important references. Reg Hill has subjected this work to searching examination, for which I am truly grateful. Rod Frehlich helped by identifying important papers from the remarkable filing system that he maintains. Jim Churnside and Gerard Ochs have been generous reviewers and have paid attention to the experiments I have cited. My good friend Robert Lawrence did all the numerical computing and I owe him a special debt.

Steven Clifford extended the hospitality of the Environmental Technology Laboratory to me and has been a strong supporter from the outset. As a result of his initiative, I have enjoyed wonderful professional relationships with the people identified above. I cannot end without acknowledging the help of Mary Alice Wheelon who edited too many versions of the manuscript and double checked all the references. The drawings and figures were designed by Andrew Davies and Peter Wheelon. Jonathan Mitchell checked and edited much of the material. Jane Watterson and her colleagues at the NOAA Library in Boulder have provided prompt and continuing reference support for my research.

At the end of the day, however, the work presented here is mine. I alone bear the responsibility for the choice of topics and the accuracy of the presentation. My reward is to have taken this journey with wonderful friends.

Albert D. Wheelon
*Santa Barbara, California*

# 1

# Introduction

The laws of geometrical optics were known from experiments long before the electromagnetic theory of light was established [1]. Today we recognize that they constitute an approximate solution for Maxwell's field equations. This solution describes the propagation of light and radio waves in media that change gradually with position [2]. The wavelength is taken to be zero in this approximation and diffraction effects are completely ignored. The field is represented by signals that travel along ray paths connecting the transmitter and receiver. In most applications these rays can be approximated by straight lines. These trajectories are uniquely determined by the dielectric constant of the medium and by the antenna pattern of the transmitter. In this approach energy flows along these ray paths and the signal acts locally like a plane wave. Geometrical optics provides a convenient description for a wide class of propagation problems when certain conditions are met.

The assumption that the medium changes gradually means that geometrical optics cannot describe the scattering by objects of dimensions comparable to a wavelength. Similarly, it cannot describe the boundary region of the shadows cast by sharp edges. A further condition is that rays launched by the transmitter must not converge too sharply – as they do for focused beams. These conditions must be refined when ray theory is used to describe propagation in random media.

Geometrical optics is widely used to describe electromagnetic propagation in the nominal atmosphere of the earth, other planets and the interstellar medium. Refractive bending of starlight and microwave signals in the troposphere is accurately described by this approximation. Standard atmospheric-profile models are used to calculate ray paths, radio horizons and angles of arrival for various elevation angles and surface conditions [3]. Ray theory is also the primary tool for describing the reflection of radio signals in the ionosphere [4]. The maximum usable frequency can be estimated for shortwave broadcast and communication services if the electron-density profile of the ionosphere is known from vertical sounding measurements or modeling. The same techniques are used to describe the transmission of acoustic waves in the ocean.

It was initially thought that geometrical optics would provide a valid description for propagation through random media and early studies all relied on this approach [5][6][7][8]. The concept assumed that the signal fluctuations are induced by small dielectric variations located close to the nominal ray trajectory. It was hoped that perturbation solutions of the ray equations would yield valid expressions for phase and amplitude variations. Only the first half of that expectation was realized.

Geometrical optics provides a good description for the phase fluctuations imposed by random media. These are caused by the random *speeding up and slowing down* of the signal as it travels along the nominal ray trajectory. Phase fluctuations computed in this way agree with experiment, even when the path is long and the fluctuations are large. For line-of-sight propagation the predicted phase variance is proportional to the path length and the first moment of the spectrum of turbulent irregularities. This means that phase fluctuations depend primarily on the largest eddies and diffraction effects can be ignored.

Geometrical optics also describes angle-of-arrival fluctuations over a wide range of propagation conditions. Angular errors at the receiver are the result of many small *random refractive bendings* along the ray path. This sets the threshold for astronomical seeing with ground-based telescopes. The angular variance is proportional to distance traveled and to the third moment of the spectrum. Angular errors depend primarily on small eddies. As a practical matter, aperture smoothing suppresses the contributions of eddies smaller than the receiver and such measurements depend primarily on the inertial range of the turbulent spectrum. Again, diffraction effects are relatively unimportant.

By contrast, this method cannot describe amplitude and intensity fluctuations in most situations of practical interest. These scintillations are due to the random *bunching and diverging* of energy-bearing rays in this approximation. The resulting expression for the logarithmic variance of the amplitude is proportional to the third power of distance and the fifth moment of the spectrum. Intensity scintillation therefore depends primarily on the smallest eddies for which diffraction effects play a dominant role. To use this approximation the influential eddies must be larger than the Fresnel length. That condition is seldom met and one cannot use this method to define scintillation levels – unless large receivers and/or transmitters are employed. The geometrical optics description of amplitude fluctuations is primarily of historical interest and one is referred to standard texts for expressions for the variance and correlation [9][10]. Amplitude and intensity fluctuations will be analyzed with diffraction theory in the next volume.

The goal of the second chapter is to describe random media. In the following chapter we adapt geometrical optics to describe propagation through random media and to establish the validity conditions for its application. The single-path phase variance is estimated in Chapter 4. In the following chapter we calculate the phase structure function as a function of the separation between receivers and compare it with results from phase-difference experiments. The temporal correlation of phase and the corresponding power spectrum are addressed in Chapter 6. In the next chapter we describe the angle-of-arrival errors induced by a random medium. We show that the random phase and phase difference are distributed as Gaussian random variables in Chapter 8. Moments of the electric field strength calculated with geometrical optics are presented in the last chapter. Problems are included at the end of each chapter to develop additional insights and to explore related topics. Helpful mathematical relations are summarized in the appendices.

## References

[1] M. Born and E. Wolf, *Principles of Optics*, 6th Ed. (Pergamon Press, New York, 1980), 109 *et seq.*

[2] Yu. A. Kravtsov and Y. I. Orlov, *Geometrical Optics of Inhomogeneous Media* (Springer-Verlag, Berlin, 1990).

[3] B. R. Bean and E. J. Dutton, *Radio Meteorology* (National Bureau of Standards Monograph 92, U.S. Government Printing Office, Washington, March 1966).

[4] J. M. Kelso, *Radio Ray Propagation in the Ionosphere* (McGraw-Hill, New York, 1964), 139 *et seq.*

[5] P. G. Bergmann, "Propagation of Radiation in a Medium with Random Inhomogeneities," *Physical Review*, **70**, Nos. 7 and 8, 486–492 (1 and 15 October 1946).

[6] V. A. Krasil'nikov, "On Fluctuations of the Angle-of-Arrival in the Phenomenon of Twinkling of Stars," *Doklady Akademii Nauk SSSR, Seriya Geofizicheskaya* **65**, No. 3, 291–294 (1949) and "On Phase Fluctuations of Ultrasonic Waves Propagating in the Layer of the Atmosphere Near the Earth," *Doklady Akademii Nauk SSSR, Seriya Geofizicheskaya*, **88**, No. 4, 657–660 (1953). (These references are in Russian and no translations are currently available.)

[7] S. Chandrasekhar, "A Statistical Basis for the Theory of Stellar Scintillation," *Monthly Notices of the Royal Astronomical Society*, **112**, No. 5, 475–483 (1952).

[8] R. B. Muchmore and A. D. Wheelon, "Line-of-Sight Propagation Phenomenon – I. Ray Treatment," *Proceedings of the IRE*, **43**, No. 10, 1437–1449 (October 1955).

[9] V. I. Tatarskii, *The Effects of the Turbulent Atmosphere on Wave Propagation* (translated from the Russian and issued by the National Technical Information Office, U.S. Department of Commerce, Springfield, VA 22161, 1971), 177–208.

[10] S. M. Rytov, Yu. A. Kravtsov and V. I. Tatarskii, *Principles of Statistical Radiophysics 4, Wave Propagation Through Random Media* (Springer-Verlag, Berlin, 1989), 21–32.

# 2

# Waves in Random Media

The first step in studying electromagnetic scintillation is to establish a firm physical foundation. This chapter attempts to do so for the entire work and it will not be repeated in subsequent volumes. We proceed cautiously because the issues are complex and the measured effects are often quite subtle. Section 2.1 explores the way in which Maxwell's equations for the electromagnetic field are modified when the dielectric constant experiences small changes. Because atmospheric fluctuations are much slower than the electromagnetic frequencies employed, their influence can be condensed into a single relationship: the *wave equation for random media*. This equation is the starting point for all developments in this field.

To proceed further one must characterize the dielectric fluctuations. We want to do so in ways that accurately reflect atmospheric conditions. Because we are dealing with a random medium, we must use statistical methods to describe them and their influence on electromagnetic signals. For instance, we want to know how dielectric fluctuations measured at a single point vary with time. Even more important, we need to describe the way in which fluctuations at separated points in the medium are correlated. There are several ways to do so and they are developed in Section 2.2. These descriptions assume that the random medium is isotropic and homogeneous. Those convenient assumptions are seldom realized in nature and we show how to remove them at the end of this section. Turbulence theory now gives an important but incomplete physi cal description of these fluctuations. Its results in the primary region of interest are expressed as a power-law scaling of the spectrum of turbulent irregularities. This approach depends on a few physical parameters, which must be found by experiment.

In Section 2.3 we describe direct measurements of these turbulence parameters in the troposphere. This region is nondispersive over broad frequency

5

bands. Fluctuations of the refractive index are related to those of temperature and humidity. We present measurements both of surface values and of
height profiles for three parameters: (*a*) the level of turbulence, (*b*) the inner
scale length and (*c*) the outer scale length. We also examine what is known
about anisotropy and how it changes with height. Scintillation experiments
now provide the most accurate way to measure these parameters.

Electromagnetic propagation through the ionized layers above 100 km is
quite different. Dielectric variations there depend on the electromagnetic
frequency and on the electron density in the plasma created by solar radiation. This dispersion made early exploration of the ionosphere by using
reflected radio signals possible. Microwave signals from artificial satellites
and radio-astronomy sources now provide a more flexible and accurate way
to probe the ionosphere. What we know about electron-density fluctuations
in its elevated layers is summarized in Section 2.4. The picture is necessarily less complete than it is for the troposphere because we can seldom make
direct measurements of the important parameters. *In situ* measurements of
ion density are possible with scientific earth satellites and infrequent rocket
flights. From microwave-transmission experiments we know that the irregularities are significantly elongated in the direction of the terrestrial magnetic
field. The spectrum of electron-density fluctuations is described by a power
law, although it may be different than the tropospheric form.

## 2.1 Maxwell's Equations in Random Media

Our first task is to establish the equations which describe electromagnetic
propagation in a random medium. We start with Maxwell's equations for
the various components of the electromagnetic field. The electric, magnetic,
displacement and induction fields are governed by four vector equations:

$$\boldsymbol{\nabla} \times \mathbf{E} = -\frac{1}{c} \frac{\partial \mathbf{B}}{\partial t} \tag{2.1}$$

$$\boldsymbol{\nabla} \times \mathbf{H} = \frac{1}{c} \frac{\partial \mathbf{D}}{\partial t} + \frac{4\pi}{c} \mathbf{J} \tag{2.2}$$

$$\boldsymbol{\nabla} \cdot \mathbf{D} = 4\pi \rho_{\mathrm{e}} \tag{2.3}$$

$$\boldsymbol{\nabla} \cdot \mathbf{B} = 0 \tag{2.4}$$

Here $\mathbf{J}$ is the current density and $\rho_{\mathrm{e}}$ is the net charge density. These equations
are verified by careful laboratory experiments. In combination they describe
the generation and propagation of electromagnetic waves. We have used

mixed Gaussian units for notational efficiency but this choice will have no effect on the final result.

The divergence of the induction field **B** vanishes because there is no magnetic charge. By taking the divergence of (2.2) and using (2.3) one establishes the continuity equation

$$\mathbf{\nabla} \cdot \mathbf{J} + \frac{\partial \rho_e}{\partial t} = 0. \tag{2.5}$$

In our applications, this balance relates to the current and charge on the transmitting antenna or laser source. Elsewhere in the transmission region, one can ignore these quantities and the divergence of the displacement field **D** also vanishes:

$$\mathbf{\nabla} \cdot \mathbf{D} = 0 \tag{2.6}$$

To this set of Maxwell's equations we must add the *constitutive equations* which characterize the propagation medium. These equations connect similar field components by employing atmospheric properties. The magnetic permeability relates the magnetic field to the induction field in a linear manner:

$$\mathbf{B} = \mu_{\mathrm{m}} \mathbf{H} \tag{2.7}$$

In our units $\mu_{\mathrm{m}} = 1$ in the earth's atmosphere so that **B** and **H** are virtually the same vector. The more important relation for our work connects the electric and displacement fields:

$$\mathbf{D} = \varepsilon(\mathbf{r}, t) \mathbf{E} \tag{2.8}$$

The dielectric constant $\varepsilon(\mathbf{r}, t)$ contains all the information we need to describe the propagation of electromagnetic waves in random media and therefore all of our attention will be focused on its consequences. It is customary to decompose this quantity into its average value and a small component that is a stochastic function of position and time:

$$\varepsilon(\mathbf{r}, t) = \varepsilon_0(\mathbf{r}) + \Delta\varepsilon(\mathbf{r}, t) \tag{2.9}$$

The average value $\varepsilon_0$ can be a function of position and it is therefore important to retain this slowly varying term in describing short-wave signals reflected by the ionosphere. In the lower atmosphere $\varepsilon_0$ is different than unity by less than 300 parts per million [1]. We concentrate here on the fluctuating component $\Delta\varepsilon(\mathbf{r}, t)$ which gives rise to electromagnetic scintillation.

To establish the wave features of electromagnetic propagation, one must combine Maxwell's equations. We apply the curl operator to the first equation and use the second to express $\nabla \times \mathbf{H}$ in terms of the current density and displacement vectors:

$$\nabla \times \nabla \times \mathbf{E} = -\frac{1}{c} \nabla \times \frac{\partial \mathbf{H}}{\partial t} = -\frac{1}{c} \frac{\partial}{\partial t} (\nabla \times \mathbf{H})$$

$$= -\frac{1}{c} \frac{\partial}{\partial t} \left( \frac{1}{c} \frac{\partial \mathbf{D}}{\partial t} + 4\pi \mathbf{J} \right)$$

$$= -\frac{1}{c^2} \frac{\partial^2}{\partial t^2} (\varepsilon \mathbf{E}) - \frac{4\pi}{c} \frac{\partial \mathbf{J}}{\partial t} \qquad (2.10)$$

The double curl operation can be simplified since the following relation holds for any vector:

$$\nabla \times \nabla \times \mathbf{E} = -\nabla^2 \mathbf{E} + \nabla(\nabla \cdot \mathbf{E})$$

We use the second constitutive relation to relate the divergence of $\mathbf{E}$ to the gradient of the dielectric constant:

$$\nabla \cdot \mathbf{D} = \nabla \cdot (\varepsilon \mathbf{E}) = \mathbf{E} \cdot \nabla \varepsilon + \varepsilon \nabla \cdot \mathbf{E} = 0$$

so that

$$\nabla \cdot \mathbf{E} = -\mathbf{E} \cdot \nabla(\log \varepsilon).$$

By combining these results we establish a general equation for the electric field vector:

$$\nabla^2 \mathbf{E} - \frac{1}{c^2} \frac{\partial^2}{\partial t^2} [(1 + \Delta\varepsilon)\mathbf{E}] = \frac{4\pi}{c} \frac{\partial \mathbf{J}}{\partial t} - \nabla\{\mathbf{E} \cdot \nabla [\log(1 + \Delta\varepsilon)]\} \qquad (2.11)$$

The last term describes polarization changes induced by scattering in the random medium. In Volume 2 we shall find that this depolarization is negligible for atmospheric propagation and this term will be carried no further.

The electric field is generated by the transmitter and modified by $\Delta\varepsilon$ as it travels through the medium. The current density and dielectric variations are functions of position and time. On the other hand, their characteristics are quite different. For a microwave system, the current density is confined to the transmitting antenna and oscillates very rapidly at microwave frequencies. By contrast, the dielectric fluctuations pervade the entire region but change very slowly with time. We can exploit this profound difference to separate their effects.

We turn first to the current density $\mathbf{J}(\mathbf{r}, t)$. Its variation with time can be a complicated function describing pulse transmissions or modulation formats used to carry information. It can also be a narrow-band signal centered on a single carrier frequency. If the dielectric fluctuations have no time dependence the wave equation (2.11) represents a linear relationship between $\mathbf{E}$ and $\mathbf{J}$. Both functions could then be Fourier analyzed and their spectra related algebraically. It is possible to do so even when $\Delta\varepsilon$ depends on time – as it does in the atmosphere. This occurs because dielectric fluctuations vary quite slowly relative to the field. In fact, their frequency components are trivial compared with the microwave or optical frequency of the transmitted field. In addition, they are usually small with respect to modulation components of the field. The frequency mixing that occurs in the term $\Delta\varepsilon\,\mathbf{E}$ is therefore not important. This means that we can consider a single frequency both for the source current and for the electric field:

$$\mathbf{J}(\mathbf{r}, t) = \mathbf{J}(\mathbf{r})\exp(-i\omega t) \quad \text{and} \quad \mathbf{E}(\mathbf{r}, t) = \mathbf{E}(\mathbf{r})\exp(-i\omega t) \qquad (2.12)$$

With this assumption we find that the wave equation depends primarily on the carrier frequency $\omega = 2\pi f$:

$$\nabla^2\mathbf{E}(\mathbf{r}) - \mathbf{E}(\mathbf{r})\frac{1}{c^2}\exp(i\omega t)\frac{\partial^2}{\partial t^2}\{[1 + \Delta\varepsilon(\mathbf{r}, t)]\exp(-i\omega t)\} = -\frac{4\pi i\omega}{c}\mathbf{J}(\mathbf{r})$$

The effects of the random medium are concentrated in the second term. We are concerned with both the spatial and the temporal fluctuations of $\Delta\varepsilon(\mathbf{r}, t)$. Let us focus first on its variability with time and write the second derivative as follows:

$$\frac{1}{c^2}e^{i\omega t}\frac{\partial^2}{\partial t^2}\Big\{[1 + \Delta\varepsilon(\mathbf{r}, t)]e^{-i\omega t}\Big\} = k^2[1 + \Delta\varepsilon(\mathbf{r}, t)]$$
$$-\frac{2ik}{c}\frac{\partial}{\partial t}\Delta\varepsilon(\mathbf{r}, t) + \frac{1}{c^2}\frac{\partial^2}{\partial t^2}\Delta\varepsilon(\mathbf{r}, t)$$

where $k = 2\pi/\lambda$ is the electromagnetic wavenumber. We must estimate each term that is proportional to $\Delta\varepsilon$. We suspect that the first term is the most influential, but we must demonstrate this by estimating the others.

The first derivative of $\Delta\varepsilon$ with respect to time can be estimated with the following qualitative argument. We shall learn later that the fluctuations are induced both by internal rearrangements of the turbulent structure and by its carriage on prevailing winds. These changes are most rapid when a horizontal wind bears the structure past a measuring point, because the prevailing wind speed is considerably greater than the turbulent velocities which it induces. The first derivative is then related to the average speed of the irregularities and the eddy of size $\ell$ by the following approximate relationship:

$$\frac{\partial}{\partial t}\Delta\varepsilon(\mathbf{r},t) \simeq \frac{v}{\ell}\,\Delta\varepsilon$$

To this we must add the Doppler shift of the moving irregularities:

$$\frac{\partial}{\partial t}\Delta\varepsilon(\mathbf{r},t) \simeq v\,\Delta\varepsilon\left(\frac{1}{\ell}+\frac{1}{\lambda}\right)$$

We want to compare this expression with the first term involving $\Delta\varepsilon$:

$$k^2\,\Delta\varepsilon \qquad \text{versus} \qquad \Delta\varepsilon\,\frac{2vk}{c}\left(\frac{1}{\ell}+\frac{1}{\lambda}\right)$$

We note that $k^2\,\Delta\varepsilon$ is substantially larger than the term on the right-hand side because the wind speed is trivial relative to the speed of light. We are therefore justified in dropping the first time derivative of $\Delta\varepsilon$. Similar reasoning shows that the second derivative is even smaller [2] and we can write the final wave equation as

$$\nabla^2\mathbf{E}(\mathbf{r}) + k^2[1+\Delta\varepsilon(\mathbf{r},t)]\mathbf{E}(\mathbf{r}) = -4\pi ik\mathbf{J}(\mathbf{r}). \qquad (2.13)$$

The vector components of the electric field are not mixed in this equation. If the source current is aligned in the $x$ direction, only the $x$ component of the field is excited. This means that we can drop the vector notation. Each of the components of $\mathbf{E}$ which is excited by the source must satisfy the following scalar wave equation:

$$\nabla^2 E(\mathbf{r}) + k^2[1+\Delta\varepsilon(\mathbf{r},t)]E(\mathbf{r}) = -4\pi ikJ(\mathbf{r}) \qquad (2.14)$$

This wave equation for random media will be the starting point for all predictions developed in these volumes. One cannot solve this equation exactly because $\Delta\varepsilon(\mathbf{r},t)$ is a stochastic function of position and time. Our challenge is to find approximate solutions that agree with experimental results.

We shall employ a succession of more capable solutions for the random wave equation. This hierarchy of increasingly sophisticated solutions has a common thread, i.e., they all depend on integrals of $\Delta\varepsilon$. In this first volume we will use the techniques of geometrical optics to express the measured quantities as line integrals of $\Delta\varepsilon$ taken along the nominal ray paths. In the second volume we use the Rytov approximation which expresses the measured quantities as volume integrals of $\Delta\varepsilon$. We describe strong scattering in the last volume using the method of path integrals in which the field strength is represented by functional integrals of $\Delta\varepsilon$. This increasingly capable and complex program means that we shall need to know a good deal about the dielectric variations. We will invariably need to know the variance $\langle \Delta \varepsilon^2 \rangle$ and the correlations of $\Delta \varepsilon$ with respect to space and time. The statistical distribution is also important in some applications.

## 2.2 Describing Random Media

Our next task is to learn how to describe $\Delta \varepsilon(\mathbf{r}, t)$ in a random medium. We want to relate these descriptions to measured properties of the real atmosphere. There is no substitute for looking at actual data when doing so. Let us begin by examining the temporal behavior of dielectric variations at a fixed point in the random medium. One can make such measurements close to the earth's surface in several ways. We shall learn later that dielectric variations at optical frequencies are sensitive to temperature fluctuations but not pressure variations. A time-series measurement of temperature variations is thus a reliable surrogate for $\Delta \varepsilon(\mathbf{r}, t)$. An experimental history of temperature fluctuations is reproduced in Figure 2.1.

This type of experiment shows that the time history of the dielectric variations has two components – and that they are quite different.

$$\Delta \varepsilon(t) = [\varepsilon(t) - \varepsilon_0] + \delta\varepsilon(t) \qquad (2.15)$$

The first term describes a change in the ambient value and is suggested by a gradual upward drift in Figure 2.1. Personal experience reminds one that temperature, humidity and wind speed vary from day to day. The first term can represent these gradual diurnal and seasonal changes observed in the atmosphere and we usually model them by a linear trend. It can also describe sudden changes. These occur when a weather front moves past a sensor in the troposphere or when a solar flare causes rapid changes in the ionosphere. The second component $\delta\varepsilon(t)$ is a random or stochastic function of time. It is characterized by the surprisingly rapid fluctuations indicated in Figure 2.1. We must develop techniques for dealing with both

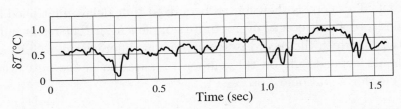

Figure 2.1: The temporal history of temperature fluctuations about a mean value measured near the earth's surface by Krechmer [3].

components. The fundamental problem in doing so is to decide which slow changes represent a gradual evolution of the ambient value and which should be regarded as low-frequency fluctuations of the random component. That question will surface often and we shall struggle with it repeatedly in these volumes.

### 2.2.1 Stationary Random Processes

The majority of our attention is devoted to the random component of the dielectric variation. If one can successfully separate out the gradual trend, $\delta\varepsilon(t)$ can be described as a *stationary process*. We must define what that means. We define a random process to be stationary if its statistical features do not depend on *when* the measurements are made. For example, the average value of a stationary process should be the same today and tomorrow and next year. We can choose the reference level to make this average value zero. The first stationary condition can be expressed as a time average:

$$\overline{\delta\varepsilon(t)} = \frac{1}{T}\int_0^T dt\,\delta\varepsilon(t) = 0 \qquad (2.16)$$

We would run the integration over an infinite interval in an ideal world. That is seldom possible and we must settle for a finite data sample. In doing so it is important to choose the integration time $T$ to be long enough to sense all the slow harmonic terms which contribute to $\delta\varepsilon(t)$.

Another important description of a random process is provided by the *temporal covariance*, which is defined by the following:

$$B_\varepsilon(t_1, t_2) = \overline{\delta\varepsilon(t_1)\,\delta\varepsilon(t_2)} \qquad (2.17)$$

It is estimated by making a copy of the signal history like that shown in Figure 2.1. This copy is then shifted by a fixed time delay, multiplied by the original record and averaged over a finite sample length. The result indicates the extent to which the record is *self-similar* at different times. The covariance should depend only on the temporal separation of the measurements $\tau = t_1 - t_2$ for a stationary process:

$$\overline{\delta\varepsilon(t)\,\delta\varepsilon(t+\tau)} = \frac{1}{T}\int_0^T dt\,\delta\varepsilon(t)\,\delta\varepsilon(t+\tau) = B_\varepsilon(\tau) \qquad (2.18)$$

The *autocorrelation function* is sometimes used instead and is the time covariance normalized by the mean square value:

$$C(\tau) = \frac{\overline{\delta\varepsilon(t)\,\delta\varepsilon(t+\tau)}}{\overline{\delta\varepsilon(t)}^2} \tag{2.19}$$

Random dielectric variations are often described by the Fourier integral transform of the time covariance:

$$B_\varepsilon(\tau) = \frac{1}{2\pi}\int_{-\infty}^{\infty} d\omega\, W_\varepsilon(\omega)\exp(i\omega\tau) \tag{2.20}$$

The transform function $W_\varepsilon(\omega)$ is called the *power spectrum*. It contains the same information as $B_\varepsilon(\tau)$ but in a different format. The power spectrum can be measured directly by passing the time history of $\delta\varepsilon(t)$ through a spectrum analyzer. These devices break the random signal into its harmonic components. The power spectrum is simply the relative energy in the various frequency components identified by $\omega = 2\pi f$.

It is significant that the gradually changing component of the dielectric constant does not satisfy the conditions for a stationary process. Because the average value of $\delta\varepsilon$ vanishes,

$$\overline{\Delta\varepsilon(\mathbf{r}_0, t)} = [\varepsilon(t) - \varepsilon_0]. \tag{2.21}$$

The first term in (2.15) is therefore identified with the *running mean value* of the dielectric variation.

### 2.2.2 Ensemble Averages

Propagation studies also depend on the statistical properties of the random fluctuations $\delta\varepsilon(r, t)$. We introduce the notion of an *ensemble average* to describe them. An ensemble is defined as *all possible configurations* of the random medium. The functions of $\delta\varepsilon$ that occur in our solutions must be averaged over that very large population and we use brackets to describe this process:

$$\text{Ensemble Average of } F(\delta\varepsilon) \quad \rightarrow \quad \langle F(\delta\varepsilon)\rangle \tag{2.22}$$

The *probability density function* describes the distribution of values that would be observed at a given point as the possible configurations of the random medium come and go. The ensemble average is intimately connected to this function. If we know this distribution, we can express the ensemble average as an integration over all possible values of $\delta\varepsilon$:

$$\langle F(\delta\varepsilon) \rangle = \int_{-\infty}^{\infty} d(\delta\varepsilon)\, F(\delta\varepsilon) \mathsf{P}(\delta\varepsilon) \tag{2.23}$$

Atmospheric measurements show that $\delta\varepsilon$ is often a Gaussian random variable with zero mean:

$$\mathsf{P}(\delta\varepsilon) = \frac{1}{\sqrt{2\pi\langle\delta\varepsilon^2\rangle}} \exp\left(-\frac{\delta\varepsilon^2}{2\langle\delta\varepsilon^2\rangle}\right) \tag{2.24}$$

On the other hand, there are many occasions when the observed distribution is quite different. Fortunately, the precise description seldom matters because our solutions of the random wave equation are represented by integrals of $\delta\varepsilon$. These integrations tend to smooth out the differences between various distributions.

Our descriptions for the electromagnetic field usually rely on the moments of $\delta\varepsilon$. The first moment vanishes:

$$\langle\delta\varepsilon\rangle = 0 \tag{2.25}$$

because gradual changes in $\Delta\varepsilon$ have presumably been absorbed in the first term of (2.15). This is the same statement as that made by (2.16), where the average value was expressed as a time integral taken over a finite interval $T$. The duality of these expressions needs to be clarified. To do so we must introduce the *ergodic theorem*. This theorem states that the ensemble average and the time average of a stationary random process must converge as the integration time goes to infinity. This makes sense since all possible configurations for the random medium should materialize if one waits long enough, but it is true only for stationary processes. We can express the condition analytically by considering the second moment of $\delta\varepsilon$ using both approaches:

$$\text{Ergodic Theorem:} \quad \langle\delta\varepsilon^2\rangle = \lim_{T\to\infty}\left(\frac{1}{T}\int_0^T dt\,(\delta\varepsilon)^2\right) \tag{2.26}$$

Because the observing time is always limited, ensemble averages are replaced by time averages estimated over a finite sample length:

$$\langle\delta\varepsilon^2\rangle \simeq \frac{1}{T}\int_0^T dt\,(\delta\varepsilon)^2 \tag{2.27}$$

We should remember that this is an *approximation* and there is an important difference for short samples.

### 2.2.3 The Spatial Covariance Description

The next step is to ask how measurements of the dielectric fluctuations vary from one point to another in the atmosphere. To do so we consider two identical sensors operating at adjacent points $\mathbf{r}_1$ and $\mathbf{r}_2$. We are primarily interested in the spatial dependence of their readings averaged over all the configurations that the medium can assume. The *spatial covariance* is a natural way to describe their similarity:

$$B_\varepsilon(\mathbf{r}_1, \mathbf{r}_2) = \langle \delta\varepsilon(\mathbf{r}_1, t)\, \delta\varepsilon(\mathbf{r}_2, t) \rangle \qquad (2.28)$$

This ensemble-averaged product occurs in almost every expression for phase and amplitude fluctuations that we will encounter. As a practical matter, the ensemble average is usually approximated by a time average over a finite data sample:

$$B_\varepsilon(\mathbf{r}_1, \mathbf{r}_2) = \frac{1}{T} \int_0^T dt\, \delta\varepsilon(\mathbf{r}_1, t)\, \delta\varepsilon(\mathbf{r}_2, t) \qquad (2.29)$$

The spatial covariance is usually simplified in two ways. The random medium can often be considered to be homogeneous over the small sensor separations usually employed in terrestrial measurements. This means that statistical averages of $\Delta\varepsilon$ do not depend on position. In other words, one should measure the same covariance if the two sensors are translated together to any other position in the region. This means analytically that the covariance depends on the separation vector $\mathbf{r}_1 - \mathbf{r}_2$ but not on the absolute end point positions:

$$\text{Homogeneous:} \quad B_\varepsilon(\mathbf{r}_1 - \mathbf{r}_2) = \langle \delta\varepsilon(\mathbf{r}_1, t)\, \delta\varepsilon(\mathbf{r}_2, t) \rangle \qquad (2.30)$$

In a sense, homogeneity of the random medium is the spatial analogy of the stationarity of $\delta\varepsilon(t)$. We will discover a connection of these characteristics in Chapter 6 when we introduce Taylor's hypothesis to describe temporal changes of the irregularities.

A second simplification is frequently made by assuming that the medium is isotropic. This means that the measured scale length is the same in the vertical direction as it is in both horizontal directions. In that case the covariance function should depend on the magnitude of the sensor-separation vector but not on its orientation:

$$\text{Isotropic:} \quad B_\varepsilon(|\mathbf{r}_1 - \mathbf{r}_2|) = \langle \delta\varepsilon(\mathbf{r}_1, t)\, \delta\varepsilon(\mathbf{r}_2, t) \rangle \qquad (2.31)$$

We can write this more succinctly in terms of the difference vector $\boldsymbol{\rho} = \mathbf{r}_1 - \mathbf{r}_2$:

Isotropic:     $B_\varepsilon(|\boldsymbol{\rho}|) = B_\varepsilon(\rho) = \langle \delta\varepsilon(\mathbf{r}, t)\, \delta\varepsilon(\mathbf{r} + \boldsymbol{\rho}, t)\rangle$     (2.32)

This assumption has been made quite consistently in previous studies of propagation in random media. It is a reasonable approximation near the surface but becomes invalid quickly as one ascends in the atmosphere. We return to this problem in Section 2.2.7 and explain the standard technique for describing anisotropic media.

The *spatial correlation function* is a normalized way of describing the variation of $\delta\varepsilon$ with position. It is simply the spatial covariance divided by the mean square value:

$$C(\boldsymbol{\rho}) = \frac{\langle \delta\varepsilon(\mathbf{r}, t)\, \delta\varepsilon(\mathbf{r} + \boldsymbol{\rho}, t)\rangle}{\langle \delta\varepsilon^2\rangle}$$     (2.33)

It begins at unity for zero separation and decreases as the distance between sensors increases. This approach isolates the spatial variation of $\delta\varepsilon$ from the level of turbulent activity. Early theoretical descriptions of propagation in random media relied on analytical models for the spatial correlation to estimate the signal phase and amplitude. These models were usually chosen for analytical convenience and bore little resemblance to atmospheric physics.

The best approach is to measure the spatial correlation directly. One can do so near the surface in several ways. Since temperature fluctuations are proportional to dielectric variations at optical wavelengths, one can investigate the spatial correlation of $\delta\varepsilon$ by comparing the fluctuations of temperature readings made with adjacent instruments:

$$C_T(\boldsymbol{\rho}) = \frac{\langle \delta T(\mathbf{r}, t)\, \delta T(\mathbf{r} + \boldsymbol{\rho}, t)\rangle}{\langle \delta T^2\rangle}$$

This was done in an early experiment using an array of bead thermistors spaced along a straight line approximately 1 m above the surface in Texas [4]. Temperature–time series were measured simultaneously with six identical instruments and the data was used to estimate the covariance and mean square value. Experimental results for the array oriented perpendicular to the wind vector are reproduced in Figure 2.2. They show that the correlation drops to 20% when the separation between sensors is approximately 2 m. This distance is called the *scale length* and is identified with the average size of turbulent irregularities. When these measurements were repeated

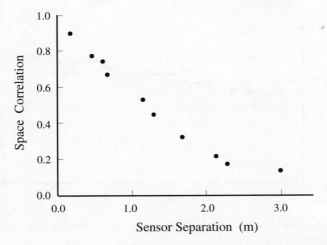

Figure 2.2: The spatial correlation of temperature fluctuations measured 1 m above the surface by Gerhardt, Crain and Smith [4]. A linear array of bead thermistors deployed perpendicular to the wind vector was used to acquire this data on 29 May 1951.

with the array oriented parallel to the wind vector, the fluctuations were correlated over a considerably longer distance in the downwind direction.

There are several problems in using spatial covariance to describe random media. Trends in the temperature readings are often present and distort estimates of the correlation function. If the temperature trend is linear with time as suggested in Figure 2.3, one can describe its influence on the correlation function by representing

$$\mathcal{T}(\mathbf{r}, t) = \langle \mathcal{T} \rangle + \delta \mathcal{T}(\mathbf{r}, t) + 2\eta \left( t - \frac{T}{2} \right) \qquad (2.34)$$

where $\langle \mathcal{T} \rangle$ is the average temperature estimated for a data sample of length $T$. The spatial covariance for fluctuation measurements

$$\delta \mathcal{T}' = \mathcal{T}(\mathbf{r}, t) - \langle \mathcal{T} \rangle$$

can be written in terms of the length of the sample in the following form:

$$\langle \delta \mathcal{T}'(\mathbf{r}, t) \, \delta \mathcal{T}'(\mathbf{r} + \boldsymbol{\rho}, t) \rangle = \frac{1}{T} \int_0^T dt \, \delta \mathcal{T}(\mathbf{r}, t) \, \delta \mathcal{T}(\mathbf{r} + \boldsymbol{\rho}, t) + \frac{1}{3}(\eta T)^2 \qquad (2.35)$$

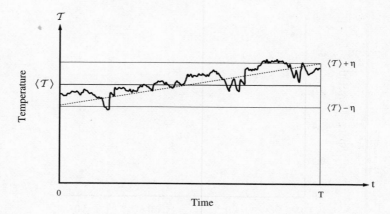

Figure 2.3: Simulated temperature–time series composed of an average value, a linear trend and a stationary random component.

This shows that the spatial correlation of temperature should approach a constant value for large separations:

$$\lim_{\rho \to \infty} [C_T(\rho)] = \left( 1 + \frac{3\langle \delta T^2 \rangle}{(\eta T)^2} \right)^{-1} \tag{2.36}$$

There is some indication of such flattening in the data of Figure 2.2. Notice that temperature offset and bias errors in the sensors correspond to a constant error in $\langle T \rangle$ and generate similar problems for the measured correlation. These complications stimulated the search for other ways to describe the spatial variation of atmosphere dielectric variations.

### 2.2.4  The Structure-function Description

Trends in temperature measurements are caused by gradual meteorological changes. By contrast, temperature sensors are remarkably stable during the relatively short data samples employed in scientific experiments. This means that the trends at nearby sensors are likely to be the same. One can remove this effect by taking the difference of temperature readings. Using the analytical description (2.34) for these readings, we see that

$$T(\mathbf{r}, t) - T(\mathbf{r} + \boldsymbol{\rho}, t) = \delta T(\mathbf{r}, t) - \delta T(\mathbf{r} + \boldsymbol{\rho}, t)$$

and the trend terms have disappeared. The remaining difference is a random function of time since $\delta T$ is a stochastic variable. We estimate its magnitude by taking the mean square value:

$$\mathcal{D}_T(\boldsymbol{\rho}) = \langle [\delta T(\mathbf{r}, t) - \delta T(\mathbf{r} + \boldsymbol{\rho}, t)]^2 \rangle \tag{2.37}$$

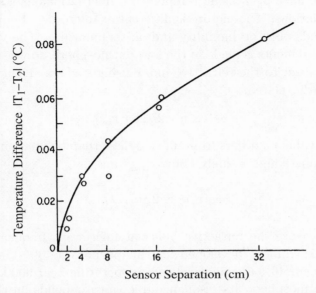

Figure 2.4: The structure function of temperature fluctuations versus separation measured near the surface by Krechmer [3]. The solid line is the 2/3 scaling law identified in (2.38).

This expression defines the *temperature structure function*. One usually replaces the ensemble average by a time average in estimating this quantity.

The original technique used to investigate the random nature of our atmosphere employed temperature-difference measurements. Results from an early experiment made near the surface are reproduced in Figure 2.4. It was found that the empirical data could be fitted by a simple scaling law over the limited range of separations employed:

$$\mathcal{D}_{\mathcal{T}}(\boldsymbol{\rho}) = C_{\mathcal{T}}^2 |\boldsymbol{\rho}|^{\frac{2}{3}} \tag{2.38}$$

where $C_{\mathcal{T}}^2$ is the *temperature structure constant*. We will learn later that this behavior is predicted by turbulence theory. It is important to note that this expression depends on the magnitude of the separation vector, so that it can describe only isotropic irregularities that have the same scale length in all directions. For large separations the structure function approaches an asymptotic value that is twice the temperature variance:

$$\lim_{\rho \to \infty} \mathcal{D}_{\mathcal{T}}(\boldsymbol{\rho}) = 2\langle \delta \mathcal{T}^2 \rangle \tag{2.39}$$

Experimental data taken with large spacings confirms that the 2/3 scaling law gives way to this constant value.

Thus far we have been using temperature fluctuations as a surrogate for dielectric variations. The picture is different at microwave frequencies since $\delta\varepsilon$ then depends both on humidity and on temperature. The vast majority of these measurements is made in the lower atmosphere, so we shift from the dielectric constant to the refractive index using $\varepsilon = n^2$. Their fluctuations are thus linearly related:

$$\delta\varepsilon(\mathbf{r}, t) = 2n_0\, \delta n(\mathbf{r}, t)$$

The ambient value $n_0$ differs from unity by less than one part in a thousand in the troposphere and is safely omitted, giving

$$\delta\varepsilon(\mathbf{r}, t) \simeq 2\, \delta n(\mathbf{r}, t). \qquad (2.40)$$

One can measure the refractive index at microwave frequencies directly by using an instrument developed for this purpose [5][6]. The *microwave refractometer* employs a cylindrical cavity operating near 9000 MHz that is open to the atmosphere. Its resonance frequency depends on the refractive index of the air sample that fills it and is compared with that of a sealed cavity that acts as a reference. Their difference in frequency provides an instantaneous measurement of $\delta n$ that is accurate to approximately one part in $10^7$. One can measure spatial differences of $\delta n$ to similar accuracy with an array of microwave refractometers. The corresponding time series allows one to estimate the *refractive-index structure function*:

$$\mathcal{D}_n(\mathbf{r}_1, \mathbf{r}_2) = \langle [\delta n(\mathbf{r}_1, t) - \delta n(\mathbf{r}_2, t)]^2 \rangle \qquad (2.41)$$

For relatively small regions of the lower atmosphere, one can assume that the random medium is homogeneous. This approximation is certainly valid over the relatively small separations employed in Figure 2.4 and one can be confident that the structure function does not depend on the end points. In this case one can write

$$\mathcal{D}_n(\boldsymbol{\rho}) = \langle [\delta n(\mathbf{r}, t) - \delta n(\mathbf{r} + \boldsymbol{\rho}, t)]^2 \rangle \qquad (2.42)$$

It is significant that this expression is relatively insensitive to irregularities that are larger that the separation since such structures are common to the two points. Their influence largely disappears on taking the difference. The assumption of homogeneity is usually valid for the structure function unless the separation is so great that entirely different atmospheric conditions characterize the locations of the two sensors. Near the surface one can also assume that the irregularities are isotropic. In that case the structure function depends only on the magnitude of the separation, not on its orientation – at least for small separations. Experimental data then suggests that the refractive-index structure function is well approximated by the 2/3 scaling law over a surprisingly wide range of separations:

$$\mathcal{D}_n(\boldsymbol{\rho}) = C_n^2 |\boldsymbol{\rho}|^{\frac{2}{3}} \tag{2.43}$$

The *refractive-index structure constant* $C_n^2$ will play a central role in our explorations of propagation through random media. Many experimental programs to measure its values near the surface and its profile with altitude have been undertaken. One finds that these values change during a twenty-four-hour period and from one day to the next. They also vary significantly from season to season. It is important to remember that $C_n^2$ is different for microwave and optical frequencies, even at the same point and time. We will review the measurements for both types of signals in Section 2.3.

### 2.2.5 *The Wavenumber-spectrum Description*

Fourier analysis is frequently used to describe time-varying electrical signals. The temporal history is replaced by a Fourier integral in which the weighting function for the harmonic components carries the same information as that contained in the original signal. Recall that we used Fourier transforms to relate the autocorrelation and power spectrum of $\delta\varepsilon(t)$ measured at a fixed point in (2.20). We can use the same technique to describe the way in which fluctuations in refractive index vary with position in the random medium. To do so we must use a three-dimensional Fourier transform for the spatial covariance:

$$\langle \delta n(\mathbf{r}_1, t)\, \delta n(\mathbf{r}_2, t) \rangle = \int d^3\kappa\, \Phi_n(\boldsymbol{\kappa}) \exp[i\boldsymbol{\kappa} \cdot (\mathbf{r}_1 - \mathbf{r}_2)] \tag{2.44}$$

The irregularities are completely described by the wavenumber spectrum of irregularities $\Phi_n(\boldsymbol{\kappa})$. In the general case of anisotropic random media, the spectrum depends on all three components of the wavenumber. The spectrum has the dimensions of length cubed since $\delta n$ is a pure number.

### 2.2.5.1 *Simplification of Propagation Calculations*

One might wonder what has been gained by this approach. The answer is a great deal. Recall our previous forecast that solutions of the random wave equation are expressed as integrals of the refractive fluctuations. To first order, the measured quantities are described by *propagation integrals* operating on the spatial covariance. For those problems which can be described with geometrical optics, the solutions are given by double line integrals of the covariance. They are expressed as repeated volume integrals of the covariance when diffraction is important. One cannot evaluate these expressions unless the covariance is modeled with an analytical expression. This means that one is *blockaded at the outset* from establishing a general description. One must repeat the propagation calculations for each model. Doing those difficult computations for different covariance models commanded much of the effort in this field for more than a decade of research.

Fortunately, we now have a superior technique for analyzing propagation in a random medium. The wavenumber representation (2.44) provides the solution. With it the propagation integrals that describe electromagnetic features of the problem can be solved quite generally by simply reversing the integrations over space and wavenumber in expressions for signal parameters. This allows the propagation integrals to operate on the universal harmonic term

$$\exp[i\boldsymbol{\kappa} \cdot (\mathbf{r}_1 - \mathbf{r}_2)].$$

The resulting propagation integrals can be done analytically in almost all cases of practical interest. They generate terms that depend only on the wavenumber vector $\boldsymbol{\kappa}$ and on the characteristics of the transmitted signal. In this way the quantities measured by experiments can be expressed as weighted averages of the wavenumber spectrum:

$$\text{Measured Quantity } = \int d^3\kappa\, \Phi_n(\boldsymbol{\kappa})[\text{Weighting function of } \boldsymbol{\kappa}] \qquad (2.45)$$

The enormous leverage of this approach is now apparent. The problem has been separated into two distinct modules. The medium is completely described by the turbulence spectrum $\Phi_n(\kappa)$. The weighting function characterizes all electromagnetic features of the propagation. In *combination* they generate a complete description of the measured quantity. The two modules are coupled through the wavenumber integration. The choice of a model for the random medium is thereby postponed to the last step. It need be made only after all other features of the propagation have been carefully analyzed.

This process is further simplified if the random medium is isotropic. The spectrum then depends only on the magnitude of the wavenumber vector, as is apparent from the inherent symmetry of (2.44):

$$\text{Isotropic:} \qquad \Phi_n(\boldsymbol{\kappa}) \rightarrow \Phi_n(\kappa)$$

The wavenumber integration is now cast in spherical coordinates centered on the difference vector $\boldsymbol{\rho} = \mathbf{r}_1 - \mathbf{r}_2$:

$$\langle \delta n(\mathbf{r},t)\, \delta n(\mathbf{r}+\boldsymbol{\rho},t)\rangle = \int_0^\infty d\kappa\, \kappa^2 \Phi_n(\kappa) \int_0^\pi d\psi \sin\psi \int_0^{2\pi} d\omega \exp(i\kappa|\boldsymbol{\rho}|\cos\omega).$$

The angular integrations are easily performed and the covariance is expressed as a single integration on the magnitude of the wavenumber vector:

$$\text{Isotropic:} \quad \langle \delta n(\mathbf{r},t)\, \delta n(\mathbf{r}+\boldsymbol{\rho},t)\rangle = 4\pi \int_0^\infty d\kappa\, \kappa^2 \Phi_n(\kappa) \left( \frac{\sin(\kappa\rho)}{\kappa\rho} \right)$$

$$(2.46)$$

We will use this result as the starting point for describing electromagnetic scintillation phenomena when the random medium is isotropic. Of course, one must use (2.44) for anisotropic irregularities.

### 2.2.5.2 Interpretation of the Wavenumber

The wavenumber representation provides the second bonus of physical insight. Turbulence theory teaches us that atmospheric irregularities are the result of a process governed by the physics of atmospheric hydrodynamics. In the standard model, large eddies are created by instabilities induced in the ambient wind field. These large eddies are sometimes kilometers in size and are often highly elongated. Their random movements are characterized by considerable kinetic energy. The large eddies break up into smaller eddies, and in doing so they slow down while gradually becoming more symmetrical. This continuing subdivision proceeds until the resulting eddies reach a size of less than 1 cm.

Our first challenge is to identify the wavenumber $\kappa$ with the eddy size in the description just introduced. There is a hierarchy of eddy sizes $\ell$ in the process of decay, which correspond to small, medium and large structures. These eddies influence the measurement of spatial covariance in markedly different ways. To explore that preferential effect, consider a spaced-sensor experiment similar to the one which produced the temperature correlation data in Figure 2.2. The data shows that the covariance decreases with the separation and that gives us an important clue. It is difficult to imagine how eddies that are smaller than the separation between sensors can influence both readings *simultaneously*. Stated differently, these eddies cannot generate correlated readings in the two sensors. By contrast, eddies that are larger than the spacing can easily influence both readings in a coordinated manner. This suggests that only those eddies which satisfy the condition

$$\rho < \ell$$

can contribute significantly to a spatial covariance measurement.

Now let us pose the same question in the context of the wavenumber description. Since we assume that the measurements are made close to the surface, the irregularities are nearly isotropic, which allows one to use the simplified spectrum description (2.46). The factor in large parentheses there shows how different wavenumbers in the spectrum contribute to the covariance estimate. The spectrum is weighted most strongly by wavenumbers that satisfy the condition

$$\kappa\rho < 1 \quad \text{or} \quad \kappa < 1/\rho$$

since the weighting function is close to unity in this case. The contributions of larger wavenumbers are suppressed by the oscillating and falling behavior of the weighting factor. On comparing the two conditions, we are led to a simple identification:

$$\kappa = 2\pi/\ell \tag{2.47}$$

In this understanding, small eddies correspond to large wavenumbers and small wavenumbers represent large eddies. In future discussions we shall move back and forth between wavenumber and eddy-size descriptions to emphasize this duality.

The inverse relationship of $\kappa$ and $\ell$ is often taken as the starting point for describing turbulence processes. In that context it is essentially a definition. We have come to this relationship by a different route, using spatial correlation experiments and their spectral representation to identify the sizes of different refractive irregularities with the inverse wavenumber.

### 2.2.5.3 Relation to Measured Quantities

The spectrum of refractive-index fluctuations was introduced in (2.44) as a mathematical substitution to give explicit form to the spatial covariance. We need now to broaden that initial definition and connect $\Phi_n(\kappa)$ to atmospheric physics. We judge that in some sense the spectrum describes the relative ability of different eddy sizes to influence measurements of refractive index in the random medium.

One way to make this connection is to estimate the mean square refractive index that would be measured at a single point. We can relate this quantity to the spectrum by setting $\mathbf{r}_1 = \mathbf{r}_2$ in definition (2.44):

$$\langle \delta n^2 \rangle = \int d^3\kappa \, \Phi_n(\kappa)$$

Near the surface one can approximate the random medium by an isotropic model and therefore use the second version (2.46):

$$\langle \delta n^2 \rangle = 4\pi \int_0^\infty d\kappa \, \kappa^2 \Phi_n(\kappa) \qquad (2.48)$$

The quantity on the left-hand side is readily measured and thus provides valuable estimates for the second moment of the spectrum.

Using the association of eddy size with wavenumber given by (2.47), we can also estimate the average eddy size as a spectrum-weighted integration:

$$\langle \ell \rangle = \frac{4\pi}{\langle \delta n^2 \rangle} \int_0^\infty d\kappa \, \kappa^2 \Phi_n(\kappa) \left( \frac{2\pi}{\kappa} \right) \qquad (2.49)$$

The average size can be identified with the *scale-length* estimate made using spatial correlation measurements. For the surface experiment summarized in Figure 2.2 we judge that

$$\langle \ell \rangle \approx 2 \text{ m}.$$

Such measurements provide useful estimates of the first moment of $\Phi_n(\kappa)$. We shall return to this relationship in Section 4.1.5 and use it to calculate the mean square phase shift for line-of-sight propagation.

### 2.2.5.4 *The Wavenumber-spectrum Expression for the Structure Function*

The structure function of refractive irregularities can also be expressed as a Fourier wavenumber integral of the turbulence spectrum. From (2.41) the following relationship emerges:

$$\mathcal{D}_n(\mathbf{r}_1, \mathbf{r}_2) = \langle \delta n^2(\mathbf{r}_1, t) \rangle + \langle \delta n^2(\mathbf{r}_2, t) \rangle - 2\langle \delta n(\mathbf{r}_1, t) \, \delta n(\mathbf{r}_2, t) \rangle$$

One can assume that the medium is homogeneous if the separation between sensors is not too great:

$$\mathcal{D}_n(\mathbf{r}_1, \mathbf{r}_2) = 2[\langle \delta n^2 \rangle - \langle \delta n(\mathbf{r}_1, t) \, \delta n(\mathbf{r}_2, t) \rangle]$$

We can relate both terms to the turbulence spectrum using (2.44):

$$\mathcal{D}_n(\mathbf{r}_1, \mathbf{r}_2) = 2 \int d^3\kappa \, \Phi_n(\boldsymbol{\kappa}) \{ 1 - \exp[i\boldsymbol{\kappa} \cdot (\mathbf{r}_1 - \mathbf{r}_2)] \} \qquad (2.50)$$

If the medium is also isotropic, the structure function is written in terms of the scalar separation $\rho = |\mathbf{r}_1 - \mathbf{r}_2|$:

$$\text{Isotropic:} \quad \mathcal{D}_n(\rho) = 8\pi \int_0^\infty d\kappa\, \kappa^2 \Phi_n(\kappa) \left( 1 - \frac{\sin(\kappa\rho)}{\kappa\rho} \right) \qquad (2.51)$$

We should ask how eddies of different sizes influence measurements of the structure function. The weighting function in large parentheses is essentially one for wavenumbers greater than the inverse spacing. This region is characterized by eddies that are smaller than the separation and they are thus the most influential. By contrast, large eddies are relatively unimportant because the weighting function goes to zero as $\kappa\rho \to 0$. This makes sense because the data is differenced in measuring $\mathcal{D}_n(\rho)$ and the influence of large eddies is common to both sensors. It also explains why inhomogeneous features of the medium can usually be ignored in such measurements.

### 2.2.5.5  *The Equivalence of the Turbulence Spectrum and the Structure Function*

Equations (2.50) and (2.51) show how the turbulence spectrum and structure function are connected. One can evidently calculate $\mathcal{D}_n(\rho)$ by integration if one knows $\Phi_n(\kappa)$. Conversely, if the structure function is measured, one can determine the spectrum by inverting these *integral equations*. We have an immediate interest in learning what spectrum corresponds to the 2/3 scaling law that has been measured frequently. The answer emerges from solving the following integral equation:

$$C_n^2 \rho^{\frac{2}{3}} = 8\pi \int_0^\infty d\kappa\, \kappa^2 \Phi_n(\kappa) \left( 1 - \frac{\sin(\kappa\rho)}{\kappa\rho} \right) \qquad (2.52)$$

It is not difficult to invert this relationship and we will suggest how to do so in Problem 1. It is even easier to guess the solution. Since power laws are usually reflected in other power laws, we try the following:

$$\Phi_n(\kappa) = Q\kappa^\gamma$$

With a change of variable in (2.52) one sees that the 2/3 power scaling law is replicated if $\gamma = -11/3$. The constant is then computed using an integral found in Appendix B:

$$Q = \frac{5}{18\pi^2} \Gamma\left(\frac{2}{3}\right) \sin\left(\frac{\pi}{3}\right) = 0.033\,005$$

and we will replace this constant by 0.033 in what follows. An important equivalence is thus established:

$$\mathcal{D}_n = C_n^2 \rho^{\frac{2}{3}} \quad \text{is equivalent to} \quad \Phi_n(\kappa) = \frac{0.033 C_n^2}{\kappa^{\frac{11}{3}}}$$

There is a valid physical basis for these relations in limited ranges of the separation and wavenumber. We turn now to their development.

### 2.2.6 *The Wavenumber Spectrum of Irregularities*

In the next few pages we will try to summarize a subtle and difficult branch of physics: *turbulence theory*. This is an extraordinary challenge. We attempt to do so only because this discipline provides an understanding of the spectrum in its most important wavenumber range. The basic problem is that of discovering how random velocity components are generated by laminar flow and how they evolve once they have been created. We believe that these velocity fluctuations are given by solutions of the Navier–Stokes equations of hydrodynamics:

$$\frac{\partial}{\partial t}\mathbf{v}(\mathbf{r}, t) + [\,\mathbf{v}(\mathbf{r}, t) \cdot \boldsymbol{\nabla}\,]\,\mathbf{v}(\mathbf{r}, t) = \nu \, \nabla^2 \mathbf{v}(\mathbf{r}, t) \tag{2.53}$$

The velocity $\mathbf{v}(\mathbf{r}, t)$ is found to be a random function of position and time. Also, it is usually assumed that the flow is incompressible:

$$\boldsymbol{\nabla} \cdot \mathbf{v}(\mathbf{r}, t) = 0 \tag{2.54}$$

A constant-velocity component has been removed from both equations with a linear transformation. Energy dissipation is described by the Laplacian term in (2.53) and is proportional to the kinematic viscosity $\nu$. It is significant that $\nu$ is the only molecular property of the fluid which enters these equations.

There is little doubt that solutions to the Navier–Stokes equation describe the chaotic behavior observed in nature. The nonlinear term prevents us from solving these equations by analytical techniques. The Reynolds number is a dimensionless parameter which describes the flow:

$$Re = \frac{LV_0}{\nu} \tag{2.55}$$

where $V_0$ is the wind speed and $L$ a relevant size scale. Numerical solutions of the Navier–Stokes equation have been constructed for small values of this parameter and exhibit many properties that are measured under controlled circumstances. The Reynolds number for atmospheric flow is between $10^6$ and $10^7$. There are between $10^{14}$ and $10^{16}$ nonlinear ordinary differential equations that must be solved numerically, a task that is clearly beyond the capability of existing computers [7]. Research in this field has turned instead to dimensional analysis. It has proven to be very productive in mapping out the general features of turbulence over a wide range of conditions.

One can represent the structure function for velocity fluctuations by a three-dimensional Fourier transform, as we did for refractive fluctuations in (2.50). In the region of primary interest we are concerned with eddies that are thought to be both isotropic and homogeneous. The structure function for different velocity components can then be expressed in terms of a single scalar function:

$$\langle |v_i(\mathbf{r}, t) - v_j(\mathbf{r} + \boldsymbol{\rho}, t)|^2 \rangle = \int d^3\kappa \, \frac{E(\kappa)}{4\pi\kappa^2} \left( \delta_{ij} - \frac{\kappa_i \kappa_j}{\kappa^2} \right) [\, 1 - \exp(i\boldsymbol{\kappa} \cdot \boldsymbol{\rho}) ]$$

$$(2.56)$$

The wavenumber $\kappa$ is again the inverse eddy size identified in (2.47). The challenge is to find the velocity spectrum $E(\kappa)$.

A physical picture of the turbulent process that was first suggested by Richardson [8] is presented in Figure 2.5. We describe that process in terms of the flow of energy. A small fraction of the kinetic energy in the ambient wind field is converted into turbulence energy at large scale lengths. This conversion is done by mechanisms that are neither specified nor understood. Fortunately the details of the external source do not matter very much to the later stages of the decay.

The interesting portion of the energy cascade begins at the *outer scale wavenumber* at which the eddy size is equal to the *outer scale length* $L_0$. The eddies begin to break up as soon as they are created. The input energy is progressively divided and redistributed to smaller and smaller scales. Their distribution then follows a universal law and they rapidly lose all statistical information about their formation – except for the mean energy-dissipation rate $\mathcal{E}$. The eddies also become symmetrical as the subdivision proceeds. The energy cascade continues until the eddies are comparable in size to the *inner scale length* $\ell_0$. Below this length their energy is dissipated by viscosity as heat.

A similar process is believed to describe the creation, subdivision and decay of fluctuations of a passive scalar. One can identify the passive scalar with temperature, humidity or the refractive index itself. The following

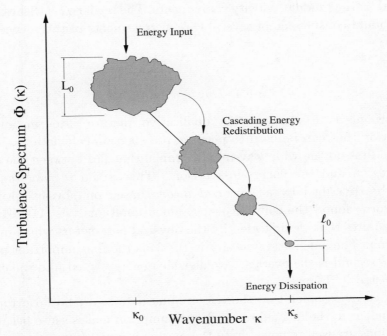

Figure 2.5: A conceptual description of the process of turbulent decay as it proceeds through an energy cascade in which eddies subdivide into progressively smaller eddies until they finally disappear.

partial differential equation describes each of these quantities or their scaled counterparts:[1]

$$\frac{\partial}{\partial t}\,\psi(\mathbf{r},t) + [\,\mathbf{v}(\mathbf{r},t)\cdot\boldsymbol{\nabla}\,]\,\psi(\mathbf{r},t) = D\,\nabla^2\psi(\mathbf{r},t) \qquad (2.57)$$

The second term describes *turbulent mixing* of the scalar quantity and depends on the random velocity vector $\mathbf{v}(\mathbf{r},t)$ generated by the Navier–Stokes equations. It represents the process by which the turbulent velocity stirs the passive scalar and thereby creates random structures. These eddies subdivide by the same process as that suggested in Figure 2.5. They evolve in response to the decaying velocity fluctuations and depend on the properties of $\mathbf{v}(\mathbf{r},t)$. Fluctuations of the scalar are dissipated through molecular diffusion, which is described by the Laplacian term. This term is similar to the kinetic-energy-removal term in (2.53) but is proportional to the diffusion constant $D$.

With these very brief remarks we have tried to set the stage for describing the velocity spectrum $E(\kappa)$ and also that of a passive scalar that

[1] For instance, *potential* temperature obeys (2.57) but temperature does not.

is mixed by the random velocity components. The reader who wishes to learn more should consult one of several careful treatments of turbulence theory [9][10].

### 2.2.6.1 *The Inertial Range*

We now need to convert this qualitative picture into an analytical description that can be used to perform propagation calculations. We concentrate first on the wide wavenumber range that lies between the energy-input region and the energy-loss region. This is the *inertial range* and it can be described by a *universal theory* based on physical principles. Kolmogorov found the solution using dimensional analysis [11]. His technique requires that we first identify the physical parameters which influence the region. The kinematic viscosity in (2.53) is the first important parameter. The second is the average rate at which energy is fed into the cascade, as suggested by Figure 2.5.

Kolmogorov assumed that the velocity fluctuations are both isotropic and homogeneous in the inertial range. His assumption makes sense because the fluctuations are well removed from the mechanism which generated the turbulence. Under this assumption, one can use (2.56) to describe the structure function. These eddies are also far from the dissipation mechanism which removes energy since the inertial range is characterized solely by interactions among the turbulent eddies. Only redistribution of energy among eddy sizes should be important. The rate at which energy enters and leaves the cascade $\mathcal{E}$ should play a unique role. The structure function for the velocity component which is parallel to the separation vector depends on this parameter and on the magnitude of the separation:

$$\text{Velocity:} \quad \mathcal{D}_v(\rho) = \langle |v(\mathbf{r}, t) - v(\mathbf{r} + \boldsymbol{\rho}, t)|^2 \rangle = F(\mathcal{E}, \rho)$$

The only combination of $\rho$ and $\mathcal{E}$ which generates a squared velocity is

$$\text{Velocity:} \quad \mathcal{D}_v(\rho) = \text{constant}(\mathcal{E}\rho)^{\frac{2}{3}} = C_v^2 \rho^{\frac{2}{3}} \quad \text{for} \quad \ell_0 < \rho < L_0 \tag{2.58}$$

since the dimensions of the energy dissipation rate $\mathcal{E}$ are m$^2$ s$^{-3}$. In this expression $C_v^2$ is called the *velocity structure constant*. The 2/3 scaling law has been verified by numerous experiments using both fluids and gases. Kraichnan noted that "*Kolmogorov's theory has achieved an embarrassment of success*" [12]. It works under a surprisingly wide range of conditions, from small-scale turbulence near the earth's surface to enormous irregularities in the interstellar plasma. It sometimes describes situations in which neither isotropy nor homogeneity seems justified.

One can find the spectrum by inverting (2.56) after setting $i = j$ and using the trick explained in solving (2.52):

$$\text{Velocity:} \quad E(\kappa) = \frac{\text{constant}}{\kappa^{\frac{5}{3}}} \quad \text{for} \quad \frac{2\pi}{L_0} < \kappa < \frac{2\pi}{\ell_0} \quad (2.59)$$

The constant is not provided by this analysis and must be established from measurements. The spectral width of the inertial range increases with the Reynolds number. This prediction was confirmed by precise meteorological measurements over more than three orders of magnitude [13].

One can also use dimensional analysis to estimate the inner scale length which marks the start of the dissipation region. It should depend on the viscosity, which is approximately $\nu = 0.148$ cm$^2$ s$^{-1}$ in air at the earth's surface. Only one combination of $\mathcal{E}$ and $\nu$ has the units of length and it is called the *Kolmogorov microscale*:

$$\eta = \left(\nu^3/\mathcal{E}\right)^{\frac{1}{4}} \quad (2.60)$$

It provides a fundamental *unit of length* for the turbulence process. The inner scale length is defined in terms of $\eta$ by the relation

$$\text{Velocity:} \quad \ell_0 = 7.4(\nu^3/\mathcal{E})^{\frac{1}{4}} = 7.4\,\eta \quad (2.61)$$

where the constant was determined experimentally. The inner scale is a few millimeters near the surface and probably increases with altitude as does viscosity. Vigorous turbulence is characterized by large values of $\mathcal{E}$ and should produce small values for the inner scale. The outer scale length is proportional to $\mathcal{E}^{\frac{1}{2}}$. This means that the inertial range should broaden as the energy rate rises, with $\ell_0$ decreasing and $L_0$ increasing.

It was soon realized that the same techniques could be applied to describe the turbulent mixing of a passive scalar [14][15][16]. An exclusive dependence on energy transfer in the inertial range again leads to the 2/3 scaling law for the refractive-index structure function:

$$\text{Scalar:} \quad \mathcal{D}_n(\rho) = C_n^2 \rho^{\frac{2}{3}}$$

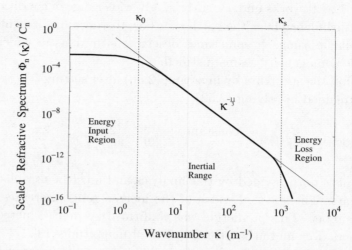

Figure 2.6: A wavenumber spectrum for refractive-index fluctuations showing three distinct stages of the turbulence process. The inertial-range behavior is the only specific prediction in this diagram. Descriptions for the other regions are only suggestive at this point.

This relationship has been confirmed by measurements of temperature and refractive index over a remarkably broad range of separations. We showed previously that it corresponds to the following refractive-index spectrum:

$$\text{Scalar:} \quad \Phi_n(\kappa) = \frac{0.033 C_n^2}{\kappa^{\frac{11}{3}}} \quad \text{for} \quad \frac{2\pi}{L_0} < \kappa < \frac{2\pi}{\ell_0} \qquad (2.62)$$

Turbulence theory has now added limits on the wavenumber range for which it is valid. This result has also been confirmed over three orders of magnitude by temperature measurements in the atmospheric boundary layer [13].

The result (2.62) provides the first and most important module needed to construct the spectrum which is plotted on logarithmic coordinates in Figure 2.6. Notice that it is valid only between the outer-scale and inner-scale wavenumbers. The inner scale is a few millimeters near the surface. The outer scale varies from a few meters near the surface to several kilometers in the free atmosphere and is quite different in the horizontal and vertical directions.

The basic result (2.62) is equivalent to the previous expression for $E(\kappa)$ since the latter is divided by $\kappa^2$ in (2.56). It is often called the *Kolmogorov model* because his conceptual breakthrough showed the way. On the other hand, it might better be called the Obukhov/Corssin spectrum to recognize their contribution in adapting this technique to passive scalars. We shall

employ the term *Kolmogorov model* in this series since it is used so widely
in other publications and because we want to avoid confusion.

### 2.2.6.2 The Energy-loss Region

The turbulence spectrum behaves quite differently in the three wavenumber
regions identified by Figure 2.6. We need to explore the energy-loss region
which lies beyond the inner-scale wavenumber cutoff:

$$\kappa > \kappa_{\mathrm{s}} = 2\pi/\ell_0$$

Energy is being removed from the cascade process at this stage and we
expect a rapid decline in the spectrum.

One might ask why we bother to characterize this region of the turbulence
spectrum. We do so because it plays an important role in describing the
fine structure of amplitude fluctuations measured with optical signals. The
Fresnel length $\sqrt{\lambda R}$ is the fundamental length scale when diffraction effects
are important. It is comparable to $\ell_0$ for short paths and optical wavelengths. The dissipation range often plays an important role in interpreting
experiments in which we measure the spatial and frequency correlation of
fluctuations in optical amplitude.[2] By contrast, the Fresnel length is always
large relative to $\ell_0$ for microwave frequencies and one need not include this
correction in their description.

Phase measurements depend primarily on large eddies and are insensitive
to the dissipation region. Angle-of-arrival and image-centroid fluctuations
lie midway between amplitude and phase measurements. We will find that
they are proportional to the spectrum's third moment,[3] which emphasizes
the dissipation region for inertial models like (2.62). Although aperture averaging overrides that dependence for large optical telescopes and microwave
receivers, the dissipation range dominates angular errors for small apertures.

To begin this exploration we write the spectrum in a form that
describes both the inertial region and the dissipation region. Kolmogorov
demonstrated that the velocity spectrum can be described as the inertial-
range result multiplied by a universal function, but did not establish an
explicit form for it [11]. The same approach can be used to describe a passive scalar in terms of the Kolmogorov microscale:

$$\text{Scalar:} \qquad \Phi_n(\kappa) = \frac{0.033C_n^2}{\kappa^{\frac{11}{3}}} \mathcal{F}(\kappa\eta) \qquad \text{for} \qquad \kappa_0 < \kappa < \infty \qquad (2.63)$$

---

[2] These effects are explored in Chapters 4 and 6 of the second volume.
[3] This dependence is established in Chapter 7 of this volume.

We refer to the function $\mathcal{F}(\kappa\eta)$ as the *dissipation function*. It evidently carries the entire burden of describing the energy-loss mechanism provided that

$$\lim_{x\to 0} [\mathcal{F}(x)] = 1$$

We are thus left to explore all analytical and physical suggestions that are compatible with this condition.

**A sharp cutoff** The first approximation employed in propagation studies ignored the dissipation region and assumed that the spectrum ends abruptly at the inner-scale wavenumber. In this case the structure function becomes

$$\text{Cutoff at } \kappa_\text{s}: \qquad \mathcal{D}_n(\rho) = 8\pi \int_0^{\kappa_\text{s}} d\kappa \, \kappa^2 \left(1 - \frac{\sin(\kappa\rho)}{\kappa\rho}\right) \left(\frac{0.033C_n^2}{\kappa^{\frac{11}{3}}}\right)$$

where we have neglected the lower limit $\kappa_0$ since it does not affect the result for small separations. By contrast, the upper-limit cutoff is influential for small spacings and some electromagnetic measurements depend on this size range, as noted above. We can expand the trigonometric terms inside the integral if the product $\kappa_\text{s}\rho$ is small:

$$\kappa_\text{s}\rho \ll 1: \qquad \mathcal{D}_n(\rho) = \rho^2 [0.1037 C_n^2 (\kappa_\text{s})^{\frac{4}{3}}] \qquad (2.64)$$

A quadratic scaling is confirmed by closely spaced-temperature experiments, but the assumption of an abrupt end to the cascade seems quite unrealistic.

**A Gaussian model** A popular description of the dissipation range is based on the Gaussian model. It represents little more than a guess at how the decay process proceeds and was patterned after a similar assumption made for the velocity field:

$$\mathcal{F}(\kappa\ell_0) = \exp(-\kappa^2/\kappa_\text{m}^2) \qquad (2.65)$$

This analytical model dominated studies of random media for three decades, primarily because the ensuing propagation integrations can usually be done in closed form. It is often referred to as the *Tatarskii model* because he was the first to use it extensively [17]. One should note that he was well aware of its limitations and observed that it did not agree with actual measurements then in hand.

The structure function for the Gaussian model is calculated from the following expression:

Gaussian: $\mathcal{D}_n(\rho) = 8\pi \int_{\kappa_0}^{\infty} d\kappa \, \kappa^2 \left(1 - \frac{\sin(\kappa\rho)}{\kappa\rho}\right)\left[\frac{0.033 C_n^2}{\kappa^{\frac{11}{3}}} \exp\left(-\frac{\kappa^2}{\kappa_m^2}\right)\right]$

The outer-scale wavenumber limit $\kappa_0$ can again be ignored and one can express the result in terms of a Kummer function using an integral found in Appendix C:

Gaussian: $\mathcal{D}_n(\rho) = 1.6865 C_n^2 \kappa_m^{-\frac{2}{3}} [M(-\frac{1}{3}, \frac{3}{2}; -\rho^2 \kappa_m^2/4) - 1]$

(2.66)

The product $\rho\kappa_m$ is very large when the separation is greater than the inner scale length. One can then use the asymptotic expansion for the Kummer functions given in Appendix G and show that the 2/3 scaling law emerges from this expression. For small spacings the power-series expansion of this function leads to a quadratic dependence:

$\kappa_m \rho \ll 1: \qquad \mathcal{D}_n(\rho) = \rho^2[0.0936 C_n^2 (\kappa_m)^{\frac{4}{3}}]$

The parameter $\kappa_m$ is usually fixed by requiring that this expression match the inertial range result at $\rho = \ell_0$:

$$C_n^2(\ell_0)^{\frac{2}{3}} = 0.0936 C_n^2 \ell_0^2 (\kappa_m)^{\frac{4}{3}}$$

This means that $\kappa_m$ and $\kappa_s$ are nearly the same:

$$\kappa_m = \frac{5.91}{\ell_0} = 0.9406 \kappa_s \qquad (2.67)$$

It became possible to probe the energy-loss region of the spectrum directly as temperature measurements became more accurate. These experiments showed convincingly that the Gaussian model is not a good model for the dissipation range. This reality stimulated a search for descriptions that are consistent with atmospheric physics.

**The Hill bump model** Redistribution comes to an end when the eddies reach the energy-loss region. We can no longer assume that the structure function depends only on the energy rate $\mathcal{E}$ and separation $\rho$ at that stage. The dissipation range is now influenced by the Laplacian term in (2.57) with the appropriate diffusion constant. However, that is not enough. One must also include the influence of the kinematic viscosity because the passive scalar is being mixed by the turbulent velocity. The spectral decline is thus caused by two effects. Viscosity drains kinetic energy from the turbulent velocity, which stirs the passive scalar, and this reduces the level of fluctuations. The irregularities are also being erased by diffusion at large wavenumbers. In air the ratio $\nu/D$ is 0.72 for temperature and 0.63 for humidity. These relationships mean that the random velocity and the passive-scalar fluctuations are attenuated by viscosity before diffusion starts to destroy the eddies.

For scalar advection Batchelor showed that the inertial-range spectrum changes from $\kappa^{-\frac{11}{3}}$ to $\kappa^{-3}$ in the range where $\nu$ becomes important but diffusion has not yet exerted control [18]. This milder decline implies a *rise* in the dissipation factor $\mathcal{F}(\kappa\eta)$ and is confirmed by careful measurements. The rise in $\mathcal{F}(\kappa\eta)$ is cut down by diffusion soon after it begins because $\nu$ and $D$ are nearly the same in air. These countervailing effects create a bump in the spectrum.

Hill placed this qualitative description on a solid analytical basis in a pioneering contribution [19]. Starting from first principles, he found Kolmogorov's universal function $\mathcal{F}(\kappa\eta)$ for temperature fluctuations. He showed that the temperature spectrum satisfies an ordinary differential equation [20]:

$$\frac{d}{d\kappa}\left(H(\kappa)\,\frac{d}{d\kappa}\Phi(\kappa)\right) = 2D\kappa^4\Phi(\kappa) \tag{2.68}$$

where

$$H(\kappa) = \frac{3}{11}\beta^{-1}\mathcal{E}^{\frac{1}{3}}\kappa^{\frac{14}{3}}[1+(13.9\kappa\eta)^{3.8}]^{-0.175}. \tag{2.69}$$

These equations were solved numerically for temperature variations in air with $\beta = 0.72$. The spectrum factor is plotted in Figure 2.7 using the inner scale length for wavenumber scaling. The following approximation is a good fit to the solution [21]:

$$\mathcal{F}(x) = \Big[1 + 0.709\,37x + 2.8235x^2 - 0.280\,86x^3 + 0.082\,77x^4\Big]e^{-1.1090x} \tag{2.70}$$

The only drawback is that one must trace the consequences of this solution through the propagation integrations but that can now be done efficiently with computer routines. Notice also that one can also solve (2.68) numerically and use the data files as a direct input to propagation calculations.

Experimental confirmations of this description have been obtained using quite different methods [22]. Careful recording of atmospheric temperature spectra confirms the prediction over three orders of magnitude in frequency

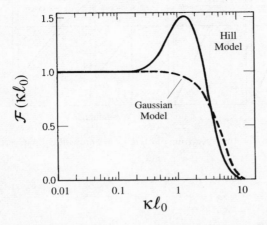

Figure 2.7: Two functions used by Hill and Clifford to characterize the dissipation-range factor in the spectrum of a passive scalar [20]. The Hill model was derived from fluid mechanics and is an accurate description. The Gaussian model is little more than an analytical assumption and is shown for comparison only.

[23][24]. Electromagnetic measurements now provide the most convenient way to explore the dissipation region of the atmosphere. The general approach is to use two optical signals that are sensitive to different ranges of the spectrum. The most effective technique uses a diverging laser signal monitored by a point detector. The amplitude scintillation experienced by this signal is sensitive to very small eddies. A second incoherent reference signal uses aperture averaging to emphasize the inertial region of the spectrum. The ratio of the scintillation imposed on these signals is independent of $C_n^2$. The results of these electromagnetic measurements were compared with *in situ* micrometeorological data and provide overwhelming evidence that the Hill model correctly describes the dissipation region [25][26]. A second method used a laser signal to measure the spatial correlation of intensity fluctuations with an array of small detectors. Its results supported the Hill solution [27]. A third method uses different types of laser to measure the wavelength correlation of intensity fluctuations and those measurements also agree with the model.

### 2.2.6.3 The Energy-input Region

The energy-input region is identified in Figure 2.6 and is defined by wavenumbers smaller than the inertial-range boundary $\kappa_0$. This is the region where the atmospheric wind field becomes unstable and creates large eddies that begin the cascading eddy-decay process. Unfortunately, we do not yet have

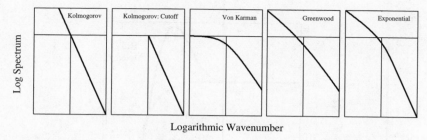

Figure 2.8: Five models for the spectrum plotted logarithmically. The vertical lines locate the outer-scale wavenumber.

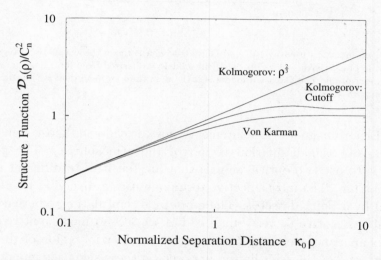

Figure 2.9: Calculated structure functions for three commonly used spectrum models plotted logarithmically versus the separation between sensors multiplied by the outer-scale wavenumber.

a physical understanding of the fluid mechanics which governs that process. We cannot ignore this region even though we have no *universal theory* with which to describe it. Estimates for the average scale length and the refractive-index variance are proportional to the first two moments of the spectrum. They depend strongly on the energy-input region. Descriptions of this region have relied on five analytical models that were constructed to join the inertial range spectrum (2.62) above the outer-scale wavenumber $\kappa_0$. They are summarized graphically in Figure 2.8, and structure functions for three of the models are plotted in Figure 2.9.

**The unrestricted Kolmogorov model** The first dissipation model employed extended the inertial-range behavior to zero wavenumber. It is illustrated in the first panel of Figure 2.8 and is described by the following expression:

$$\text{Kolmogorov:} \qquad \Phi_n(\kappa) = 0.033 C_n^2 \, \kappa^{-\frac{11}{3}} \qquad \text{for} \qquad 0 < \kappa < \kappa_{\text{s}} \qquad (2.71)$$

This choice facilitates many propagation calculations but generates serious problems in some applications. For instance, it leads to infinite results for the mean square fluctuation $\langle \delta n^2 \rangle$ and average eddy size $\langle \ell \rangle$. We shall learn later that it gives infinite values for the phase of electromagnetic signals. On the other hand, the signal amplitude and intensity are not sensitive to the outer-scale range and we can use this model with confidence to describe these important quantities.

The corresponding structure function is estimated by combining this model with (2.51):

$$\mathcal{D}_n(\rho) = 0.033 C_n^2 \times 8\pi \int_0^{\kappa_{\text{s}}} \frac{d\kappa}{\kappa^{\frac{5}{3}}} \left( 1 - \frac{\sin(\kappa\rho)}{\kappa\rho} \right)$$

When $\rho$ is smaller than the inner scale it begins quadratically as noted in (2.64). One must evaluate this expression numerically if the spacing is close to the inner scale length $\ell_0$. The separation is usually much greater than the inner scale length, in which case the upper limit is infinite and one recovers the 2/3 scaling law:

$$\mathcal{D}_n(\rho) = C_n^2 \rho^{\frac{2}{3}} \qquad \text{for} \qquad \ell_0 < \rho < \infty \qquad (2.72)$$

This differs from (2.43) only in that the separation range is now restricted by the inner scale. One can usually ignore that refinement and rely on the unconstrained version which is plotted in Figure 2.9. It is significant that this function does not bend over and reach an asymptotic value – a feature that is observed in terrestrial experiments. That flaw results from the unwarranted assumption made in running the Kolmogorov model down to zero wavenumber.

**The Kolmogorov model with a cutoff** To remedy the problems with the first model, we admit that we do not know what happens in the energy-input region and place a lower bound on the wavenumbers to which the spectrum applies. Doing so does not deny the existence of large energies. It simply acknowledges that we cannot describe their influence on refractive-index irregularities. It is important to note that this lower bound influences the prediction of some properties of the signal and by using it we cannot avoid the consequences of our ignorance:

$$\text{Sharp Cutoff:} \qquad \Phi_n(\kappa) = 0.033C_n^2\,\kappa^{-\frac{11}{3}} \qquad \text{for} \qquad \kappa_0 < \kappa < \kappa_s$$

$$(2.73)$$

This model is illustrated in the second panel of Figure 2.8. It has the virtue of providing finite expressions for the first two moments of the spectrum in terms of the outer-scale wavenumber $\kappa_0$. The variance of refractive-index fluctuations is calculated by combining this model with (2.48):

$$\langle \delta n^2 \rangle = 4\pi \int_{\kappa_0}^{\kappa_s} d\kappa\,\kappa^2 (0.033C_n^2\,\kappa^{-\frac{11}{3}})$$

Since the upper limit $\kappa_s$ is much greater than the outer-scale wavenumber in the atmosphere, we find

$$\text{Cutoff:} \qquad \langle \delta n^2 \rangle = 0.622C_n^2(\kappa_0)^{-\frac{2}{3}} \qquad\qquad (2.74)$$

One can measure this quantity at radio frequencies using microwave refractometers [5][6] and thereby provide experimental values for the parameter combination on the right-hand side. In a similar way one can estimate the average eddy size using (2.49) and the same approximations:

$$\text{Cutoff:} \qquad \langle \ell \rangle = 0.40\left(\frac{2\pi}{\kappa_0}\right) = 0.40L_0 \qquad\qquad (2.75)$$

This provides a way of estimating the outer-scale wavenumber, to the extent that one can identify $\langle \ell \rangle$ with the scale length identified in spatial correlation measurements. By combining these two measurements, one can estimate both $C_n^2$ and $\kappa_0$ with fair accuracy.

We have noted before that the structure function is the objective of many terrestrial measurements. It can be estimated by introducing our model into the defining relationship (2.51):

$$\mathcal{D}_n(\rho) = 0.033C_n^2 \times 8\pi \int_{\kappa_0}^{\kappa_s} \frac{d\kappa}{\kappa^{\frac{5}{3}}}\left(1 - \frac{\sin(\kappa\rho)}{\kappa\rho}\right) \qquad (2.76)$$

This expression yields a quadratic relationship when the product $\kappa_s\rho$ is small. Most measurements employ values of $\rho$ considerably larger than the inner scale, which allows one to send the upper limit to infinity. With a change of variables we find the following result:

$$\text{Cutoff:} \qquad \mathcal{D}_n(\rho) = 0.8294 C_n^2 \rho^{\frac{2}{3}} \int_{\kappa_0\rho}^{\infty} \frac{dx}{x^{\frac{5}{3}}} \left(1 - \frac{\sin x}{x}\right) \qquad (2.77)$$

When the combination $\kappa_0\rho$ is less than unity this reduces to the 2/3 scaling law established with the unrestricted Kolmogorov model. When the separation increases beyond the outer scale length one must do the integral numerically. The result is plotted in Figure 2.9 and exhibits the desired asymptotic behavior. That agrees generally with experimental results but one is left with an awkward situation. For this cutoff to work, one must assume that the spectrum *vanishes below* $\kappa_0$. This assumption denies the *influence* of irregularities in the energy-input region, which is quite different than saying that we do not know how to describe their effect. That problem has motivated the search for more satisfying descriptions.

**The von Karman model** This model *does* make a statement about large eddies while preserving the behavior of the spectrum in the inertial range. It was first proposed by von Karman [28] to describe the turbulence generated by a fluid flowing in a circular pipe. The diameter of the pipe provides a unique spatial scale in that application. That is a far cry from atmospheric turbulence, which is bounded only by the earth's surface and contains many scale lengths. On the other hand, there is some experimental support for its applicability to the troposphere and we will review such data later.

The von Karman model is the most widely used description for the energy-input region. It is easy to handle analytically and provides finite estimates for virtually all quantities of practical interest. This model is defined by

$$\Phi_n(\kappa) = 0.033 C_n^2 (\kappa^2 + \kappa_0^2)^{-\frac{11}{6}} \qquad \text{for} \qquad 0 < \kappa < \kappa_{\text{s}} \qquad (2.78)$$

which undergoes a smooth transition to the inertial-range result (2.62) for wavenumbers larger than $\kappa_0$. It is presented graphically in the third panel of Figure 2.8. It goes flat below $\kappa_0$ and approaches a constant value:

$$\lim_{\kappa \to 0} [\Phi_n(\kappa)] = 0.033 C_n^2 (\kappa_0)^{-\frac{11}{3}}$$

Notice that this is just the value of the previous models at the outer-scale wavenumber.

The first two moments of the von Karman model are finite and give the following estimates for the variance of refractive-index fluctuations and average eddy size:

$$\text{von Karman}: \qquad \langle \delta n^2 \rangle = 0.523 C_n^2 (\kappa_o)^{-\frac{2}{3}}$$

$$\langle \ell \rangle = 0.476 \left( \frac{2\pi}{\kappa_0} \right) = 0.476 L_0$$

The numerical constants differ very little from those established with the sharp-cutoff model and measurement uncertainties probably mask their small differences.

For a tighter test of the von Karman model we turn to the structure function, which is expressed as follows:

$$\mathcal{D}_n(\rho) = 8\pi \int_0^{\kappa_s} d\kappa \, \kappa^2 \left( 1 - \frac{\sin(\kappa\rho)}{\kappa\rho} \right) \left( \frac{0.033 C_n^2}{\left( \kappa^2 + \kappa_0^2 \right)^{\frac{11}{6}}} \right)$$

The spacing is usually greater than the inner scale length and one can let the upper limit go to infinity. The result is expressed in terms of the MacDonald function discussed in Appendix D:

$$\text{von Karman}: \qquad \mathcal{D}_n(\rho) = 1.0468 \, C_n^2 (\kappa_0)^{-\frac{2}{3}} \left( 1 - \frac{2^{\frac{2}{3}}}{\Gamma\left(\frac{1}{3}\right)} (\kappa_0 \rho)^{\frac{1}{3}} K_{\frac{1}{3}}(\kappa_0 \rho) \right)$$

$$(2.79)$$

This result is also plotted in Figure 2.9. When the separation is less than the outer scale $\kappa_0 \rho < 1$ the curve follows the traditional 2/3 scaling law.[4] The expression (2.79) approaches a constant asymptotic value when the spacing is larger than the outer scale length:

$$\kappa_0 \rho \gg 1: \qquad \mathcal{D}_n(\rho) = 1.0468 \, C_n^2 (\kappa_0)^{-\frac{2}{3}} \qquad (2.80)$$

Although the von Karman model has the general behavior that we consider desirable, we need to test it more carefully before embracing it.

---

[4] One can also use the small-argument expansion for the MacDonald function given in Appendix D to verify that

$$\mathcal{D}_n(\rho) = C_n^2 \rho^{\frac{2}{3}} \qquad \text{for} \qquad \kappa_0 \rho \ll 1$$

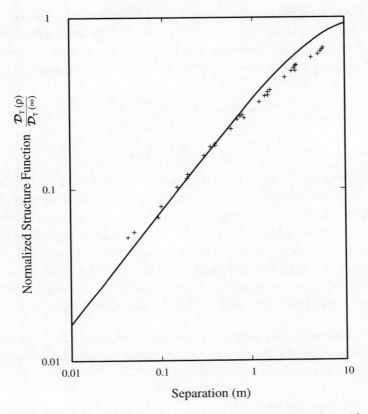

Figure 2.10: A temperature structure function measured 3 m above the surface with nine sensors deployed horizontally by Greenwood and Tarazano [29]. The prediction for a von Karman model with $\kappa_0 = 4.12$ m$^{-1}$ is overplotted.

A careful measurement of the temperature structure function was made near Rome, New York [29]. Nine temperature sensors were mounted on a beam 2.26 m above a grassy field with separations in the range

$$4 \text{ cm} < \rho < 6 \text{ m}.$$

Judicious choice of the sensors' locations provided simultaneous measurements of the structure function for thirty-two separations. The experiment used sensors composed of silver plating on 2.5-μm platinum wire, which produced responses between 300 and 900 μs, depending on wind conditions. The noise level and drift were established by immersing one sensor in a room-temperature oil bath, which showed that atmospheric temperature fluctuations were 20 dB above the sensor noise level. The actual data from one of these runs is reproduced in Figure 2.10.

The outer-scale wavenumber is estimated from such data in the following manner. One first extrapolates the 2/3 portion of the actual data curve until it intercepts the asymptotic value and notes the separation $\rho_0$ for which this occurs. Equating this to the predicted value for the asymptote (2.80) gives

$$C_n^2(\rho_0)^{\frac{2}{3}} = 1.0468\, C_n^2(\kappa_0)^{-\frac{2}{3}}$$

or

$$\kappa_0 = 1.0710/\rho_0. \qquad (2.81)$$

This identification requires that one measure the asymptotic value of the structure function $\mathcal{D}_n(\infty)$ quite carefully. A fitted value of $\kappa_0 = 4.12\ \mathrm{m}^{-1}$ was estimated in this way for the data in Figure 2.10 and the corresponding theoretical prediction from (2.79) is overplotted on the experimental points. The prediction departs from the data for spacings greater than 1 m, suggesting that the von Karman model is not a tight fit to the data and other runs confirmed that observation. Notice also that the deviation for small separations cannot be explained by inner-scale effects since $\ell_0 < 1$ cm.

**The Greenwood–Tarazano model** Data like that presented in Figure 2.10 prompted Greenwood and Tarazano to consider other analytical models [29]. They found that the spectrum

$$\Phi_n(\kappa) = 0.033 C_n^2 (\kappa^2 + \kappa_0 \kappa)^{-\frac{11}{6}} \qquad \text{for} \qquad 0 < \kappa < \kappa_s \qquad (2.82)$$

gave considerably better fits to their structure-function and power-spectrum measurements. This model is portrayed in the fourth panel of Figure 2.8 and suggests that there is a rising influence for large eddies, which behave as $\kappa^{-\frac{11}{6}}$ below the outer-scale wavenumber. The refractive-index variance and average scale length are given by the following expressions:

$$\text{Greenwood} - \text{Tarazano:} \qquad \langle \delta n^2 \rangle = 0.554 C_n^2 (\kappa_0)^{-\frac{2}{3}}$$

$$\langle \ell \rangle = 4 \left( \frac{2\pi}{\kappa_0} \right) = 4 L_0$$

The integral that defines the average eddy size is barely convergent, which is reflected in the relatively large coefficient. The structure function for this model is described by the hypergeometric function $_2F_3(a, b,\ c, d, e; z)$ [30]. Other propagation computations also involve these functions which are difficult for most engineers and scientists to use. That analytical challenge might explain why the model is often cited but seldom used in propagation studies.

**The exponential model** A different analytical form was used frequently in research in the USSR, primarily for analytical convenience [31]:

$$\Phi_n(\kappa) = 0.033 C_n^2 \kappa^{-\frac{11}{3}}[1 - \exp(-\kappa^2/\kappa_0^2)] \quad \text{for} \quad 0 < \kappa < \kappa_s$$

$$(2.83)$$

This model is portrayed in the fifth panel of Figure 2.8 and exhibits an increasing influence of large eddies following $\kappa^{-\frac{5}{3}}$ below $\kappa_0$. It is close to the Greenwood–Tarazano model. Many theoretical calculations were performed both with this form and with the von Karman model. Experiments in the USSR using collimated laser signals gave better agreement with the von Karman model than with the exponential form [32].

The variance and scale size for this model are given by the following expressions:

$$\text{Exponential:} \quad \langle \delta n^2 \rangle = 0.842 C_n^2 (\kappa_0)^{-\frac{2}{3}}$$

$$\langle \ell \rangle = 1.644 \left( \frac{2\pi}{\kappa_0} \right) = 1.644 L_0$$

The structure function is expressed in terms of Kummer functions:

$$\text{Exponential:} \quad \mathcal{D}_n(\rho) = C_n^2 \rho^{\frac{2}{3}} - 1.6846 C_n^2 \kappa_0^{-\frac{2}{3}}[M(-\tfrac{1}{3}, \tfrac{3}{2}; -\rho^2 \kappa_0^2/4) - 1]$$

$$(2.84)$$

Only the first term is important when the spacing is less than the outer scale length and yields the usual 2/3 scaling law. One must use the asymptotic expansion for the Kummer function when the separation is relatively large:

$$\mathcal{D}_n(\rho) = 1.6846 C_n^2 \kappa_0^{-\frac{2}{3}}$$

This is similar to the expression for the von Karman model. Other properties of the signal are expressed in terms of Kummer functions [30].

**A perspective** The reader might well ask where we are in this forest of formulas. At the outset, we were frank enough to note that such models were little more than analytical conjectures – at least when they were first introduced. They were used to make propagation estimates in the absence of a universal theory for this region. A good deal of effort has been expended to work out their consequences.

In subsequent years they have been tested in two ways. Measurements of temperature and refractive-index fluctuations near the surface provide structure functions for these quantities. The von Karman and Greenwood–Tarazano models agree reasonably well with such data. It would be helpful to extend those measurements to larger separations since such data should be more sensitive to the energy-bearing eddies.

The second important class of tests comes from measurements made on electromagnetic signals that pass through the atmosphere. The signal phase is well suited to this purpose because it is very sensitive to large eddies. To exploit this possibility, one must trace through the consequences of the various spectrum models and compare them with propagation data. Much of the effort in this first volume is directed to that end.

Optical signals propagating near the surface probably give one the best chance of discovering hints about the refractive-index spectrum in the energy-input range. The power spectrum of single-path phase fluctuations provides a sensitive test for different turbulence models. It was measured with a laser signal in a single *published* experiment [33], which is examined in Section 6.2.3. Those results suggest a model that goes flat below $\kappa_0$ and are consistent with the von Karman model. Phase- difference experiments provide a less sensitive test but represent a simpler measurement challenge. We noted above that results of collimated laser sources monitored by phase-locked receivers agreed better with the von Karman model than they did with the exponential model [32]. Referring to their plots in Figure 2.8, this suggests that a model that goes flat below $\kappa_0$ is preferable to one that keeps rising.

These limited experiments provide a hint about the spectrum's behavior in the energy-input region. There is a good deal of flexibility in the data and it can accommodate several interpretations. It is clear that large eddies are present and are quite influential for some measurements. We shall often use the von Karman spectrum to estimate signal parameters that are sensitive to this region in our work. We do so because we judge that it is a *reasonable* model and because it is analytically tractable.

Describing the energy-input region of the turbulence spectrum evidently remains a major challenge. It requires a fundamental examination of the processes by which the eddies are created. One can be sure that it will be necessary to recognize the anisotropic nature of large eddies. It is a problem for experts in fluid dynamics and atmospheric physics to solve. Their objective should be to establish a *universal theory* for the range $\kappa < \kappa_0$ but it is not clear that this is possible. A great deal of propagation data is available for its validation when such a theory emerges.

### 2.2.7 Describing Anisotropic Irregularities

Our description of irregularities presented thus far has assumed that they are isotropic. In the troposphere this means that the correlation of refractive-index fluctuations in the vertical and horizontal directions should be the same. It is important to determine whether experimental data confirms that assumption.

The spatial correlation of refractive-index variations was measured with microwave refractometers close to the ground in conjunction with a series of propagation tests conducted on Maui [34]. Five identicalinstruments were used, each oriented so that the prevailing wind carried atmospheric samples through their cylindrical cavities. They were first deployed vertically, then horizontally and normal to the wind direction, and finally parallel to the wind vector. Unfortunately these measurements could not be made simultaneously since it took a considerable amount of time to redeploy the instruments. The spatial correlations were measured in three orthogonal directions for spacings in the range 16 cm $< \rho <$ 456 cm and are reproduced in Figure 2.11.

The correlations measured normal and parallel to the wind direction are quite different. This is not necessarily an indication of anisotropy. The wind

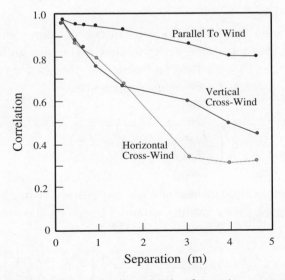

Figure 2.11: The spatial correlation of refractive-index fluctuations measured with microwave refractometers 1.7 m above the surface by Thompson, Kirkpatrick and Grant [34]. These data sets were taken at different times with five 3.2-cm-cavity refractometers deployed along the orthogonal directions noted.

carries refractive-index irregularities along with it and similar features are measured by successive instruments until they have time to rearrange themselves. This horizontal transport is called *advection* and will be analyzed in Chapter 6. Downwind stretching distorts the correlation measurement in the manner suggested by Figure 2.11.

We look for evidence of anisotropy in the two cross-wind measurements. The vertical and horizontal cross-wind correlations are similar for spacings less than 1.5 m and diverge beyond that point. Because these measurements were not made simultaneously, one can only conclude that the vertical and horizontal correlation lengths are about the same. This suggests that the eddies which are important to such measurements are reasonably isotropic. In these volumes we will often assume isotropy for propagation near the surface.

Anisotropy of the refractive index was measured directly in a series of optical experiments [35][36]. A collimated laser beam was transmitted over a short distance near the surface. Image motion was measured simultaneously in the vertical and horizontal directions. These measurements were sensitive to the scale length in that direction. Their ratios showed that the horizontal scale length is approximately twice the vertical value. This suggests that the eddies can be visualized as the compressed *circular pillows* suggested in Figure 2.12. If one ignores downwind stretching, the surfaces of constant correlation are oblate spheroids. This observation suggests a way to characterize them.

Anisotropic media were first analyzed in conjunction with studies of the ionosphere [37]. Plasma irregularities in that region are highly elongated in the direction of the magnetic field and were described by a spatial correlation that depends on *stretched rectangular coordinates*:

$$C\left(\sqrt{\left(\frac{x}{a}\right)^2 + \left(\frac{y}{b}\right)^2 + \left(\frac{z}{c}\right)^2}\right) \tag{2.85}$$

Surfaces of constant spatial correlation are ellipsoids in this formulation. One can learn more by examining a model in which the outer scale length appears explicitly:

$$C(x, y, z) = \exp\left\{-\frac{1}{L_0^2}\left[\left(\frac{x}{a}\right)^2 + \left(\frac{y}{b}\right)^2 + \left(\frac{z}{c}\right)^2\right]\right\}$$

This Gaussian model *does not* describe ionospheric irregularities accurately and is used here only for illustration. The dimensionless constants $a, b$ and $c$

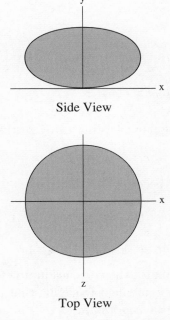

Side View

Top View

Figure 2.12: A graphical description of anisotropic irregularities near the surface suggested by image motion measured by Lukin with collimated laser beams in orthogonal directions [36].

are called *scaling parameters.* They carry the burden of representing the anisotropy in this format by assuming different values. These parameters are proportional to the correlation lengths in each direction:

$$L_x = aL_0 \qquad L_y = bL_0 \qquad L_z = cL_0 \qquad (2.86)$$

It was soon realized that the ellipsoidal model (2.85) could also describe tropospheric structures. To represent the horizontal symmetry suggested in Figure 2.12 we take $a = b \neq c$. This formulation seems to catch the *essence* of anisotropy and is well suited to analytical operations. It is not unique, but is consistently employed to analyze anisotropic media [38] [39][40].

Most propagation calculations depend on the spectrum description of irregularities. We need to know how the ellipsoidal model (2.85) is reflected in wavenumber coordinates. To answer that question we combine it with the spatial covariance expression (2.44) and take the inverse Fourier transformation:

$$\Phi_n(\boldsymbol{\kappa}) = \frac{1}{(2\pi)^3} \int_{-\infty}^{\infty} dx \int_{-\infty}^{\infty} dy \int_{-\infty}^{\infty} dz \, \langle \delta n^2 \rangle C\left(\sqrt{\left(\frac{x}{a}\right)^2 + \left(\frac{y}{b}\right)^2 + \left(\frac{z}{c}\right)^2}\right)$$
$$\times \exp[i(\kappa_x x + \kappa_y y + \kappa_z z)]$$

If one rescales the rectangular coordinates by setting

$$x = ax' \qquad y = by' \qquad z = cz'$$

the anisotropic spectrum becomes

$$\Phi_n(\boldsymbol{\kappa}) = \frac{1}{(2\pi)^3} \int_{-\infty}^{\infty} dx' \int_{-\infty}^{\infty} dy' \int_{-\infty}^{\infty} dz' \, \langle \delta n^2 \rangle C\left(\sqrt{(x')^2 + (y')^2 + (z')^2}\right)$$
$$\times abc \, \exp[i(a\kappa_x x' + b\kappa_y y' + c\kappa_z z')].$$

In the transformed coordinates the correlation function depends only on the scalar separation. That is equivalent to saying that the spectrum is isotropic in the *stretched wavenumber coordinates*

$$\boldsymbol{\kappa}' = \mathbf{i}_x(a\kappa_x) + \mathbf{i}_y(b\kappa_y) + \mathbf{i}_z(c\kappa_z)$$

and the anisotropic spectrum can be written in the following form:

$$\Phi_n(\boldsymbol{\kappa}) = abc\Phi_n\left(\sqrt{a^2\kappa_x^2 + b^2\kappa_y^2 + c^2\kappa_z^2}\right) \tag{2.87}$$

To see how it should be applied, suppose that the von Karman model (2.78) describes the eddies when they are isotropic. The stretched-wavenumber transformation suggests that the anisotropic version of the same model is

$$\Phi_n(\boldsymbol{\kappa}) = 0.033C_n^2 abc\left(a^2\kappa_x^2 + b^2\kappa_y^2 + c^2\kappa_z^2 + \kappa_0^2\right)^{-\frac{11}{6}}. \tag{2.88}$$

This evidently reduces to the isotropic form when $a = b = c = 1$. It helps to identify the outer-scale wavenumbers that might be measured in the three directions:

$$\kappa_{0_x} = a\kappa_0 \qquad \kappa_{0_y} = b\kappa_0 \qquad \kappa_{0_z} = c\kappa_0 \tag{2.89}$$

We believe that the large eddies in the energy-input range are elongated, reflecting the directional properties of the mean wind field which generates the turbulence. The relations (2.86) and (2.89) give us a way to recognize the likelihood of this hypothesis. In doing so, we must remember that we still have no universal physical description for this region. All we have done here is to embroider anisotropic features on analytical assumptions about the large-eddy region. It is all we now have and we will work with it quite effectively. Nonetheless, we should not forget the fragile foundation on which it rests.

### 2.2.8 Inhomogeneous Random Media

Our descriptions thus far have assumed that the random medium is homogeneous. That is a reasonable assumption for links near the surface, at least for the path lengths usually employed. On the other hand, homogeneity is not a valid assumption for the very large spacings often exploited by microwave interferometry. Turbulent conditions can be quite different at their widely spaced receivers, which sometimes span entire continents. The argument that inhomogeneous conditions do not influence the structure function clearly breaks down for these large baselines. The level of turbulent activity also changes with altitude, reflecting the different rates of input and dissipation of energy. Optical and microwave signals that travel through the atmosphere are influenced differently by the varying conditions they encounter. For these important applications we cannot describe the irregularities by using the homogeneous model (2.30).

To enlarge the description of random media, we return to the covariance expression (2.28). The inhomogeneity notwithstanding, we can represent the covariance as the product of a local variance and the spatial correlation:

$$\langle \delta\varepsilon(\mathbf{r}_1)\, \delta\varepsilon(\mathbf{r}_2)\rangle = \langle \delta\varepsilon^2\rangle C(\mathbf{r}_1, \mathbf{r}_2) \tag{2.90}$$

We appeal to experimental data to gauge the relative variability of the two terms. In the troposphere one can measure them directly with refractometers or indirectly using temperature and humidity sensors. The refractometer data in Figure 2.11 shows that the correlation function changes significantly within just a few meters. By contrast, the variance term changes quite slowly in the horizontal direction. In the vertical direction it can change by one or two orders of magnitude within a kilometer. That rate of change is large but still quite modest relative to the correlation function and indicates that the variance is relatively constant over separations for which the correlation goes to zero. The variance can be expressed as a function of the average position, which is usually the mean altitude of the points. The correlation depends primarily on the distance between the sensors and seems to have the same form at widely separated points. This indicates that we can represent the covariance as the product of a slowly changing variance term evaluated at the midpoint and a correlation that changes rapidly with the separation:

$$\langle \delta\varepsilon(\mathbf{r}_1)\, \delta\varepsilon(\mathbf{r}_2)\rangle = \left\langle \left[\delta\varepsilon\left(\frac{\mathbf{r}_1+\mathbf{r}_2}{2}\right)\right]^2\right\rangle C(\mathbf{r}_1-\mathbf{r}_2) \qquad (2.91)$$

This type of atmosphere is called a *locally homogeneous random medium with smoothly varying characteristics* [41]. It is also labeled *quasi- homogeneous* in some treatments [42]. The analytical approach suggested by (2.91) is widely used to describe atmospheric conditions.

We need to see how the assumption (2.91) is reflected in the wavenumber description of irregularities since most propagation studies depend on that approach. Tatarskii showed how to do this by absorbing the gradual change of the variance into the turbulence spectrum [41]:

$$\langle \delta\varepsilon(\mathbf{r}_1)\, \delta\varepsilon(\mathbf{r}_2)\rangle = \int d^3\kappa \, \Phi_n\!\left(\boldsymbol{\kappa}, \frac{\mathbf{r}_1+\mathbf{r}_2}{2}\right) \exp[i\boldsymbol{\kappa}\cdot(\mathbf{r}_1-\mathbf{r}_2)] \qquad (2.92)$$

No approximations have been made up to this point. We shall frequently use this expression as our starting point. Notice that the propagation integrals over the spatial coordinates of $\mathbf{r}_1$ and $\mathbf{r}_2$ must now operate both on the traditional exponential term *and* on the position-dependent spectrum. That additional burden is manageable in most problems using numerical methods, although it depends on knowing a good deal about the atmosphere.

The basic description (2.92) is often simplified by assuming that the entire effect of the inhomogeneous behavior is reflected in a changing refractive-index structure constant. In this approximation one would write the von Karman spectrum with an inner scale factor in the following way:

$$\Phi_n\!\left(\boldsymbol{\kappa}, \frac{\mathbf{r}_1+\mathbf{r}_2}{2}\right) = C_n^2\!\left(\frac{\mathbf{r}_1+\mathbf{r}_2}{2}\right)\frac{0.033\mathcal{F}(\kappa\ell_0)}{\left(\kappa^2+\kappa_0^2\right)^{\frac{11}{6}}} \qquad (2.93)$$

The greatest variation in $C_n^2$ occurs with altitude and the following replacement is usually made:

$$C_n^2\!\left(\frac{\mathbf{r}_1+\mathbf{r}_2}{2}\right) \to C_n^2(z) \qquad (2.94)$$

A good deal of effort has been expended on measuring the profile of $C_n^2(z)$ and we will review the results of those experiments in Section 2.3.3.

The concentration of effort on $C_n^2$ profiles has deflected attention from the potential variation of the inner scale length $\ell_0$ and outer scale length $L_0$.

The outer scale changes from a few meters near the surface to several kilometers in the free atmosphere. Moreover, the eddies undergo a transition from nearly isotropic to highly elongated as one ascends. In some problems involving transmission through the atmosphere, the outer-scale variability $L_0(z)$ exerts an important influence on the measured quantities in conjunction with $C_n^2(z)$. In using models like (2.93) it is well to keep in mind that all three parameters as well as the anisotropy depend on height.

## 2.3 Tropospheric Parameters

To establish the properties of turbulent irregularities in the lower atmosphere we can rely on direct measurements. Temperature, humidity and refractive index have been measured near the surface at many locations. These have been extended to considerable height by routine instrumented meteorological balloon flights, occasional airborne sampling programs and ground-based reflective probing. As our understanding of propagation in random media has matured, the process has been reversed. Microwave and optical techniques now provide the most reliable way to study turbulent conditions in the troposphere.

### 2.3.1 Refractive-index Expressions

One can measure the refractive index at radio frequencies with the microwave refractometer [5][6]. These instruments are relatively expensive and are seldom employed. The usual procedure is to exploit standard meteorological measurements [43] and convert them into the refractive index by using basic relationships established by laboratory experiments. From 1 MHz to 30 GHz one finds that the refractive index is independent of frequency and given by

$$\text{Microwave:} \qquad n = 1 + \frac{77.6}{T}\left(p + 4810\frac{e}{T}\right)10^{-6} \qquad (2.95)$$

where $T$ is the temperature measured in kelvins [44][45]. The atmospheric pressure $p$ is measured in millibars. The partial pressure of water vapor which describes humidity is denoted by $e$ and is also expressed in millibars. These quantities are routinely measured as functions of altitude by atmospheric sounding. One can take differentials of this expression to relate fluctuations in refractive index to those in $p$, $T$ and $e$:

$$\delta n \times 10^6 = \delta T\left[-\frac{77.6}{T^2}\left(p + 9,620\frac{e}{T}\right)\right] + \delta p\left(\frac{77.6}{T}\right) + \delta e\left(\frac{373,256}{T^2}\right)$$

As a practical matter, one can disregard pressure variations, which leaves the refractive index solely dependent on temperature and humidity.

At optical and most infrared frequencies one can ignore the humidity term in (2.95). On the other hand, molecular resonances generate a small dispersion term that multiplies the previous result:

$$\text{Optical:} \quad n = 1 + 77.6\,\frac{p}{T}\left(1 + \frac{7.52 \times 10^{-3}}{\lambda^2}\right)10^{-6} \qquad (2.96)$$

Here $\lambda$ is measured in micrometers [46][47]. Notice that the microwave result (2.95) is inherently larger because it contains the water-vapor term. That is reflected in significantly larger ground-level values for the refractive index. Since water vapor disappears rapidly with height, one expects the microwave and optical results to merge as one ascends in the troposphere.

Fluctuations in the refractive index at optical wavelengths are related primarily to temperature variations, which are large relative to pressure changes:

$$\delta n = \delta T\left[-77.6\,\frac{p}{T^2}\left(1 + \frac{7.52 \times 10^{-3}}{\lambda^2}\right)\right]10^{-6}$$

This permits one to connect the structure constants for temperature and refractive index:

$$C_n^2 = C_T^2\left[77.6\,\frac{p}{T^2}\left(1 + \frac{7.52 \times 10^{-3}}{\lambda^2}\right)\right]^2 10^{-12} \qquad (2.97)$$

Small additional wavelength-dependent corrections have been developed but are not required in our work [48].

### 2.3.2  *Values of $C_n^2$ Near the Surface*

The structure constant $C_n^2$ is the most important turbulence parameter for tropospheric propagation. We begin by reviewing measurements of $C_n^2$ near the surface. Microwave values are important for the worldwide network of microwave communication links – especially as they move to higher frequency – that propagate from one tower to another along line-of-sight paths. This parameter defines the ultimate accuracy of distance-measuring systems used for geodetic control. It is also responsible for the scintillation experienced by millimeter-wave systems used for target acquisition and missile guidance. The corresponding optical values for $C_n^2$ influence the resolution of infrared thermal imaging systems, which are now widely used in military operations and increasingly for civilian applications. They also define the accuracy of laser range finders and laser-guided short-range missiles.

High-powered lasers are limited by atmospheric turbulence and the ground-level value of $C_n^2$ is uniquely important for horizontal transmission of these signals.

### 2.3.2.1 Microwave Values of $C_n^2$ Near the Surface

Refractometer readings taken at a single resonance frequency can be used over a wide range of microwave frequencies since the index of refraction is independent of wavelength from approximately 1 MHz to 30 GHz. These instruments yield time histories of the random function $\delta n(t)$. The problem is that of how to estimate $C_n^2$ from such readings. With two refractometers this is a simple task. The following expression gives $C_n^2$ directly from spaced refractometer readings provided that the separation is intermediate between the inner and outer scale lengths:

$$\langle [\delta n(\mathbf{r} + \boldsymbol{\rho}, t) - \delta n(\mathbf{r}, t)]^2 \rangle = C_n^2 |\boldsymbol{\rho}|^{\frac{2}{3}} \qquad \text{for} \qquad \ell_0 < \rho < L_0 \qquad (2.98)$$

Microwave refractometers are more expensive than meteorological instruments and this motivated the search for ways to measure $C_n^2$ with a single refractometer. One can do so with expressions like (2.74) if one knows the outer scale length $L_0$. The outer scale is approximately the same as the height of the instrument near the surface. This gives only a crude estimate of $C_n^2$ but provides a good way to monitor its variability with time since $L_0$ is relatively constant.

A more accurate technique uses a single refractometer in combination with a wind-speed indicator. This approach uses Taylor's hypothesis, which assumes that the eddies are frozen during the measurement and are carried past the instrument at constant speed $v$ by the wind.[5] The power spectrum of refractive-index fluctuations can be expressed in terms of the wavenumber spectrum by the following relationship, as explained in Problem 2:

$$W_n(\omega) = \frac{4\pi^2}{v} \int_{\omega/v}^{\infty} d\kappa \, \kappa \Phi_n(\kappa) \qquad (2.99)$$

For the Kolmogorov model with a sharp outer-scale cutoff defined by (2.73) the power spectrum assumes different forms depending on the ratio $\omega/(\kappa_0 v)$:

$$\omega < \kappa_0 v: \qquad W_n(\omega) = 0.782 C_n^2 \frac{1}{v} (\kappa_0)^{-\frac{5}{3}}$$

$$\omega > \kappa_0 v: \qquad W_n(\omega) = 0.782 C_n^2 v^{\frac{2}{3}} \omega^{-\frac{5}{3}} \qquad (2.100)$$

---

[5] This approximation is examined in Chapter 6 of this volume.

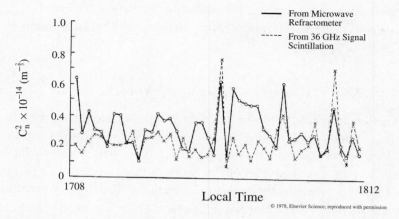

Figure 2.13: A history of $C_n^2$ values measured with a microwave refractometer and from intensity scintillation of a 36-GHz signal by Ho, Mavrokoukoulakis and Cole [49]. The microwave signal was transmitted over a 4.1-km path near London with an average height of 50 m. The refractometer data was taken near ground level and combined with wind-speed data.

This approach to constructing the power spectrum was taken in an experiment using refractometer data taken near the surface [49]. The results agreed with the predicted $-5/3$ scaling in the high-frequency range. Independent wind-speed measurements were then used in the second expression to provide the instantaneous values of $C_n^2$. They are reproduced in Figure 2.13 and indicate that there is a good deal of variability during the hour of sampling. This data gives the following average value for the structure constant for microwave frequencies:

$$\text{Microwave} \qquad \overline{C_n^2} = 0.32 \times 10^{-14} \text{ m}^{-\frac{2}{3}} \qquad \text{Refractometer} \qquad (2.101)$$

This estimate was confirmed by a completely independent measurement. Fluctuations of field strength were measured for a microwave signal at 36 GHz transmitted over a 4.1-km path near London [49]. The variations of logarithmic amplitude for a plane wave are related to the structure constant, path length and electromagnetic wavenumber by the following formula:[6]

$$\langle \chi^2 \rangle = 0.307 \, C_n^2 \, k^{\frac{7}{6}} R^{\frac{11}{6}} \qquad (2.102)$$

This determines the instantaneous value of $C_n^2$ since the other parameters can be measured quite accurately. A simultaneous history of the structure constant generated in this way is compared with refractometer data in

---

[6] This relationship is established in Chapter 3 of the second volume.

Figure 2.13. They agree quite well. The average value determined for the radio data was close to the average value of the refractometer data:

$$\text{Microwave:} \qquad \overline{C_n^2} = 0.25 \times 10^{-14} \text{ m}^{-\frac{2}{3}} \qquad \text{Radio Link} \qquad (2.103)$$

These measurements cover only a short sample. Their variability suggests that even larger variations should be expected over periods of days or months. This question gave rise to an experiment using the same 36-GHz link that lasted five months [50]. The refractometer readings were replaced by less expensive temperature and humidity sensors, which provided estimates of the refractive index through (2.95). The short-term radio and meteorological values for $C_n^2$ again agreed remarkably well. Over this longer period it was possible to measure $C_n^2$ under both stable and convective conditions. The radio link gave the following average values:

$$\text{Microwave:} \qquad \overline{C_n^2} = 1.2 \times 10^{-14} \text{ m}^{-\frac{2}{3}} \text{ Stable Conditions}$$
$$\overline{C_n^2} = 6.6 \times 10^{-14} \text{ m}^{-\frac{2}{3}} \text{ Convective Conditions}$$
$$(2.104)$$

These results indicate that there is a substantial difference in the structure constant for the two types of meteorological conditions.

### 2.3.2.2 Optical Values of $C_n^2$ Near the Surface

As yet there is no direct analog of the microwave refractometer for optical wavelengths. The preceding discussion suggests using amplitude fluctuations of a laser signal to determine the instantaneous values of $C_n^2$. The simple formula (2.102) is valid only for weak scattering, which restricts one to path lengths less than several hundred meters. A new problem arises for these short paths. When diffraction is important the basic unit of length is the Fresnel length $\sqrt{\lambda R}$. For optical signals and a path length of 200 m the Fresnel length is about 1 cm, which is comparable to the inner scale length $\ell_0$. This means that one must include the dissipation-range spectrum factor $\mathcal{F}(\kappa\ell_0)$ introduced in (2.63) when one is estimating the logarithmic amplitude variance. That requirement invalidates the simple expression (2.102).

**Optical scintillometer measurements of** $C_n^2$ One can resolve this dilemma by using an optical scintillometer [51]. These instruments employ incoherent transmitting and receiving apertures that are considerably larger than the inner scale length. The influence of small eddies is removed by a process called *aperture averaging*.[7] The dissipation spectrum factor $\mathcal{F}(\kappa\ell_0)$

---

[7] This topic is discussed in Chapter 3 of the second volume.

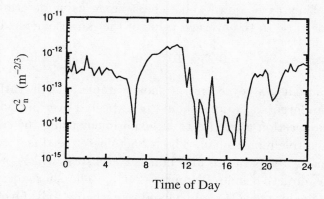

Figure 2.14: Diurnal variations of $C_n^2$ measured near the earth's surface with an optical scintillometer by Churnside [46]. The points here are 15-min averages taken near Boulder, Colorado in the summer.

is then replaced by a wavenumber-weighting term that depends on the diameter of the aperture. For a spherical wave signal and the Kolmogorov models one finds that the logarithmic amplitude variance is

$$\langle \chi^2 \rangle = 0.124 C_n^2 k^{\frac{7}{6}} R^{\frac{11}{6}} G\left(\frac{D_{\mathrm{r}}}{\sqrt{\lambda R}}\right) \tag{2.105}$$

where $D_{\mathrm{r}}$ is the diameter of the large apertures and $G(\eta)$ is a known function. One can measure all the quantities in this expression except $C_n^2$. A time history of $C_n^2$ established in this way is reproduced in Figure 2.14 and indicates that it changes by *three orders of magnitude* in a twenty-four hour cycle.

**The distribution of synoptic changes in $C_n^2$** The considerable variability evident during a single day encouraged measurements over longer time scales to gauge the influence of changing weather conditions. This was first done in the USSR during July and August [52]. Temperature differences among six sensors deployed vertically on a meteorological tower at heights ranging from 0.5 to 12 m were measured. This data gave $C_T^2$, which was converted into the refractive-index structure constant using the relationship (2.97). Thirty-minute data averaging was used to construct distributions for $C_n^2$ during three different periods: midday, late afternoon and evening. The largest values were found at midday and were spread over less than two orders of magnitude. The distribution broadened in late afternoon to cover three orders of magnitude and its most likely value dropped by a factor of ten. The evening distribution is the broadest of all, spanning four orders of

Table 2.1: Monthly average values of $\log(C_n^2)$ presented by Fante [54], based on temperature measurements made by Spencer in New York, 2 m above the surface [53]

| Month | Dawn | Day | Dusk | Night |
|---|---|---|---|---|
| February | $-14.26$ | $-12.76$ | $-13.59$ | $-14.13$ |
| March | $-14.34$ | $-12.70$ | $-13.86$ | $-14.33$ |
| April | $-13.79$ | $-12.79$ | $-13.59$ | $-13.67$ |
| May | $-13.66$ | $-12.56$ | $-13.53$ | $-13.49$ |

magnitude in $C_n^2$ with essentially the same average value as that found in late afternoon. This wide range indicates that the level of refractive-index fluctuations is as variable as the meteorology which generates the turbulent conditions.

Similar temperature measurements were made in New York over a four-month period [53]. Temperature differences among three probes deployed horizontally 2 m above the ground were taken. Instantaneous values of $C_T^2$ were estimated from the measured structure functions and converted to $C_n^2$ [54]. Monthly averages formed from this data are reproduced in Table 2.1. These values agree generally with average values of the Russian data [52]. Notice that the diurnal variation shown in Figure 2.14 is substantially larger than the values indicated in the table. The reason is fairly obvious; monthly averages smooth the swings evident in instantaneous histories. One should be sensitive to both sets of data in estimating $C_n^2$. The important thing to recognize is that electromagnetic measurements are determined by the level of $C_n^2$ which exists during the time they are being made. To achieve agreement between theory and experiment one should measure $C_n^2$ independently, as was done in the experiment of Figure 2.13.

It is significant that few of these long-term distributions are fitted by a Gaussian model. That is not surprising when one thinks about it. The conditions which generate atmospheric turbulence are changing *sequentially* as the weather itself evolves with time. A Gaussian distribution usually results from a large number of independent influences acting *simultaneously* on the measured quantity. That is evidently not the situation for temperature-difference fluctuations.

**The distribution of short-term fluctuations of $\delta n$** In view of the foregoing discussion one should examine the distribution for *rapid* fluctuations in the random medium. The averaging time exerts a powerful influence on estimates of the probability density function (PDF) and we expect the distribution of rapid temperature fluctuations to be quite different than the 15- and 30-min averages that were used above. The short-term PDF of the temperature difference was measured 1.5 m above the surface for two sensors 3 cm apart [55]. The sensors had frequency responses of 200–500 Hz depending on wind conditions and negligible averaging was performed. The result was completely different than a Gaussian distribution, perhaps for the same reasons as those noted above. Moreover, scintillation experiments done on short paths show that $C_n^2$ changes in an intermittent manner and that its distribution is log normal[8] [21]. In view of these observations, we must question the familiar assumption made in (2.24), namely that $\delta\varepsilon$ is a Gaussian random variable with zero mean.

Fortunately, the precise description for the PDF does not matter a great deal. The electromagnetic signals are represented by integrals of $\delta n$. These integrations tend to smooth out the differences among the various possible distributions for $\delta n$ in most applications. The quantities measured are determined almost entirely by the level of $C_n^2$ experienced during the time when the experiment is performed.

**Intermittency** Our descriptions of tropospheric properties thus far have assumed that $C_n^2$ does not change *during* the course of an electromagnetic experiment. The actual situation is often quite different and one observes significant changes in $C_n^2$ during the measurement interval. This variability manifests itself in two ways. Recall that we considered gradual temperature changes in Figure 2.3. We learn from direct measurements that $C_n^2$ also changes *suddenly*.

Thermal plumes occur in the troposphere near the surface in a spontaneous manner. They are best represented by cloud-like structures whose horizontal dimension is measured in tens or even hundreds of meters. These buoyant thermal plumes can extend to considerable heights and are often felt by low-flying aircraft. Large-scale temperature inhomogeneities were studied carefully in Russia with six sensors mounted on a vertical mast [56]. Sudden changes in the temperature profiles, which occurred randomly in time, were noted. This is called *intermittency* and is an intrinsic property of turbulent media. The surges were observed as temperature jumps that appeared sequentially at the sensors. This was interpreted as a *temperature ramp* moving through the vertical array. These ramps are imagined to be responsible for transferring heat from the surface to the atmosphere. The same variability was observed by airborne measurements made on a clear

---

[8] The log-normal distribution is described in Appendix E of the second volume.

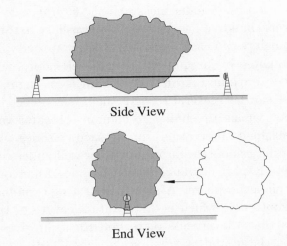

Side View

End View

Figure 2.15: An intermittent structure is shown moving through an electromagnetic path. In some cases these structures are identified with thermal plumes or temperature ramps. At microwave frequencies they are associated with columns of water vapor.

day at 400 m. The data exhibited *temperature pulsations* and demonstrated that intermittency is not limited to the surface [57].

These transient structures cause rapid changes in refractivity when they move through a propagation path as suggested in Figure 2.15. That change is rapidly reflected in the phase and amplitude of a received signal. Ground-based laser experiments show that $C_n^2$, $\ell_0$ and $L_0$ can change rapidly under such conditions [21]. These fluctuations are quite different than the diurnal pattern of Figure 2.14 and the synoptic variations suggested in Table 2.1. A statistical distribution of the turbulence parameters is therefore required in order to characterize the short-term behavior of electromagnetic signals.[9] Short data samples are often used in order to minimize the influence of nonstationary conditions but they are especially sensitive to this variability. In addition, short path lengths are frequently used to ensure that weak scattering conditions pertain, but they are strongly influenced by a large structure that can dominate the propagation physics. Its influence is reduced roughly by the ratio $\ell/R$ when the horizontal link is long relative to the size of the structure.

Microwave signals are influenced by *columns of water vapor* in the convective boundary layer in an analogous way. Microwave interferometers see sudden shifts in phase difference when such a structure intersects the signal path to one receiver but not the other. This results in a brief change in

[9] We examine these distributions in Chapter 10 of the second volume.

apparent source position. Hinder measured the lateral scale of these irregularities to be several hundred meters [58]. The corollary to this inconvenience is that interferometers can easily monitor their occurrence [59]. Anomalous refraction is also noted by large-aperture radio telescopes, which experience brief, unexpected jumps in pointing. The result is an angular shift of a few arc seconds that can last several seconds [60]. This effect is attributed to moist air packets, which are similar to columns of water vapor. The best approach is to suspend observations during such episodes.

The physics of these spontaneous events is not well understood. Some hold that they are simply part of the confusing mosaic of meteorological conditions in a dynamic atmosphere. A more penetrating examination identifies them with intermittency, which is an essential feature of turbulence. The basic concept is that the process of turbulence itself gives rise to bursts of energy both at large and at small scale sizes [7][61]. These events are considered to be a natural consequence of the Navier–Stokes equations and are overlaid on the routine eddy-subdivision process described by Figure 2.5. Their influence on electromagnetic propagation can be quite surprising and accounts for the sudden *improvement* of image quality and bursts of scattering intensity that are sometimes observed [62][63]. This field is an active area of research that is still evolving, but it is beyond our horizon.

### 2.3.3 *Vertical Profiles of* $C_n^2$

Electromagnetic scintillation was first addressed in connection with astronomical observations made at optical wavelengths. This concern is shared today by a large community of astronomers working at microwave, millimeter-wave, infrared and optical wavelengths. These signals are used to create images of distant galaxies and to perform spectroscopy on their constituent molecules. The electromagnetic waves emitted must travel through the terrestrial atmosphere before reaching ground-based receivers. This means that each level in the troposphere can influence the result and is dominated by the values of $C_n^2$ at successive levels. A good deal of effort has therefore gone into establishing its height profile both for optical and for microwave frequencies.

#### 2.3.3.1 *Vertical Profiles of Microwave Values for* $C_n^2$

In retrospect it was clear from tropospheric scatter communication links that $C_n^2$ must decrease rapidly with altitude [64]. Field-strength measurements showed that there is a steady decline with distance as the common scattering volume was forced to greater heights by radio-horizon limitations. The specific height profile of $C_n^2$ is needed in order to interpret several types of microwave measurement. Radar tracking of silent missiles and spacecraft is limited by angular errors imposed in the troposphere if good range and Doppler data is not available. In Chapter 7 we shall learn that angular errors are proportional to the integral of $C_n^2$ taken over all heights. The resolution of large radio telescopes is limited in the same way. Ground-based radars are occasionally used to image spacecraft in earth orbit and they are limited by the atmosphere in two ways. The quality of their images is set by the angular error and by amplitude fluctuations introduced by tropospheric irregularities. In the second volume we will find that the amplitude variance is proportional to $C_n^2$ times the altitude raised to the power 5/6 and this product integrated over all heights. As communication-satellite signals move to higher frequencies and wider information bandwidths, amplitude fluctuations are becoming important for their signals. Residual range errors for GPS position location depend on the profile of $C_n^2$ and the outer scale length through a more complicated expression, which is developed in Chapter 4. Synthetic-aperture radars are now used widely for military missions and often for scientific purposes. These radars are carried in aircraft, by the space shuttle and in specially designed earth satellites. As these carriers move along their flight paths, the radar views the target through different combinations of atmospheric eddies. That variability blurs the radar images and their ultimate quality is set by the microwave profile.

**Airborne refractometer measurements** The vertical profile of $C_n^2$ was first measured directly by a single 9.4-GHz microwave refractometer carried aloft in a light aircraft [65]. The airplane was often flown on a level flight path at constant speed and successive refractometer readings provided the structure function of $\delta n$ as a function of the delay time. One can ignore changes in the random medium between such readings because the airspeed is so much greater than the eddy velocity. This allows the time delay to be directly related to the horizontal separation and airspeed. The basic relation (2.43) for a fixed altitude then becomes

$$\mathcal{D}_n(\tau, z) = C_n^2(z)|v\tau|^{\frac{2}{3}} \tag{2.106}$$

and path-averaged values of $C_n^2$ were determined for each flight altitude. Measurements were made from ground level to $29\,000$ ft over Colorado and Florida, generally under clear-air conditions. Some flights employed level runs 5 km long made at height intervals of 1000 ft. A second type of flight pattern used gradual climbing and descents to sample the profile. A large range of values was encountered in doing so:

$$10^{-17} \mathrm{m}^{-\frac{2}{3}} < C_n^2 < 3 \times 10^{-13} \ \mathrm{m}^{-\frac{2}{3}} \tag{2.107}$$

These measurements are approximately two orders of magnitude greater than the optical values below 10 000 ft because of the strong contribution of water vapor to the microwave refractivity.

The smallest values were measured at the highest altitude sampled and a general increase was observed as the plane descended. However, the most striking feature of these profiles was the significant variation with altitude and conspicuous layering that was observed. Maximum values were usually encountered in one or more elevated layers. Not surprisingly, these layers often coincided with temperature inversions or jumps in concentration of water vapor. The layer heights changed from day to day, although the maximum values were invariably found below 10 000 ft. The minimum values occurred above that level. The measured profiles changed significantly from one day to the next, reminding us that turbulence processes are governed by atmospheric meteorology.

These results indicate that one must actually measure the local profile at the time of observation to achieve close agreement bet- ween predictions and electromagnetic transmission measurements. Since aircraft operations are relatively expensive, it is not feasible to make airborne refractometer flights in conjunction with the numerous experiments that are conducted around the world. That reality stimulated the search for techniques that could measure the profile with surface-based instruments.

**Radar backscatter measurements** By 1935 radio pulses reflected from ionized regions above 80 km were being used on a regular basis to investigate ionospheric structure and to predict high-frequency propagation conditions. When microwave radars were developed during the Second World War, it was assumed that only metallic aircraft could reflect their much higher frequencies. Radar reflections from rain were observed somewhat later and soon became a useful tool for meteorologists.

As the power of pulsed radars increased, reports of echoes from apparently clear atmospheric regions became more frequent. In some cases, these radar detections coincided with aircraft reports of significant clear-air turbulence (CAT). It seemed logical to identify the mechanical turbulence of CAT with refractive irregularities because of the strong coupling between turbulent velocity and irregularities of a passive scalar suggested by (2.57). Microwave transmissions well beyond the optical horizon were observed in 1950 and tropospheric scatter propagation was developed to provide secure military communications during the following decade. This type of propagation was attributed to electromagnetic scattering by turbulent refractive irregularities in the common volume of the transmitter and receiver beams, in large part because of the random nature of the received signals. Scientific interest in this phenomenon and concern about the air-safety implications of CAT stimulated a vigorous examination of the phenomenon.

Initial theoretical predictions were based on a number of simplifications. The irregularities were assumed to be isotropic. They were also assumed to be in the far field of the radar and their largest scale was assumed to be small relative to the available scattering volume. This meant that many eddies could contribute to the backscattered signal. The ratio of the received to the transmitted power for a fixed radar range $R$ is expressed in terms of the antenna's effective capture area $A_e$ and the range-gate distance $\delta R$ by the following [66]:

$$\frac{P_R}{P_T} = \frac{A_e \, \delta R}{4\pi R^2} [8\pi^2 k^4 \Phi_n(2k, R)] \qquad (2.108)$$

The term in square brackets is the radar cross section per unit volume and is proportional to the spectrum of refractive irregularities. It is specialized in two important ways. The pulse range gate selects values of the spectrum that lie at a distance $R$ from the radar. This allows one to focus the measurements on one slice of the atmosphere at a time. More fundamentally, the cross section depends on the spectrum evaluated at the *Bragg wavenumber*, which is uniquely important for backscattering:

$$\kappa = 2k = 4\pi/\lambda$$

This selection of a single wavenumber by the scattering process is portrayed in Figure 2.16 and can be understood on physical grounds.

The outgoing radar signal encounters a wide variety of eddy sizes in the atmosphere. Consider first the eddies whose size is exactly half the wavelength of the radar signal. Waves that are scattered back to the radar by these eddies will all be in phase with one another and generate a strong return. By contrast, waves scattered by other eddies that are smaller or larger than half the wavelength will interfere with one another and result in negligible reflected signal. The actual range of eddy sizes that satisfy this resonance criterion is quite narrow and the backscatter wavenumber filter indicated in Figure 2.16 reflects the eddy selection made by the radar signal.

It is desirable that the backscattering filter fall within the inertial range since one has a confident model for the turbulence spectrum in that region of the spectrum. The candidate frequencies chosen for sounding radars thus

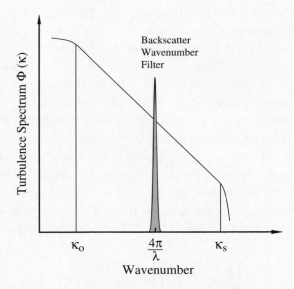

Figure 2.16: A typical spectrum of refractive irregularities illustrating the narrow-band filter at the Bragg wavenumber which characterizes backscattering. The wavelength of the sounding radar is usually chosen to ensure that this filter falls within the inertial range of the spectrum.

fall in the range 50–3000 MHz. The corresponding resonant eddy sizes lie in the range

$$5 \text{ cm} < \ell < 3 \text{ m}$$

and are therefore comfortably between the inner and outer scale lengths – at least near the surface.[10] This means that one can confidently use the Kolmogorov model (2.62) in most practical situations and the power ratio is given by the following scaling law:

$$\frac{P_R}{P_T} = 0.379 \frac{A_e \, \delta R}{4\pi R^2} \, C_n^2(R) \lambda^{-\frac{1}{3}} \tag{2.109}$$

This wavelength dependence was confirmed in a pioneering experiment that compared backscattered signals measured simultaneously at 10.7 and 71.5 cm [67]. This was further confirmed in three experiments in which FM/CW radar returns were correlated to independent measurements of $C_n^2$. The first comparison used data taken on a 300-m meteorological tower [68]. The second compared radar-derived values of $C_n^2$ with those measured by a tethered balloon [69]. The third used a refractometer suspended from a helicopter

---

[10]  We shall learn in Section 2.3.4 that the inner scale length increases with altitude to values approaching 10 cm. To probe the upper atmosphere, sounding radars usually operate at the lower frequencies for this and other reasons.

Table 2.2: Characteristics of sounding radars used to measure winds aloft and profiles of $C_n^2$ at microwave frequencies (the heights are measured relative to local ground level and $P$ is the peak pulse power)

| Radar | Location | $f$ (MHz) | $P$ (kw) | $H_{\min}$ (km) | $H_{\max}$ (km) | $\Delta H$ (m) | Ref. |
|---|---|---|---|---|---|---|---|
| Millstone | Westford, MA | 1295 | 4000 | 5 | 30 | 1800 | 72 |
| Sunset | Boulder, CO | 40 | 125 | 5 | 15 | 1000 | 73 |
| Jicamarca | Peru | 50 | 1000 | 15 | 85 | 5000 | 74 |
| Arecibo | Puerto Rico | 430 | 2000 | 10 | 30 | 150 | 75 |
| Arecibo | Puerto Rico | 2380 | (400) | 14 | 19 | 30 | 76 |
| Millstone | Westford, MA | 440 | 1000 | 7 | 20 | 1500 | 77, 78 |
| Poker Flat | Alaska | 50 | 15 | 4 | 20 | 750 | 79 |
| Platteville | Colorado | 50 | 15 | 4 | 20 | 750 | 79 |
| SOUSY | Mobile, NM | 49 | 200 | 2 | 22 | 150 | 80 |
| SOUSY | West Germany | 53 | 600 | | | 150 | 81 |
| MU Radar | Japan | 46 | 1000 | | | | 82 |
| Chung-Li | Taiwan | 52 | 180 | 2 | 12 | 150 | 83 |
| Weather | Denver, CO | 915 | 6 | 0.5 | 10 | 150 | 84 |

which flew toward the radar which measured its range-gated return before the instrument reached that point [70]. The agreement was satisfactory in each case.

With these confirmations one can confidently use the relationship (2.109) to infer profiles of microwave values for $C_n^2$ from radar power measurements for different slant ranges. These are easily converted to vertical profiles by using concurrent pointing information from the radar. This means that one can make continuous measurements of the vertical profile at a chosen site. With modern technology the system can often be automated and this made comprehensive studies of $C_n^2$ profiles economically feasible [71]. Existing VHF and UHF radars with adequate power and antenna area were first exploited for such studies.[11] Special-purpose sounding radars were then built to explore atmospheric irregularities. The characteristics of radars used in this program are summarized in Table 2.2.

A typical profile of microwave values for $C_n^2$ is reproduced in Figure 2.17. This data was taken in New Mexico with the mobile SOUSY radar which had a range resolution of 150 m [80]. The fine-grain profile measured in this way exhibits considerable variation with altitude. Notice that $C_n^2$ changes by two orders of magnitude between 2 and 8 km. For heights between 2 and

---

[11] Birds and insects generate measurable radar echoes at wavelengths less than 10 cm. Although these observations are useful for entomology and ornithology, the primary concern with CAT detection suggested that synoptic studies should be made at VHF and UHF frequencies in order to avoid these reflections.

*Waves in Random Media*

22 km the measured values fell in the range

$$10^{-19} \text{ m}^{-\frac{2}{3}} < C_n^2 < 5 \times 10^{-14} \text{ m}^{-\frac{2}{3}}. \qquad (2.110)$$

These values agree generally with the airborne refractometer measurements (2.107), which necessarily covered a smaller height range.

Radar profiles measured at the same site exhibit significant diurnal and seasonal change. As the network of sounding radars expanded it became clear that there is also considerable geographical variability, probably due to the variability of the concentration of water vapor at different locations. Pulsed Doppler pulse radars were built at Platteville, Colorado and Poker Flats, Alaska to investigate the climatology of microwave $C_n^2$ profiles [79]. Monthly averages of $C_n^2$ were obtained from 3 to 15 km in Colorado and from 4 to 19 km in Alaska. Profiles were taken at each site for approximately one year and established a synoptic guide to refractive variability in the lower atmosphere. Monthly mean values varied by an order of magnitude

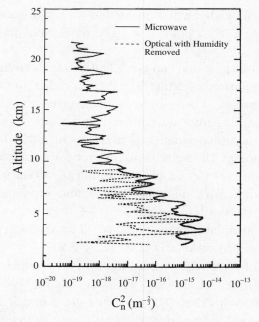

Figure 2.17: Vertical profiles of $C_n^2$ measured by Eaton, Peterson, Hines, Peterman, Good, Beland and Brown [80]. The solid curve is the microwave profile measured with a pulse Doppler radar operating at 49.25 MHz and represents the combined influence of humidity and temperature fluctuations. The effect of water vapor has been removed in the dotted curve using sounding data and should represent the $C_n^2$ profile for optical wavelengths. The two profiles are quite different below 10 km but merge above that level. This data was taken in New Mexico on 3 August 1985 and represents 15-min averages. Altitude is measured relative to local ground level.

during the course of a year. The largest values were found in winter, with a secondary peak in the summer. The maximum profile value usually occurred between 5 and 6 km, with an average decline of 2 dB km$^{-1}$ above that level. Monthly means give only a general guide and Figure 2.17 demonstrates that the instantaneous profile is a good deal more complicated.

Short-term variability is also an important feature of profile measurements. Radar soundings show that $C_n^2$ sometimes changes by a factor of ten in less than an hour [73]. The probability distribution of these variations is found to be log normal at all heights and times [79], in agreement with the surface measurements noted earlier.

Comparison of vertical and oblique radar soundings indicates that there is a substantial aspect sensitivity of the returns [83][85]. There are two explanations for these observations – and they may really be the same. One view holds that turbulent refractive irregularities coexist with relatively thin stable layer structures that are coherent in the horizontal plane. Vertical probing should sense reflections from both types of structure. Oblique signals would be returned primarily by the turbulent eddies because of the strong aspect sensitivity of reflections from stable layers or sharp refractive gradients. The second explanation accepts anisotropy as a fundamental feature of the troposphere, with the eddies becoming horizontally elongated as one ascends. There is considerable experimental data taken *in situ* that supports this description, as noted in Section 2.3.5. Backscattering would come primarily from highly elongated irregularities in this model. Viewed vertically, one would see the short outer scale lengths which characterize the eddies in that direction. Oblique echoes would be more sensitive to the much larger horizontal outer scale lengths. In this connection, it is significant that radar soundings have provided convincing evidence of acoustic gravity waves traveling horizontally in the troposphere [73].

As the transmitter power and antenna size of sounding radars increased, it became possible to explore ever higher regions of the atmosphere. The large fixed dish at Arecibo was used with both 430 and 2380 transmitters to measure refractive scattering to 30 km [75][76]. With high power and a large array, the Jicamarca radar in Peru has obtained reflections from the region between 10 and 35 km and also that between 55 and 85 km [74]. This powerful radar is able to probe even higher regions using incoherent Thomson scattering by free electrons in the ionosphere.

Theoretical research has also provided more realistic descriptions of the process of backscattering. The crude description (2.108) has been extended to include the actual radar-antenna pattern and signal waveform [86]. In establishing the cross-section approximation (2.108) it was assumed that the

Fresnel length $\sqrt{\lambda R}$ is large relative to the correlation length of the eddies. While this may be true for vertical scale lengths, it is probably not valid for the large horizontal correlation lengths which are measured at sounding altitudes. That assumption was avoided in a more general description, which was also able to treat backscattering by anisotropic irregularities [87]. This refinement is important for wind profilers that operate at relatively small slant ranges.

The introduction of pulse Doppler radars provided the capability for measuring winds aloft from the surface. One can measure the power spectrum of the backscattered signal with this technology. The first moment of the power spectrum gives the wind speed at the selected altitude. By varying the height one can measure the vertical profile of the wind speed. With a dual-beam pulse Doppler radar one can also measure the direction of the wind vector. This development formed the basis for a network of automated wind-measuring radars in the USA. In 1990 NOAA established the Wind Profiler Demonstration Network of 32 radars operating at 404 MHz. That system is gradually being replaced by an operational network of unattended wind-profiler radars operating at 449 MHz.

### 2.3.3.2 *Vertical Profiles of Optical Values for $C_n^2$*

The profile of optical $C_n^2$ values is important for the vast majority of signals passing through the atmosphere. Light from distant stars is influenced by tropospheric temperature fluctuations in a fundamental manner. Variances for angle-of-arrival and intensity fluctuations are important parameters for optical astronomy and both depend on weighted height averages of the $C_n^2$ profile at night. Similar problems arise when ground-based telescopes are used to image the small moons of neighboring planets using reflected sunlight. Ground-based telescopes are occasionally used to image artificial satellites in earth orbit using either sunlight or laser illumination.

Similar concerns were addressed for optical systems that look down from earth orbit. The most important group is strategic reconnaissance satellites that employ large optical telescopes operating primarily in the visible band [88]. Satellite systems are less sensitive to atmospheric scintillation than are surface-based telescopes looking up, for reasons that are not immediately apparent. This conclusion might change as their resolution improves still further. A second group of scientific satellites monitors surface and atmospheric conditions for environmental studies. Their sensors operate primarily in the infrared band. Warning satellites in geosynchronous orbit also rely on infrared signatures of missile launching but are not now limited by propagation considerations.

These applications account for only part of the intense activity that has been generated in order to model $C_n^2$ profiles at optical wavelengths. A variety of experimental programs to measure the profile throughout the day for different seasons, especially in New Mexico, has been sponsored. The analytical community has been engaged for several decades on working to convert these measurements into models that can be used to simulate system performance. The applications which explain this extensive effort are the high-power laser pointer/tracker systems and laser weapons being developed for military purposes. Both ground and airborne laser systems are being developed. Their ability to hold a laser beam steady on a moving target is limited by refractive irregularities in the atmosphere. A great deal of scientific work has therefore been sponsored in order to investigate those limitations.

If the microwave data on $C_n^2$ provides a good guide, the optical profiles are both complicated and variable. This means that one must find ways to actually measure the vertical profiles of $C_n^2$ at optical wavelengths. There are three ways to do so. The first depends on temperature-difference measurements and the conversion established in (2.97). The second uses optical instruments sighted on stellar point sources and tries to invert atmospheric diffraction patterns to recover the profile. The third approach converts microwave sounding data to optical profiles by subtracting the contribution of water vapor to $C_n^2$. To forecast diurnal and seasonal variability, the third method was used to convert a five-year data set of radar soundings [89].

**Pioneering experiments** Experiments to measure optical values directly were first done in the USSR. To explore the convective boundary layer, temperature was measured at various heights on a 300-m meteorological tower [90]. To explore the free atmosphere, a temperature sensor was mounted externally on an airplane that flew level courses at various altitudes [91]. Horizontal aircraft motion created the required separation for the temperature readings. The data showed that the structure constant decreased by a factor of three or four as the altitude increased.

**Thermosonde balloons** Instrumented balloons provided a major improvement in the quality and quantity of profile data. The first experiment used a simple arrangement that has been emulated in all subsequent measurements

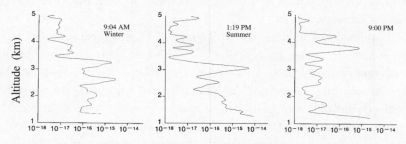

Vertical Profiles of Refractive Index Structure Function $C_n^2$

Figure 2.18: Profiles of $C_n^2$ derived from data taken by a balloon-borne instrument that measured the temperature structure constant $C_T^2$. The two profiles on the left were measured on clear sunny days over the New Mexico desert. The profile on the right-hand side was measured on a clear calm night in the same location. This data was assembled by Beland [48].

[92]. Two platinum-wire temperature sensors were mounted 1 m apart on an instrument module that was suspended 300 m beneath the balloon to avoid wake effects. Temperature difference was measured continuously as the balloon rose gradually from 1 to 15 km. The data was returned by a ave link and then converted to $C_n^2$. This system measured profiles over New Mexico for three days. The results exhibited significant structure and layering, although their major features were similar. The same technique was then used over Australia to 8 km [93], and later over France, Portugal and the Canary Islands to heights of 25 km [94].

These dual-sensor balloon payloads are called *thermosondes* and are flown with some regularity at several locations. They often reach altitudes of 30 km and provide height resolution as fine as 20 m. Typical profiles of optical values for $C_n^2$ measured in this way for different times and seasons are reproduced in Figure 2.18. The data exhibits the same complexity as that noted in microwave profiles.

Thermosonde data provides the most accurate and widespread optical profiles for $C_n^2$ and is widely used for modeling. Despite their detail and accuracy, balloon soundings pose several operating problems. By their very nature, they provide occasional profiles and can miss significant diurnal variations. Notice that values at a fixed level can change by a factor of 100 from one night to the next. Balloon launching and retrieval necessarily requires a skilled team. The payloads are sometimes lost. These considerations stimulated a search for a ground-based instrument that could perform the same task more often and more cheaply.

**Profiles measured with stellar scintillation** The optical analog of a radar is the pulsed laser or lidar. It seemed that one could aim the lidar upward and measure the radiation reflected from various heights. The return would be sensitive to the optical values of $C_n^2$ and thus require no correction. The problem with this idea is apparent from Figure 2.16. Because the optical wavelength is so small, the Bragg wavenumber falls well beyond $\kappa_s$ and there is not enough energy in the dissipation spectrum to provide a measurable return.

Ancient peoples observed shadow patterns of alternating dark and light bands that are cast on white walls during solar eclipses. This is one manifestation of the diffraction patterns created by atmospheric irregularities. This suggested that the spatial correlation of amplitude fluctuations could be measured with two or more telescopes using a bright star. Analysis showed that the result should depend on the height integral of the turbulence spectrum, weighted by a function that depends on altitude and telescope separation in a known way. It was hoped that one could invert this relationship to establish the profile from the measured correlation. Prolonged efforts to do so have been frustrated by the mathematical sensitivity of the inversion process.[12]

The group led by Ochs in Boulder then made an important discovery [95]. They reasoned that, by placing a *spatial filter* in front of a telescope, they could isolate the origin of the speckle pattern measured at the focal plane to a particular level in the atmosphere. One could then shift the emphasis from one level to another by changing the spatial filter. They demonstrated this concept with a 36-cm Schmidt–Cassegrain telescope. The spatial filters consisted of 0.5-mm aluminized stripes with 1-mm spacings. With a strong point source they were able to measure $C_n^2$ for seven broad altitude bins between 2 and 15 km.

This remarkable instrument is called a *stellar scintillometer*. The profiles provided by it were compared with simultaneous profiles generated by radars and balloons [77][80]. It was found that the scintillometer correctly determines the general trend with altitude but often misses layered structures. This meant that one can provide *rough-cut* estimates for the profile. On the other hand, one can do so on a continuous basis and monitor profile changes as the night proceeds. The results were sufficiently encouraging to deploy the first instrument to New Mexico, where it was used operationally for seven months [96]. Over 1000 profiles were taken with sources close to the zenith. Since a good deal of variability was observed, the cumulative

---

[12] This subject is explored in Chapter 4 of the second volume.

probability of $C_n^2$ was assembled for each altitude bin. The distribution was log normal at each height, in agreement with the radar observations.

Other techniques for determining the $C_n^2$ profile have been demonstrated [97]. A double star provides two sets of ray paths through the atmospheric irregularities. Crossed paths give the mathematical leverage required to invert the speckle patterns. This *Scidar* method was tested with two telescopes carried aloft by a balloon and the results agreed fairly well with thermosonde data. The problem with Scidar is that pairs of stars are seldom visible at the desired site when the profile is needed. Another method exploits the leverage offered by two-color observations of speckle patterns. This method works best at small elevation angles, for which strong scattering is often a problem.

**Profile modeling** The measured profiles invite scientific explanation. Simirity considerations explain conditions in the convective boundary layer with some confidence [98]. The scaling of $C_n^2$ with height should change from $z^{-\frac{2}{3}}$ to $z^{-\frac{4}{3}}$ near the surface as the meteorological conditions go from stable to neutral to unstable. That prediction is confirmed by experiments [90]. These results are of limited practical value because system designers must invariably describe a long propagation path through the atmosphere.

To meet the need of large laser systems, a variety of analytical models and computer routines has been developed in order to describe the $C_n^2$ profile from sea level to 30 km. This is a daunting task and the program is far from complete. Beland provides a good description of progress in this area and the interested reader is encouraged to consult his review [48].

### 2.3.4 Inner-scale Measurements

Since Kolmogorov's breakthrough, considerable importance has been attached to measuring the inner scale length for atmospheric turbulence. It leads to estimates for the energy-dissipation rate of temperature luctuations through the following expression:

$$\ell_0 = 7.4 \left( \frac{\nu^3}{\mathcal{E}} \right)^{\frac{1}{4}} \tag{2.111}$$

The kinematic viscosity $\nu$ is measured independently and the proportionality constant is taken from a careful review of many determinations [22]. We judge that the inner scale is only a few millimeters near the surface. It is challenging to make temperature sensors that are small enough and sensitive enough to measure the structure function for separations that small. o complicate matters, one must make such measurements without disturbing the tiny, weak eddies one is examining. Intermittent conditions exert a strong influence on single-point measurements and we shall find that these are an important consideration.

Electromagnetic scintillation measurements provide a welcome solution for this problem. sing collimated optical signals propagating at constant height, they can probe the atmosphere on exceedingly small scales without disturbing the random medium. This method averages the readings spatially over the entire path and provides more representative values for irregular terrain. Several techniques have been devised to do so [22].

The first practical method exploited two optical signals transmitted over the same 260-m path near Boulder [25]. The signals were designed to probe different portions of the turbulence spectrum. The first signal was a coherent spherical wave generated by a He–Ne laser. It was monitored by a point-like receiver 1 mm in diameter, which is less than both the inner scale and the Fresnel length. The logarithmic amplitude fluctuations of this signal were especially sensitive to the atmospheric fine structure. A second incoherent signal at 0.94 μm used 4.4-cm circular lenses both for transmission and for reception. These diameters are substantially larger than the Fresnel length for the short optical paths employed in this experiment. The second signal was responsive to larger scale sizes in the inertial range because the influence of the fine structure was suppressed by aperture averaging. The ratio of their scintillation levels is independent of $C_n^2$ and should depend only on the ratios of the aperture diameter and inner scale to the Fresnel length:[13]

$$\ell_0/\sqrt{\lambda R} \quad \text{and} \quad D_\mathrm{r}/\sqrt{\lambda R}$$

A time history of $\ell_0$ measured with this technique is reproduced in Figure 2.19 and exhibits significant diurnal variation:

$$\mathrm{h} = 1.5 \text{ m:} \quad 2.5 \text{ mm} < \ell_0 < 10 \text{ mm} \tag{2.112}$$

The data also exhibits rapid variations that are quite large. Our notion of a constant parameter must be discarded and $\ell_0$ regarded as a variable, in the same way as that in which we have come to view the structure constant. This means that we can only specify an average spectrum when describing atmospheric turbulence.

[13] The estimation of scintillation levels for various types of signal, with and without aperture averaging, is explored in Chapter 3 of the next volume.

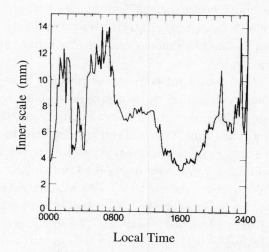

Figure 2.19: The diurnal variation of the inner scale length measured by Ochs and Hill with two optical signals near Boulder [25]. The propagation path was 1.5 m above the surface and the data points are 10-min averages.

This measurement was repeated using a 150-m path at a height of 4 m [26][99]. The shorter path length allowed the experiment to probe somewhat deeper into the dissipation range. The laser data was augmented with four different types of *in situ* micrometeorological measurements. The results of this highly redundant experiment were self-consistent and gave inner scale lengths that are essentially the same as those in Figure 2.19:

$$\mathsf{h} = 4 \text{ m:} \qquad 3.5 \text{ mm} < \ell_0 < 12 \text{ mm} \tag{2.113}$$

A second technique used the spatial correlation of laser signals to measure the inner scale length [21]. Two diverged He–Ne laser signals traveled along the same 50-m path at a height of 1.5 m. The spatial covariance of one signal was monitored by an array of eight photodiodes. For short path lengths and small diode spacings this covariance should be uniquely sensitive to the energy-loss region. The second laser signal was monitored by four relatively large circular apertures and the variance of its scintillation was influenced primarily by the inertial range. Comparing these measurements gave values in the range $2 \text{ mm} < \ell_0 < 9 \text{ mm}$.

An important conclusion emerged from this short-path experiment. It was observed that the inner scale length is an intermittent function of time and the distribution of $\ell_0$ is log normal [21]. This is not too surprising. In the absence of significant spatial averaging along the path, we know that scintillation can be dominated by intermittent phenomena. That increases the challenge for describing the turbulence spectrum since we now know that both of its defining parameters are intermittent random functions.

A third approach is based on comparing the scintillation experienced by laser signals operating at *different* wavelengths [22]. There are two ways to do so and both use short paths to ensure that weak scattering conditions pertain. The first method takes the ratio of scintillation variances measured with the two signals after averaging over a suitable interval to minimize the effects of intermittency. This ratio is independent of $C_n^2$ and depends only on

$$\frac{\lambda_1}{\lambda_2} \quad \text{and} \quad \frac{\ell_0}{\sqrt{R\lambda_{\mathrm{m}}}}$$

where $\lambda_{\mathrm{m}}$ is the mean wavelength. The inner scale length is easily extracted from that expression. The second version of this approach uses instantaneous amplitude values for both signals to form the frequency correlation function, which should also depend on the dimensionless parameters indicated above.[14] These bichromatic methods are most accurate when the wavelengths are widely separated, suggesting that one use a visible laser in combination with an infrared laser. This strategy is compromised by the occurrence of significant dispersion between widely separated wavelengths [27].

One sometimes needs values for the inner scale length measured at specific heights above the surface. The inner scale depends on the kinematic viscosity

$$\nu = \mu/\rho \tag{2.114}$$

where $\mu$ is the coefficient of viscosity and $\rho$ is the local air density. If the energy-dissipation rate is held constant, the inner scale length should vary almost inversely with density:

$$\ell_0 = \frac{\text{constant}}{\rho^{\frac{3}{4}}} \tag{2.115}$$

This suggests that the inner scale length should increase with altitude [72]. A five-month study of the altitude profile for $\ell_0$ was done using a VHF radar at White Sands, New Mexico [100]. The inner scale length was extracted from its backscattered signals and increased exponentially from $\ell_0 = 1$ cm

---

14 This expression is established in Chapter 6 of the second volume.

at 5 km to $\ell_0 = 8$ cm at 19 km. The standard exponential model of the atmosphere would predict this behavior with the above expression. The measured profiles exhibited little seasonal variation.

### 2.3.5 Anisotropy and the Outer Scale Length

We come now to the most elusive parameter which is used to characterize atmospheric turbulence. The outer scale length is imprecisely defined for several reasons. In discussing the wavenumber spectrum of irregularities, we noted that there is no universal physical theory available to describe the energy-input region. The outer-scale wavenumber $\kappa_0 = 2\pi/L_0$ is simply the value below which the well-established inertial-range result gives way to ignorance. This is more a warning than a definition. In actual spatial coordinates, the outer scale length is identified with the spacing beyond which the structure function departs from the 2/3 law. That departure is apparent in the temperature-difference data reproduced in Figure 2.10. This negative definition is often quite helpful in using direct measurements of temperature and refractivity to suggest values for $L_0$.

The outer scale length is sometimes identified with the distance at which the spatial correlation falls to half its initial value. By that standard, $L_0$ appears to be about 1 or 2 m in the temperature and refractivity measurements made near the earth, whose results were reproduced in Figures 2.2 and 2.11.

A third approach is to trace the consequences of the 2/3 law through the propagation integrations to see how properties of the received signal should scale with independent variables. For example, the 2/3 law implies that the variance of the phase difference between two receivers should scale as the 5/3 power of their separation. Departures from these derived expressions thus provide another way to estimate the outer scale length, although it is not as sensitive to $L_0$ as are direct refractivity measurements.

The larger problem is that the atmosphere is not isotropic. This means that one will often measure different outer scale lengths in the vertical and horizontal directions. Although the anisotropy is usually not significant near the surface, the horizontal scale can be a hundred times larger than the vertical component in the free atmosphere. It is for this reason that we combine our discussion of anisotropy and the outer scale length.

An additional complication is that the value for the outer scale length depends very much on the way one measures it. Sounding radars and interferometers provide quite different estimates for $L_0$. They do so because one measures the vertical component and the other the horizontal component.

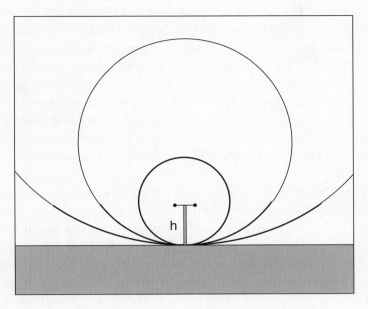

Figure 2.20: A refractivity measurement made near the surface, showing the selective influence of various eddy sizes. The influence of eddies with diameters larger than the measurement height is considerably reduced.

The type of measurement can therefore emphasize completely different features of the large eddy structure. That often prompts us to use the term *effective outer scale* to remind ourselves that large irregularities can have quite different influences on various electromagnetic measurements. This brief discussion cautions us that our search for the outer scale length will be both difficult and frustrating.[15]

### 2.3.5.1 The Outer Scale Length Near the Surface

Measurements near the surface are strongly influenced by the surface boundary condition and its irregularities. Tatarskii showed that the outer scale length in the surface layer is proportional to the height at which the experiment is done [101]:

$$\text{Near the surface:} \quad L_0 \simeq 0.4h \qquad (2.116)$$

This result emerges from an intuitive analysis of Figure 2.20. Imagine two probes used to measure the refractive-index structure function at a height h

---

[15] Proximity to natural or man-made features also complicates the problem. Telescope enclosures and nearby buildings generate turbulent eddies that are similar in size to the size of the structure, rather than the inherent atmospheric scale length.

above the surface. The probes are immersed in a fluid with many eddy sizes. From each probe's perspective, most of the large eddies are sliced off and it sees only a small portion of the change in refractive index that they present in a larger context. By contrast, eddies comparable in size to the probes' height can completely surround the probes. In doing so, they present their complete range of change in refractive index within a volume defined by the probes' height. Smaller eddies also influence the measurement according to their relative strengths. The entire range of eddy sizes is present but the geometry of the experiment self-selects eddies that are smaller than the probes' height.

A considerable body of experimental data supports this intuitive result. The temperature measurements reflected in Figures 2.2 and 2.10 suggest that $L_0 = 1.5$ m compared with the respective probe heights of 1 and 3 m. The refractometer experiment of Figure 2.11 used h $= 1.7$ m and indicates a scale length of about 2 m. Similar reasoning suggests that electromagnetic signals traveling close to the surface should emphasize eddies smaller than the path height. The phase variance was measured with laser signals 1.5 m above the surface over a considerable time and gave a distribution of outer scale lengths concentrated in the range $1 < L_0 < 2$ m [31]. Four optical phase-difference experiments[16] were each performed with h $= 1.5$ m and their measurements suggested outer scale lengths in the same range. The power spectrum of microwave amplitude fluctuations was measured with 36- and 110-GHz signals on a 4.1-km path over London [102] and the inferred outer scale lengths were about half the average path height. The agreement of (2.116) with experimental data is reasonable.

### 2.3.5.2 The Outer Scale Length in the Free Atmosphere

Now let us jump from the convective boundary layer to the *free atmosphere*. Signals that reach the ground from distant sources are strongly influenced by this turbulent region for two reasons. The first is that it is deep and the signals travel a relatively long distance through it. The second reason emerges from an examination of the microwave profile in Figure 2.17 and the optical profiles of Figure 2.18. The different features of electromagnetic signals are each proportional to weighted integrals of $C_n^2(z)$. Large values of $C_n^2$ evidently occur up to 5 km and indicate that the free atmosphere is quite influential for transmission experiments.

---

[16] These experiments are identified and reviewed in Chapter 5.

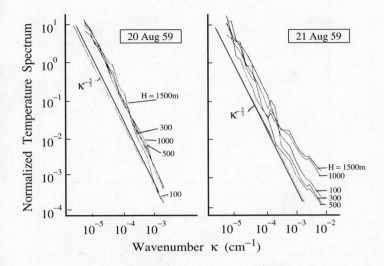

Figure 2.21: One-dimensional wavenumber spectra of temperature measured in the horizontal direction by Tsvang using an airborne sensor flown at the five constant altitudes which are indicated [91].

**The horizontal outer scale length** The features of the free atmosphere were first measured by Tsvang in the USSR [91]. He used an external temperature sensor on an aircraft that flew level courses at altitudes of 100, 300, 500, 1000 and 1500 m. The time history of the measured temperature was converted to a one-dimensional power spectrum for each altitude. The spectral frequency is related to the air speed and spatial wavenumber by $\kappa v = f$ and the one-dimensional wavenumber spectrum should be

$$S(\kappa) = \text{constant} \times \kappa^{-\frac{5}{3}} \qquad (2.117)$$

so long as $\kappa$ is in the inertial range. Experimental flight data taken on two days in 1959 is reproduced in Figure 2.21.

The striking feature of the data is the extent to which it agrees with the $-5/3$ scaling law over the entire wavenumber range. More fundamentally, it does so down to $\kappa = 10^{-5}$ cm$^{-1}$, which corresponds to

$$L_0 \geq 6 \text{ km}. \qquad (2.118)$$

This indicates that the 2/3 law is valid over very large horizontal distances at such altitudes. Those observations were greatly extended by the Global

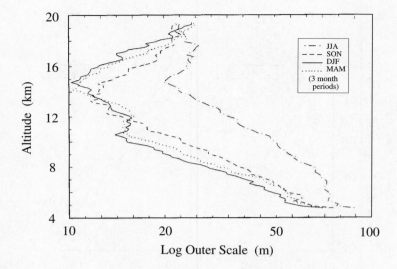

Figure 2.22: Profiles of the vertical outer scale length measured over five years above White Sands, New Mexico by Eaton and Nastrom with a pulse Doppler radar [100].

Atmospheric Sampling Program, which placed temperature and velocity sensors on commercial aircraft. During countless routine airline flights, these sensors confirmed the 2/3 law for large separations – sometimes apparently as great as 300 and 400 km [103].

Electromagnetic transmission experiments also confirm the existence of large horizontal scale lengths in the free atmosphere. Microwave interferometers exhibit surprisingly large phase correlations for signals received by antennas 5 or 10 km apart. Examinations of the temporal variability of the phase for line-of-sight microwave signals on elevated paths give the same estimate as (2.118). There can be little doubt that the horizontal outer scale length is several kilometers in the free atmosphere.

**The vertical outer scale length** The profiles of $C_n^2(z)$ provided in Figures 2.17 and 2.18 contain a good deal of fine structure and layering. They suggest that the outer scale length measured in the vertical direction might be quite small. That suspicion is confirmed by radar sounding of the troposphere. These measurements are based on a relationship between $L_0$ and the vertical gradient of wind speed established by Tatarskii [104],

$$L_0 = \sqrt{\mathcal{E}} \left( \frac{\partial v}{\partial z} \right)^{-\frac{3}{2}} \tag{2.119}$$

where $\mathcal{E}$ is the energy-dissipation rate. It is important to note that this formula assumes isotropy and was established for heights less than 100 m.

A five-year campaign was sponsored to measure outer scale profiles over White Sands, New Mexico [100]. A pulse Doppler sounding radar operating at 49.25 MHz was used to measure the wind speed as a function of height from 5 to 20 km. The vertical wind shear was estimated by differencing the readings in adjacent range bins. Average profiles were established for the four seasons and are reproduced in Figure 2.22. They exhibit a gradual decline with altitude up to 15 km and a modest increase above that level. The seasonal variation is modest and the range of values is not great:

$$\text{Vertical:} \qquad 10\text{ m} < L_0 < 70\text{ m} \qquad\qquad (2.120)$$

These profiles are the best and most comprehensive data now available for the vertical outer scale length.

### 2.3.5.3 A Composite Picture of Atmospheric Anisotropy

From the measurements reproduced here, it is clear that irregularities in the free atmosphere are highly anisotropic. By contrast, the eddies next to the earth's surface which influence many measurements are not far from symmetrical. These observations suggest the model illustrated in Figure 2.23. Irregularities of all sizes are presumably present at each level, subject only to the largest size limit which is set by the turbulence- initiating process. Eddies in each size class become more elongated as the altitude increases. This description is consistent with many measurements and reconciles a number of apparently contradictory observations. At the very least, it is a very helpful way to think about the troposphere.

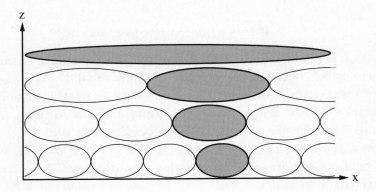

Figure 2.23: A suggested model of how the elongation of turbulent eddies changes with altitude in the troposphere.

This model has important consequences for electromagnetic propagation. The outer scale length one must anticipate depends on the *direction in which one looks*. If the signal is transmitted horizontally near the surface, one should see a spectrum that is relatively uniform, nearly isotropic and bounded in size by the path height. If one looks straight up one sees an inhomogeneous structure with eddies of relatively small thickness – which may also be described as layers. Oblique paths evidently encounter both the horizontal breadth and the small vertical thickness of the elongated irregularities. The signal experiences an interesting combination of their separate influences.

By contrast, parallel rays coming down to the separated receivers of an interferometer each encounter a variety of horizontal eddy sizes. Comparisons of signals for these dual paths will be influenced most strongly by those eddies whose horizontal dimensions are comparable to the spacing between receivers.

This examination compels us to make a strong effort to include anisotropy in our descriptions of optical and microwave transmission experiments. We introduced the ellipsoidal spectrum model (2.87) earlier and will use it often in this series. That model is probably accurate for describing modest anisotropy. We do not know whether it can also represent the highly elongated eddies which fill the free atmosphere. Moreover, acoustic gravity waves travel horizontally in the upper atmosphere and we have made no provision for their inclusion. We mention these points to emphasize the complexity of this problem. The difficult task of including anisotropy using the ellipsoidal model will dominate studies in these volumes and will provide revealing insights into atmospheric propagation.

## 2.4 Ionospheric Properties

The situation is fundamentally different in the ionized regions of the earth's atmosphere which begin at an altitude of approximately 80 km. Plasma layers are formed there by solar radiation that ionizes the tenuous gases. These layers were first discovered by recording radio reflections, for which Sir Edward Appleton received the 1947 Nobel Prize for physics. This technique stimulated development of a worldwide network of radio sounders used for predicting short-wave propagation conditions and for studying the ionosphere [105].

The second chapter in ionospheric research began when radio signals arriving from distant galaxies were first measured. That discovery created the new field of radio astronomy which now exploits electromagnetic waves ranging from the short-wave band to submillimeter wavelengths. These signals are pursued primarily to study the sources from which they originate. They are also important for ionospheric research because they provide signals at many frequencies that pass completely through the ionized regions, in contrast to the traditional reflected signals. Their passages provide a powerful new research tool and much of what we know today has come from radio astronomy. Changes in polarization of these signals reflect the total electron content along the path. Angle-of-arrival variations mirror the tilting of ionized layers and the horizontal movement of large-scale irregularities. Intensity fluctuations are sensitive to smaller scale lengths and provide a method for exploring turbulent conditions in the plasma. These opportunities emphasize the importance of understanding amplitude and phase fluctuations for microwave signals that pass through the ionosphere.

Entirely new possibilities for studying the ionosphere emerged when earth-orbiting satellites became a reality. It was soon realized that the controlled microwave transmissions from satellites provide a powerful tool for exploring the ionosphere. The passage of these signals through the ionosphere can provide a tomographic portrait of its structure, in the same way as that whereby X-rays reveal the skeleton of a human body. It seemed that communication satellites in geosynchronous orbits would be particularly useful because their positions do not change relative to ground receivers. Navigation-satellite signals are stabilized very accurately in frequency and provide another valuable opportunity for research. Several special-purpose satellites were launched to study the ionosphere. One series carried instruments to measure electron density. A second type of satellite transmitted phase-coherent signals at several frequencies, which allowed accurate measurements both of phase and of amplitude to be made – an option that is not available with incoherent galactic signals. These experiments were sponsored both to learn about the atmosphere and to address problems of concern to the military services.

Experiments to examine scintillation imposed by the atmosphere were also stimulated by technical questions related to communication satellites. Satellites provide a service to mobile users at 400–1500 MHz. GPS satellites that provide location and navigation services also operate in the L band. Electron-density variations have a strong influence on these frequencies and the signals exhibit undesirable amplitude and phase fluctuations.[17] Civilian

---

[17] To avoid large range errors which would be caused by ionospheric irregularities, the GPS transmits two signals at different frequencies and exploits their dispersion difference to remove the plasma effect. This duplication option is not possible for communication satellites since they are severely limited by the allocated bandwidths.

communication satellites were first assigned frequencies at 4000 and 6000 MHz. Both the ionosphere and the troposphere impose measurable scintillations on those signals. As these services moved to higher frequencies, the ionospheric influence declined and tropospheric effects became more important. These concerns have led to the provision of additional governmental and commercial support for measurement campaigns.

### 2.4.1 The Ambient Ionosphere

The ionosphere is created by solar radiation falling on the earth's atmosphere. The sun bombards the sunlit face of the atmosphere with ultraviolet radiation and soft X-rays. These photons have enough energy to separate electrons from the atoms and molecules which form the upper atmosphere. This creates a sea of free electrons and ions. Fortunately, the flux of radiation is rapidly reduced as this ionization proceeds and only a small fraction of the original flux reaches the earth's surface.

As we examine this process more carefully, we see that there is likely to be a height at which the plasma's density is a maximum so that a layer will be formed. One can understand this by following the solar radiation as it descends through the atmosphere. We note first that the atmospheric density decreases by thirteen orders of magnitude from a surface value of 1.22 kg m$^{-3}$ to approximately $2.3 \times 10^{-13}$ kg m$^{-3}$ at 700 km. Even though the solar radiation is strongest when it first reaches the atmosphere, there are very few atoms or molecules to be ionized. As the radiation continues to penetrate it loses its effectiveness because many of its photons have been consumed in ionizing the overlying regions. The number of atoms and molecules is rising, thereby presenting more ionization targets for the photons. At even lower altitudes, the targets become more plentiful but the radiation is further weakened. This suggests that there is a height at which the radiation is strong enough and the targets are sufficiently abundant to produce a maximum amount of ionization. That is apparent in the profile data reproduced in Figure 2.24 which was measured by an ion probe carried to an altitude of 700 km by a sounding rocket [106]. This experiment indicates that there is a peak ionization of $2.5 \times 10^5$ electrons cm$^{-3}$ at about 350 km and this is called the *F layer*. Electromagnetic scintillation is induced primarily in the F region, which is approximately 200 km thick.

One should also ask what happens to the ionized layers when they are shadowed from solar radiation rather than meeting its full force. The data in Figure 2.24 gives the answer since it was taken at night; some of the ionization persists during the night and is ready to be recharged

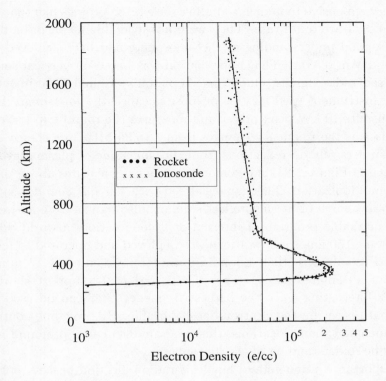

Figure 2.24: Measurements of the electron-density profile from 240 to 1875 km made by Sagalyn and Smiddy [106]. The crosses denote ionosonde measurements using reflected HF signals that cannot extend past the F-layer maximum at 370 km. The points noted by dots were measured with an ion-trap probe carried by a Blue Scout rocket launched from Florida. These simultaneous profiles were measured at night.

the next day. That introduces the concept of electron–ion recombination, which usually proceeds by electrons and ions finding one another to recreate the original type of atom or molecule. This also happens by attachment of electrons to neutral atoms. These processes proceed rapidly at lower altitudes where the D and E layers form at 85 and 105 km, respectively, and these layers largely disappear at night as a result of recombination.[18] Recombination also reduces the level of ionization of the F layer and one often finds daytime levels that are ten times greater than those evident in Figure 2.24. Electron-density levels rise and fall as the sun goes through its eleven-year solar cycle. They also respond to solar flares that subject the atmosphere to bursts of radiation. These changes are reflected in the scintillation levels experienced by microwave signals passing through the ionosphere.

---

[18] Notice that these layers are not evident in the nighttime rocket measurement, although that is probably due here to altitude and instrumentation cutoffs.

It is very expensive to launch sounding rockets, so profiles like that shown in Figure 2.24 are quite rare. They were not available at all prior to 1950 and yet a good understanding of the ionosphere had been achieved before the Second World War. That appreciation was based on vertical sounding with pulsed radio signals, emulating Appleton's original experiment. The standard instrument used for this purpose is called the *ionosonde*. It typically transmits 10 kW peak power and measures the round-trip travel time of its pulses as the frequency changes from 1 to 25 MHz in a steady sweep. These sounders rely on reflections from the *underside* of plasma layers like that shown in Figure 2.24 and cannot measure layer features above the level of maximum ionization. This gap was filled with *top-side sounder* satellites placed in low earth orbit that reflect similar pulse signals from above.

In a remarkable program of international cooperation, a worldwide network of some 150 ionosonde stations was deployed and operated to measure ionospheric conditions [105]. Soundings are often taken every 15 min, usually on an automated basis. This was first done to support short-wave frequency forecasting since the highest frequency that can be used on an oblique path is set by the electron-density profile. By providing continuous measurements at many locations, this program also had a profound impact on scientific understanding of the ionosphere.

It is important to understand how this radio-reflection process works. In doing so, we will lay a foundation for describing scintillation of signals that pass completely through the ionosphere. We first need to describe the dielectric constant for an electromagnetic wave traveling in an ionized plasma. We focus attention on the electrons, since they are much lighter than the ions and respond more readily to the incident wave. The electric field strength of the arriving signal drives each free electron with a force $eE(t)$, which generates harmonic motion of the electron at the same frequency as that of the arriving wave. Since the electron is accelerated by this oscillating force, it radiates a tiny secondary field that is frequency synchronized with the arriving field. The composite secondary field should depend on the total number of electrons per unit volume $N$. It also depends on the electron charge $e$ and on the electron mass $m$ which describes the inertia of the individual electrons. A relatively simple analysis gives the following expression for the dielectric constant and refractive index:[19]

$$\epsilon = \epsilon_0 \left( 1 - \frac{e^2 N}{\epsilon_0 m \omega^2} \right) = n^2 \qquad (2.121)$$

---

[19]  For short-wave and low-frequency communication signals that are reflected obliquely by the ionosphere, one often needs to include the influence of the terrestrial magnetic field and collisions with neutral gases. This leads to a very complicated replacement for (2.121), which is provided by *magnetoionic theory*.

This indicates that the ionosphere is a dispersive medium, since the dielectric constant depends on the probing frequency $\omega$. We often write it in terms of the radiation wavelength

$$\epsilon = \epsilon_0 - r_e \lambda^2 N \tag{2.122}$$

where the *classical electron radius* is defined by fundamental physical constants and has the following numerical value:

$$r_e = 2.818 \times 10^{-15} \text{ m} \tag{2.123}$$

The next step is to use the dispersion relation to understand how pulsed sounding can establish ionospheric plasma profiles with reflected signals. We first notice that the dielectric constant goes to zero at the *plasma frequency* which identifies the value of electron density for which the wave is reflected:

$$\omega_p^2 = \frac{e^2 N(H)}{\epsilon_0 m} \tag{2.124}$$

In writing this we remind ourselves that the electron density depends on altitude in a way that we do not yet know. It is not hard to show that the group velocity with which a pulse travels is given by the following expression:

$$v_g = \frac{c}{\sqrt{1 - \dfrac{e^2 N(z)}{\epsilon_0 m \omega^2}}} \tag{2.125}$$

The time that it takes a pulse to travel to the reflection height $H$ and back is therefore given by an integral involving the unknown profile function:

$$\tau(\omega) = 2c \int_0^H dz \left( 1 - \frac{e^2 N(z)}{\epsilon_0 m \omega^2} \right)^{-\frac{1}{2}} \tag{2.126}$$

This travel time increases as the carrier frequency does because the waves can penetrate farther into the layers. With numerical methods one can invert this relationship to establish the electron density $N(z)$ as a function of height if accurate measurements of $\tau(\omega)$ are available. This is the method that was used to determine the ionosonde data points plotted in Figure 2.24. Notice that there are no ionosonde points above the layer maximum since the pulse penetrates the F layer for frequencies greater than $\omega_p$.

The worldwide network of ionosonde stations has provided electron-density profiles with regularity at many stations for the past half century. This remarkable data bank provides the basis for studying the significant diurnal and seasonal variations of the ionosphere. This data has also been used to explore the correlation with magnetic latitude and thereby define the influential role played by the terrestrial magnetic field in configuring the layers.

### 2.4.2  Observations of Ionospheric Irregularities

The ionospheric sounding program had given occasional hints of random irregularities superimposed on the normal electron-density profiles. Reflections are sometimes received simultaneously from many heights. This is called *spread F* and is now attributed to backscattering by plasma irregularities.

Radio astronomy provided the first clear evidence that plasma irregularities are a consistent feature of the ionosphere. Hey, Parsons and Phillips made an important discovery in 1946 while measuring the spatial distribution of galactic radiation [107]. They noted *"short period irregular fluctuations of the (64 MHz) signal coming from the direction of Cygnus."* This marked the beginning of radio-scintillation research, which now provides a unique tool with which we can explore the ionosphere, the solar wind and the interstellar plasma.

### 2.4.2.1  Locating the Source of Scintillations

It was initially believed that the random intensity fluctuations of galactic signals were generated at the source. Widely separated observations questioned that assumption and two coordinated experiments were mounted. One compared the short-term variations of 3.7- and 6.7-m signals from Cassiopeia [108]. Intensity fluctuations on these wavelengths were remarkably well correlated, indicating a common cause. The second experiment was a collaboration between Jodrell Bank and Cambridge to make simultaneous measurements of the 3.7-m signal [109]. No correlation between these signals measured 210 km apart was found. This question was next pursued using receivers separated by 100 m near Jodrell Bank and these signals exhibited complete correlation. The correlation was 0.5–0.95 when the separation was extended to 3900 m.

These measurements demonstrated that the scintillation was imposed by a random medium through which the waves had passed, rather than at the source. High correlations at nearby receivers indicated that the region responsible was probably terrestrial. Their scaling with frequency showed that the random medium was ionized. The terrestrial ionosphere must therefore be the primary source of such scintillation. We have since learned that variations in the interstellar plasma and solar wind also influence signals from distant galaxies. It was soon realized that a comparison of scintillation at adjacent receivers or at different frequencies could provide important information about these regions. In some cases, that infortion could be obtained in no other way.

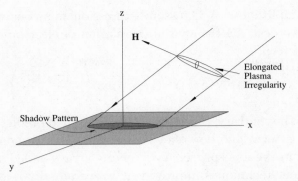

Figure 2.25: Illustrating the technique used to study shadow patterns cast by field-aligned plasma irregularities.

### 2.4.2.2 Magnetic Field Alignment

Radio astronomy also provided the second building block needed to understand plasma irregularities in the ionosphere. It was known that a plasma irregularity casts a radio shadow pattern on the ground as suggested in Figure 2.25. Spencer examined the shape and orientation of these patterns by comparing the intensity fluctuations at three nearby receivers using a 38-MHz signal from Cassiopeia at high elevation angles [110]. Three receirs were placed at the vertices of a right triangle with an East–West separation of 700 m and a separation of 1200 m in the North–South direction. Instantaneous correlation coefficients ranged from 0.41 to 0.88 on a typical day, with the *largest* value corresponding to the *greatest* separation. Further exploration showed that the shadow pattern was elliptical and considerably stretched. More fundamentally, the pattern's long axis always coincided with the direction of the magnetic field. These measurements suggested that the plasma irregularities responsible have two characteristics: (*a*) they are aligned with the terrestrial field, and (*b*) they are highly elongated with axial ratios of 20 or more. That makes sense because the collision frequency is only $10^3$ s$^{-1}$ at such heights and the free electrons should circle the magnetic field vector in classical orbits unimpeded by the ambient gas. Subsequent experiments using both satellite and radio star signals confirmed this description at many locations with a variety of magnetic field orientations [111].

### 2.4.2.3 Measuring the Structure of Plasma Irregularities

Radio astronomers have traditionally modeled the ionosphere as a phase screen that imparts phase fluctuations to the passing waves but does not change their amplitudes. A Gaussian expression was consistently used in early research to describe the spatial correlation of electron-density fluctuations in this screen:

$$\langle \delta N(\mathbf{r}, t)\, \delta N(\mathbf{r}', t)\rangle = \langle \delta N^2\rangle \exp\left(-\frac{|\mathbf{r} - \mathbf{r}'|^2}{\ell^2}\right) \qquad (2.127)$$

Absent data to the contrary, this assumption simplified predictions about the radio shadow patterns and how they should change in response to horizontal movements in the ionosphere. As receiving arrays added more antennas and measurement techniques improved, it was gradually realized that this single-scale model is not a good representation. Scintillation measurements suggested that a power law would better describe the data which was being gathered [112][113]. These power-law models are related to fractals and represent a self-similar structure that is continuously dividing itself according to a defined rule.[20]

As the discrepancy between theory and measurements grew, it became important to measure the electron-density irregularities *in situ*. Satellites provided the means to do so. In a timely experiment, an ion-concentration probe was installed on the OGO 6 spacecraft and launched into an earth orbit that dipped into the F region [114]. The movement of the spacecraft generated a history of $\delta N(\mathbf{r}, t)$ measured with a spatial resolution of about 35 m along the satellite's trajectory. This time series was detrended and used to generate a one-dimensional power spectrum. In terms of the wavenumber defined by $f = v\kappa$ the measured spectrum was well represented by

$$\text{One-dimensional:} \quad W_N(\kappa) = \frac{\text{constant}}{\kappa^\mu} \qquad (2.128)$$

The scaling exponent was measured over two orders of magnitude in scale size. During the repeated passes the satellite made through the ionosphere, the results fell within a small range of values and seemed to change very little with the magnetic dip latitude:

$$\mu = 1.9 \pm 0.1 \quad \text{for} \quad 70\,\text{m} < 2\pi/\kappa < 7\,\text{km} \qquad (2.129)$$

This result was expanded by the Atmospheric Explorer E spacecraft which also sampled the ionospheric plasma repeatedly [115]. Its measurements revealed a substantially larger distribution with $\mu$ ranging from 1.0 to 2.8. It also noticed for the first time a nonlinear effect in which the scaling exponent seemed to depend on the level of fluctuation itself.

---

[20] The Kolmogorov spectrum (2.62) for tropospheric temperature fluctuations is a good example of a power-law model. It presents all scale sizes down to the dissipation cutoff. While we doubt that it describes ionospheric irregularities, we cannot rule out that possibility at this point.

A vertical slice through the ionosphere was provided by a rocket flight launched from Kwajalein to a height of 600 km. This complemented the horizontal cuts taken by orbiting spacecraft. The rocket carried a Langmuir probe for measuring the density of charged particles and four phase-coherent microwave transmitters whose phase and amplitude were monitored on the ground [116]. Because the probe sampled many altitudes at the same time, it provided a wider picture than the constant-altitude satellites could gather. Evaluation of the probe and microwave data suggested that $\mu$ varied between 0.9 and 2.3, with different values associated with different heights and with different scale-size ranges. From these *in situ* experiments it is clear that a power-law scaling model is the right answer. There is considerable uncertainty in identifying a unique value for the exponent and we will treat it as a parameter to be defined by scintillation experiments.

### 2.4.3   A Description of Plasma Irregularities

With these basic observations in hand, we can construct a description of the plasma irregularities. We are concerned primarily with the influence of the dielectric constant on electromagnetic waves, yet most of our information relates to electron-density structures. We can connect the two by taking differentials of the basic relation (2.122):

$$\Delta\epsilon = -r_e\lambda^2\,\delta N \tag{2.130}$$

Because the features of electromagnetic fields are expressed as ray-path or volume integrations of $\Delta\epsilon$, we need the spatial covariance of $\Delta\epsilon$ to estimate the usual signal moments. This brings us back to the spatial covariance of electron-density fluctuations:

$$\langle\Delta\epsilon(\mathbf{r},t)\,\Delta\epsilon(\mathbf{r}',t)\rangle = r_e^2\lambda^4\langle\delta N(\mathbf{r},t)\,\delta N(\mathbf{r}',t)\rangle \tag{2.131}$$

#### 2.4.3.1  The Wavenumber-spectrum Description

The wavenumber spectrum provides significant advantages for describing the electron-density irregularities, just as it does in the tropospheric case:

$$\langle\delta N(\mathbf{r},t)\,\delta N(\mathbf{r}',t)\rangle = \int d^3\kappa\,\Psi_N(\boldsymbol{\kappa})\exp[\,i\boldsymbol{\kappa}\cdot(\mathbf{r}-\mathbf{r}')] \tag{2.132}$$

The spectrum depends on all three components of the wavenumber because the eddies are elongated in the direction of the magnetic field. It is important to remember that the *spectrum of electron-density fluctuations* $\Psi_N(\boldsymbol{\kappa})$ is entirely different than the spectrum of refractive-index variations $\Phi_n(\boldsymbol{\kappa})$.

They have completely different dimensions. More to the point, they describe phenomena controlled by quite different physics.

We have discovered several hints about the electron-density spectrum already. Shadow-pattern measurements show that it is highly anisotropic in the direction of the terrestrial magnetic field. The ellipsoidal representation of the spectrum (2.87) is ideally suited to describing these elongated plasma structures. Since they are apparently symmetrical around the field lines, we adopt wavenumber coordinates that use the magnetic field vector as their polar axis and set $a = b$:

$$\boldsymbol{\kappa}: \qquad \kappa \sin \psi \cos \omega, \ \kappa \sin \psi \sin \omega, \ \kappa \cos \psi$$

The spectrum is independent of the azimuth angle $\omega$ because the eddies are symmetrical about the field lines and a general expression for the plasma spectrum emerges:

$$\Psi_N(\boldsymbol{\kappa}) = a^2 c \, \Psi_N \left( \kappa \sqrt{a^2 \sin^2 \psi + c^2 \cos^2 \psi} \right) \qquad (2.133)$$

The large problem we face here is that of finding a description for the spectrum $\Psi_N(\kappa)$ that is consistent with ionospheric measurements.

### 2.4.3.2  *The Spectrum of Electron-density Fluctuations*

Booker carefully reviewed the enormous body of ionospheric data which had been accumulated [117]. From this data he generated a *global model* of the spectrum $\Psi_N(\kappa)$ in graphical form that emphasized its qualitative nature. It starts with scale lengths comparable to the earth's radius and ends with the electron's gyroradius, which is a few centimeters. His model is flat from the largest scales to about 40 km and then follows a power law down to the *ion gyroradius* at a few meters. On this graphical presentation, Booker identified the regions which account for different types of ionospheric phenomena and noted considerable overlap among them. This model was intended to draw attention to the common origins of different phenomena. As a result, many important connections were identified and a good deal of collaboration among groups working on different ionospheric problems was stimulated.

This qualitative description prompts us to identify the portion of the spectrum which is important for our work. From correlations with ionosonde data, it is known that scintillation is influenced primarily by the F layer at an altitude of approximately 350 km [118]. The corresponding Fresnel length for microwave transmissions shows that plasma irregularities in the scale range

$$5 \text{ m} < \ell < 40 \text{ m} \tag{2.134}$$

should be responsible for phase and amplitude fluctuations. Large structures are most influential for phase variations, whereas amplitude fluctuations are sensitive to the smaller scales.

**The power-law model** Booker's global model suggests that the spectrum is a simple power law in the scale range that is important for our work [117]. More to the point, the OGO 6 measurements found that the spectrum *is a power law* from 70 m to 7 km, which covers most of our concerns [114]. We will use the following form since it is easily related to these *in situ* measurements:[21]

$$\Psi_N(\boldsymbol{\kappa}) = \frac{Q_\nu}{2\pi} \frac{\langle \delta N^2 \rangle \kappa_0^{\nu-3}}{\kappa^\nu [1 + (\mathcal{A}^2 - 1) \cos^2 \psi]^{\frac{\nu}{2}}} \qquad \kappa_0 < \kappa < \frac{2\pi}{\ell_0} \tag{2.135}$$

This form is widely used for ionospheric studies, although some measurements suggest that a break between two such models that have different slopes might occur. Notice that this is a three-dimensional spectrum, in contrast to the one-dimensional result (2.128) that was measured along a straight line. The exponent $\mu$ is related to the exponent in that scaling law by $\nu = 2 + \mu$. It is important to try to estimate the other parameters in this model as best we can before proceeding to use it for propagation calculations.

**The spectral index** We do not yet have a universal theory of plasma turbulence that provides a unique value of $\nu$. The values of $\mu$ measured by probes *in situ* contain a good deal of variability. We will look primarily to satellite and radio-astronomy transmission measurements to establish the scaling exponent and two types of experiments are especially helpful in this regard.[22] The wavelength dependence of amplitude scintillations is sensitive to $\nu$ and was measured by comparing several signals transmitted from the ATS 6 geosynchronous satellite [120][121]. The second method exploits the temporal variability of amplitude scintillations imposed on satellite and radio-astronomy signals [112]. The scintillation power spectrum is also given

---

[21] A more complicated and capable analytical model was proposed [119] but is not required in our work.

[22] These experiments are discussed in Chapters 3 and 5 of the second volume.

by a scaling law whose exponent is $\nu + 1$. The values of $\nu$ measured by these quite different techniques fall in a small range:

$$3 < \nu < 4 \qquad (2.136)$$

In using (2.135) we will keep our options open with respect to $\nu$ to accommodate these possibilities.

**The outer scale length** The outer scale length of plasma irregularities is difficult to define and difficult to estimate. In the troposphere we can measure $L_0$ as the separation beyond which the 2/3 scaling law breaks down. In the ionosphere we cannot measure $L_0$ directly and must rely on electromagnetic signals that are sensitive to it. We do not know how plasma irregularities begin, just as we are not sure how the tropospheric spectrum behaves in the energy-input range. We are further handicapped in the ionosphere because we do not have a universal theory for the power-law region of the plasma spectrum. This means that we cannot use its boundary as a plausible definition for the outer scale length.

Rino provided the following appreciation: "*The effective outer scale emerges as a somewhat arbitrary division between the well-developed structures that can be modeled by a statistically random process and the evolving structures that depend on the initial configuration of the plasma at the time of instability onset. The available evidence shows that ionospheric outer scale, if indeed a verifiable outer scale exists at all, is at least several tens of kilometers*" [122]. Booker identified *traveling ionospheric disturbances* as a likely source of the energy required to generate plasma turbulence [117]. He equates $L_0$ to the scale size at which these traveling waves undergo a transition to field-aligned turbulent eddies and suggested that this break should occur at about 10% of the local scale height. The scale height is 50 km in the F region [105], which led him to use $L_0 = 5$ km. Other authors have adopted $L_0 = 10$ km but we will use a range of values:

$$5 \,\text{km} < L_0 < 20 \,\text{km} \qquad (2.137)$$

These estimates are important for estimating phase fluctuations that are measured with coherent satellite transmitters.

**The inner scale length** The inner scale length of plasma irregularities is not well defined, primarily because there is no analog of the Kolmogorov microscale to use as a length standard. Booker suggested that the spectrum changes slope at an eddy size of several meters, which is approximately equal to the ion gyroradius [117]. In his view, the gyroradius *"is the minimum scale to which one might expect field-aligned turbulence to extend."* We therefore use the following range of values:

$$2 \text{ m} < \ell_0 < 5 \text{ m} \tag{2.138}$$

**The axial ratio** Elongation of plasma irregularities was first measured by Spencer, who found large values – greater than 5 and often more than 20 [110]. His measurements were made at night with 38-MHz signals from Cassiopeia at high elevation angles. That experiment was repeated and enlarged with a variety of source positions using the same arrangement of signal and receiver [123]. The apparent axial ratio was measured for $20° < E < 85°$ and found to decrease with elevation angle, sometimes giving values as low as 2. This is consistent with a projection of field-aligned irregularities on the surface at Cambridge. We will recognize a range of values of the axial ratio but try to remember that the actual value depends on the propagation geometry:

$$2 < \mathcal{A} = a/c < 25 \tag{2.139}$$

**Spectrum normalization** There is no structure constant in the assumed spectrum model (2.135). This reflects the absence of a universal inertial model like that proposed by Kolmogorov for the lower atmosphere. Instead, we have the variance of electron-density fluctuations $\langle \delta N^2 \rangle$ and the normalization constant $Q_\nu$. Since the covariance expression (2.132) must reduce to $\langle \delta N^2 \rangle$ when the two measurement points are the same, we write

$$\langle \delta N^2 \rangle = \int_{\kappa_0}^{\kappa_s} d\kappa \ \kappa^2 \int_0^\pi d\psi \sin\psi \int_0^{2\pi} d\omega \left( \frac{Q_\nu}{2\pi} \frac{\langle \delta N^2 \rangle \kappa_0^{\nu-3}}{\kappa^\nu [1 + (\mathcal{A}^2 - 1)\cos^2\psi]^{\frac{\nu}{2}}} \right)$$

and the normalization is defined as follows:

$$Q_\nu^{-1} = \int_{\kappa_0}^{\kappa_s} d\kappa \ \frac{d\kappa}{\kappa^{\nu-2}} \int_0^\pi d\psi \sin\psi \left( \frac{\kappa_0^{\nu-3}}{[1 + (\mathcal{A}^2 - 1)\cos^2\psi]^{\frac{\nu}{2}}} \right) \tag{2.140}$$

When the power-law index is larger than 3 we can exploit the fact that the inner-scale wavenumber is much greater than the outer-scale wavenumber:

$$\nu > 3: \quad Q_\nu = \frac{\nu - 3}{2} \left( \int_0^1 \frac{du}{[1 + (\mathcal{A}^2 - 1)u^2]^{\frac{\nu}{2}}} \right)^{-1} \tag{2.141}$$

One can evaluate this exactly when $\nu = 4$,

$$Q_4 = \frac{\mathcal{A}^2 \sqrt{\mathcal{A}^2 - 1}}{\sqrt{\mathcal{A}^2 - 1} + \mathcal{A}^2 \tan^{-1}(\sqrt{\mathcal{A}^2 - 1})}$$

but it is well approximated as follows because the axial ratio is usually large:

$$\mathcal{A} \gg 1: \qquad Q_4 = (2/\pi)\,\mathcal{A}$$

The ionospheric spectrum changes its behavior markedly when the scaling index approaches 3. In this region, one must respect the wavenumber limits and the normalization depends on their ratio:

$$Q_3 = \frac{\mathcal{A}}{2\log\!\left(\dfrac{\kappa_s}{\kappa_0}\right)} \tag{2.142}$$

This sensitivity near $\nu = 3$ is reflected in predictions of properties of signals for microwave transmission through the ionosphere [124][125].

**The electron-density variance** With the normalization constant $Q_\nu$ defined by the other parameters, we see that the level of plasma irregularities is set primarily by the variance of electron-density fluctuations $\langle \delta N^2 \rangle$. This can be inferred both from *in situ* and from scintillation measurements. The OGO 6 satellite measured the ratio of electron-density variations to the mean value at an altitude of 446 km, giving values from 0.3% to 10% [114]. From Figure 2.24 we see that the mean electron density is approximately $10^5$ cm$^{-3}$ at this height, giving the following estimates:

$$10^9 \text{ m}^{-3} < \delta N_{\text{rms}} < 10^{10} \text{ m}^{-3} \tag{2.143}$$

One must question combining these measurements. Both were taken at night near the equator but they were made ten years apart in different seasons, meaning that (2.143) is only suggestive. Scintillation measured with radio star and satellite signals gave similar estimates for similar conditions. We present this range of values only to give an impression of the values that are likely to be encountered.

## 2.5 Problems

### Problem 1

For homogeneous and isotropic random media we found in (2.51) that the structure function of $\delta n$ is related to the turbulence spectrum by the following equation:

$$\mathcal{D}_n(\rho) = 8\pi \int_0^\infty d\kappa \; \kappa^2 \Phi_n(\kappa)\left(1 - \frac{\sin(\kappa\rho)}{\kappa\rho}\right)$$

Show that this integral equation can be solved exactly if $\mathcal{D}_n(\rho)$ is known from temperature or refractometer measurements [126][127][128]:

$$\Phi_n(\kappa) = \frac{1}{4\pi^2\kappa^2} \int_0^\infty d\rho \left(\frac{\sin(\kappa\rho)}{\kappa\rho}\right) \frac{\partial}{\partial\rho}\left(\rho^2 \frac{\partial}{\partial\rho}\mathcal{D}_n(\rho)\right)$$

Use this solution to calculate the turbulence spectrum which corresponds to the familiar power law:

$$\mathcal{D}_n(\rho) = C_n^2 \rho^{\frac{2}{3}}$$

This procedure requires that the structure function be known for all values of $\rho$. Although the 2/3 law is valid for a wide range of spacings between sensors, it is modified for very large and very small separations. How does that reality restrict the inversion process?

### Problem 2

Establish an explicit relationship between the power spectrum of refractive-index fluctuations $\delta n(t)$ and the wavenumber spectrum of irregularities that generate them. Notice that the Fourier inverse of (2.20) gives

$$W_n(\omega) = \int_{-\infty}^\infty d\tau \langle \delta n(t)\, \delta n(t+\tau)\rangle.$$

Assume that the eddies are frozen during the measurements and carried past the sensor by a constant wind so that the time delay is equivalent to a spatial separation:

$$\boldsymbol{\rho} = \mathbf{v}_0\tau$$

This is Taylor's hypothesis discussed in Chapter 6. Combine this approximation with the spatial covariance expression (2.44) and show that, for isotropic irregularities,

$$W_n(\omega) = \frac{4\pi^2}{v} \int_{\omega/v}^\infty d\kappa\, \kappa\Phi_n(\kappa).$$

Complete the evaluation for the Kolmogorov model with an outer-scale cut-off defined by (2.73) and show that, for

$$\omega > \kappa_0 v: \qquad W_n(\omega) = 0.782 C_n^2 v^{\frac{2}{3}}\omega^{-\frac{5}{3}}$$

whereas it approaches a constant value when $\omega < \kappa_0 v$.

*Waves in Random Media*

## Problem 3

To first order the electromagnetic field is represented by integrals of the dielectric variation. The instantaneous phase and amplitude of the signal are therefore related to double integrals of the product of dielectric variations measured at various points in the random medium:

$$\delta\varepsilon(\mathbf{r}_1)\,\delta\varepsilon(\mathbf{r}_2) = \delta\varepsilon_1\,\delta\varepsilon_2$$

The phase and amplitude of doubly scattered waves are also represented by multiple integrals of this product. We should therefore ask how this random variable is distributed. Show that its probability density function is not Gaussian – even when the individual $\delta\varepsilon$ are Gaussian random variables. To establish this important point, use the joint distribution for two different Gaussian random variables from Appendix E,

$$\mathsf{P}[\,\delta\varepsilon_1, \delta\varepsilon_2; C(\rho)] = \frac{1}{2\pi\langle\delta\varepsilon^2\rangle\sqrt{1-C^2(\rho)}} \exp\left(-\frac{\delta\varepsilon_1^2 + \delta\varepsilon_2^2 - 2\,C(\rho)\,\delta\varepsilon_1\,\delta\varepsilon_2}{2\langle\delta\varepsilon^2\rangle(1-C^2(\rho))}\right)$$

where $C(\rho)$ is their spatial correlation. Introduce their product and quotient as new random variables

$$p = \delta\varepsilon_1\delta\varepsilon_2 \qquad \text{and} \qquad q = \frac{\delta\varepsilon_1}{\delta\varepsilon_2}$$

and integrate over all possible values of $q$ to show that

$$\mathsf{P}[\,p\,; C(\rho)] = \frac{1}{\pi\langle\delta\varepsilon^2\rangle\sqrt{1-C^2(\rho)}}\exp\left(p\,\frac{C(\rho)}{\langle\delta\varepsilon^2\rangle(1-C^2(\rho))}\right)$$
$$\times\,K_0\left(\frac{|p|}{\langle\delta\varepsilon^2\rangle(1-C^2(\rho))}\right).$$

Plot this function for several values of the spatial correlation and notice that it is not symmetrical about the origin unless the correlation vanishes. Show that this distribution diverges logarithmically for small values but is normalized to unity over the entire range of $p$.

## References

[1] B. R. Bean and E. J. Dutton, *Radio Meteorology* (National Bureau of Standards Monograph 92, U.S. Government Printing Office, Washington, 1 March 1966), 89 *et seq.*

[2] V. I. Tatarskii, *The Effects of the Turbulent Atmosphere on Wave Propagation* (translated from the Russian and issued by the National Technical Information Office, U.S. Department of Commerce, Springfield, VA 22161, 1971), 181–208.

[3] S. I. Krechmer, "Investigations of microfluctuations of the temperature field in the atmosphere," *Doklady Akademii Nauk SSSR, Seriya Geofizicheskaya*, **84**, No. 1, 55–58 (1952). (This is the Russian reference; no translation is available.)

[4] J. R. Gerhardt, C. M. Crain and H. W. Smith, "Fluctuations of Atmospheric Temperature as a Measure of the Scale and Intensity of Turbulence near the Earth's Surface," *Journal of Meteorology*, **9**, No. 5, 299–311 (October 1952).

[5] G. Birnbaum, "A Recording Microwave Refractometer," *The Review of Scientific Instruments*, **21**, No. 2, 169–176 (February 1950).

[6] C. M. Crain, "Apparatus for Recording Fluctuations in the Refractive Index of the Atmosphere at 3.2 Centimeters Wave-Length," *The Review of Scientific Instruments*, **21**, No. 5, 456–457 (May 1950).

[7] M. Nelkin, "Universality and Scaling in Fully Developed Turbulence," *Advances in Physics*, **43**, No. 2, 143–181 (1994). A less challenging description is given in the following article by the same author: "In What Sense Is Turbulence an Unsolved Problem?" *Science*, **255**, 566–570 (31 January 1992).

[8] L. F. Richardson, *Weather Prediction by Numerical Process* (Cambridge University Press, Cambridge, 1922).

[9] See [2], pages 46–76.

[10] G. K. Batchelor, *The Theory of Homogeneous Turbulence* (Cambridge University Press, Cambridge, 1953).

[11] A. N. Kolmogorov, "The Local Structure of Turbulence in Incompressible Viscous Fluid for Very Large Reynolds' Numbers," *Comptes Rendus (Doklady) de l'Academie des Sciences de l'URSS*, **30**, 301–305 (1941). (This journal has an English-language version.)

[12] R. H. Kraichnan, "On Kolmogorov's Inertial-Range Theories," *Journal of Fluid Mechanics*, **62**, part 2, 305–330 (23 January 1974).

[13] S. Pond, S. D. Smith, P. F. Hamblin and R. W. Burling, "Spectra of Velocity and Temperature Fluctuations in the Atmospheric Boundary Layer Over the Sea," *Journal of the Atmospheric Sciences*, **23**, No. 4, 376–386 (July 1966).

[14] S. Corrsin, "On the Spectrum of Isotropic Temperature Fluctuations in an Isotropic Turbulence," *Journal of Applied Physics*, **22**, No. 4, 469–473 (April 1951).

[15] A. M. Obukhov, "Structure of the Temperature Field in a Turbulent Flow," *Izvestiya Akademii Nauk SSSR, Seriya Geograficheskaya i Geofizicheskaya* (Bulletin of the Academy of Sciences of the USSR, Geographical and Geophysical Series), **13**, No. 1, 58–69 (1949). (English translation available from National Research Council of Canada, Ottawa, Ontario, Canada K1A OS2.)

[16] A. S. Monin and A. M. Yaglom, *Statistical Fluid Mechanics*, vol. 2 (MIT Press, Cambridge, MA, 1975).

[17] See [2], pages 65–67.

[18] G. K. Batchelor, "Small-Scale Variation of Convected Quantities Like Temperature in Turbulent Fluid. Part I. General Discussion and the Case of Small Conductivity," *Journal of Fluid Mechanics*, **5**, Part 1, 113–133 (January 1959).

[19] R. J. Hill, "Models of the Scalar Spectrum for Turbulent Advection," *Journal of Fluid Mechanics*, **88**, Part 3, 541–562 (13 October 1978).

[20] R. J. Hill and S. F. Clifford, "Modified Spectrum of Atmospheric Temperature Fluctuations and its Application to Optical Propagation," *Journal of the Optical Society of America*, **68**, No. 7, 892–899 (July 1978).

[21] R. Frehlich, "Laser Scintillation Measurements of the Temperature Spectrum in the Atmospheric Surface Layer," *Journal of the Atmospheric Sciences*, **49**, No. 16, 1494–1509 (15 August 1992).

[22] R. J. Hill, "Review of Optical Scintillation Methods of Measuring the Refractive-Index Spectrum, Inner Scale and Surface Fluxes," *Waves in Random Media*, **2**, No. 3, 179–201 (July 1992).

[23] F. H. Champagne, C. A. Friehe, J. C. LaRue and J. C. Wyngaard, "Flux Measurements, Flux-Estimation Techniques, and Fine-Scale Turbulence Measurements in the Unstable Surface Layer Over Land," *Journal of the Atmospheric Sciences*, **34**, No. 3, 515–530 (March 1977).

[24] R. M. Williams and C. A. Paulson, "Microscale Temperature and Velocity Spectra in the Atmospheric Boundary Layer," *Journal of Fluid Mechanics*, **83**, Part 3, 547–567 (5 December 1977).

[25] G. R. Ochs and R. J. Hill, "Optical-Scintillation Method of Measuring Turbulence Inner Scale," *Applied Optics*, **24**, No. 15, 2430–2432 (1 August 1985).

[26] R. J. Hill and G. R. Ochs, "Inner-Scale Dependence of Scintillation Variances Measured in Weak Scintillation," *Journal of the Optical Society of America A*, **9**, No. 8, 1406–1411 (August 1992).

[27] R. J. Hill, "Comparison of Scintillation Methods for Measuring the Inner Scale of Turbulence," *Applied Optics*, **27**, No. 11, 2187–2193 (1 June 1988).

[28] Th. von Karman, "Progress in the Statistical Theory of Turbulence," *Proceedings of National Academy of Science, U.S.*, **34**, 530–539 (1948). See also J. O. Hinze, *Turbulence*, Second ed (McGraw-Hill, New York, 1975), 244–249 and 298–301.

[29] D. P. Greenwood and D. O. Tarazano, "A Proposed Form for the Atmospheric Microtemperature Spatial Spectrum in the Input Range," USAF Rome Air Development Center, Technical Report RADC-TR-74-19 (February 1974).

[30] V. V. Voitsekhovich, "Outer Scale of Turbulence: Comparison of Different Models," *Journal of the Optical Society of America A*, **12**, No. 6, 1346–1352 (June 1995).

[31] V. P. Lukin, *Atmospheric Adaptive Optics* (SPIE Optical Engineering Press, Bellingham, Washington, 1995; originally published in Russian in 1986), 63 and 85.

[32] V. P. Lukin and V. V. Pokasov, "Optical Wave Phase Fluctuations," *Applied Optics*, **20**, No. 1, 121–135 (1 January 1981).

[33] V. P. Lukin, V. V. Pokasov and S. S. Khmelevtsov, "Investigation of the Time Characteristics of Fluctuations of the Phases of Optical Waves Propagating in the Bottom Layer of the Atmosphere," *Izvestiya Vysshikh Uchebnykh Zavedenii, Radiofizika (Soviet Radiophysics)*, **15**, No. 12, 1426–1430 (December 1972).

[34] M. C. Thompson, A. W. Kirkpatrick and W. B. Grant, "Measurements of Radio Refractive Index Microstructure of the Near-Ground Atmosphere," *Journal of Geophysical Research*, **73**, No. 20, 6425–6433 (15 October 1968).

[35] A. S. Drofa, "Determination of Certain Parameters of Atmospheric Turbulence from Optical Measurements," *Izvestiya Akademii Nauk, Fizika*

*Atmosfery i Okeana (Bulletin of the Academy of Sciences of the USSR, Atmospheric and Oceanic Physics)*, **15**, No. 5, 358–363 (1979).

[36] See [31], pages 99–101.

[37] K. C. Yeh and C. H. Liu, "Radio Wave Scintillations in the Ionosphere," *Proceedings of the IEEE*, **70**, No. 4, 324–360 (April 1982).

[38] A. Ishimaru, *Wave Propagation and Scattering in Random Media*, vol. 2 (Academic Press, San Diego, CA, 1978), 338–339.

[39] S. M. Rytov, Yu. A. Kravtsov and V. I. Tatarskii, *Principles of Statistical Radiophysics 3, Elements of Random Fields* (Springer-Verlag, Berlin, 1989), 9.

[40] A. I. Kon, "Qualitative Theory of Amplitude and Phase Fluctuations in a Medium with Anisotropic Turbulent Irregularities," *Waves in Random Media*, **4**, No. 3, 297–306 (July 1994).

[41] See [2], pages 35–38.

[42] See [39], pages 17–19.

[43] "The Global Observing System of the World Weather Watch," World Meteorological Organization, WMO No. 872 (1998). (This document can be obtained from World Weather Watch Department, WMO/OMM, Case Postale No. 2300, CH-1211 Genève 2, Switzerland.)

[44] E. K. Smith and S. Weintraub, "The Constants in the Equation for Atmospheric Refractive Index at Radio Frequencies," *Proceedings of the IRE*, **41**, No. 8, 1035–1037 (August 1953).

[45] See [1], pages 4–9.

[46] J. H. Churnside, "Optical Remote Sensing," in *Wave Propagation in Random Media (Scintillation)*, edited by V. I. Tatarskii, A. Ishimaru and V. U. Zavorotny (SPIE and Institute of Physics Publishing, Bristol, 1993), 235–247.

[47] L. C. Andrews and R. L. Phillips, *Laser Beam Propagation through Random Media* (SPIE Optical Engineering Press, Bellingham, WA, 1998), 49.

[48] R. R. Beland, "Propagation through Atmospheric Optical Turbulence," Chapter 2 in *Atmospheric Propagation of Radiation*, edited by F. R. Smith, vol. 2 of the *Infrared and Electro-Optical Systems Handbook* (SPIE Optical Engineering Press, Bellingham, WA, 1993), 171–174.

[49] K. L. Ho, N. D. Mavrokoukoulakis and R. S. Cole, "Determination of the Atmospheric Refractive Index Structure Parameter from Refractivity Measurements and Amplitude Scintillation Measurements at 36 GHz," *Journal of Atmospheric and Terrestrial Physics*, **40**, No. 6, 745–747 (June 1978).

[50] C. G. Helmis, D. N. Asimakopoulos, C. A. Caroubalos, R. S. Cole, F. C. Medeiros Filho and D. A. R. Jayasuriya, "A Quantitative Comparison of the Refractive Index Structure Parameter Determined from Refractivity Measurements and Amplitude Scintillation at 36 GHz," *IEEE Transactions on Geoscience and Remote Sensing*, **GE-21**, No. 2, 221–224 (April 1983).

[51] Ting-i Wang, G. R. Ochs and S. F. Clifford, "A Saturation-Resistant Optical Scintillometer to Measure $C_n^2$," *Journal of the Optical Society of America*, **68**, No. 3, 334–338 (March 1978).

[52] M. A. Kallistratova and D. F. Timanovskiy, "The Distribution of the Structure Constant of Refractive Index Fluctuations in the Atmospheric Surface Layer," *Izvestiya Akademii Nauk, Fizika Atmosfery i Okeana*

*(Bulletin of the Academy of Sciences of the USSR, Atmospheric and Oceanic Physics)*, **7**, No. 1, 46–48 (January 1971).

[53] J. L. Spencer, "Long-Term Statistics of Atmospheric Turbulence near the Ground," USAF Rome Air Development Center, Technical Report RADC-TR-78-182 (August 1978).

[54] R. L. Fante, "Electromagnetic Beam Propagation in Turbulent Media: An Update," *Proceedings of the IEEE*, **68**, No. 11, 1424–1443 (November 1980).

[55] R. S. Lawrence, G. R. Ochs and S. F. Clifford, "Measurements of Atmospheric Turbulence Relevant to Optical Propagation," *Journal of the Optical Society of America*, **60**, No. 6, 826–830 (June 1970).

[56] B. M. Koprov, S. L. Zubkovsky, V. M. Koprov, M. I. Fortus and T. I. Makarova, "Statistics of Air Temperature Spatial Variability in the Atmospheric Surface Layer," *Boundary Layer Meteorology*, **88**, No. 3, 399–423 (September 1998).

[57] A. S. Gurvich and V. P. Kukharets, "The Influence of Intermittence of Atmospheric Turbulence on the Scattering of Radio Waves," *Radiotekhnika i Electronika (Radio Engineering and Electronic Physics)*, **30**, No. 8, 1531–1537 (1985).

[58] R. Hinder, "Fluctuations of Water Vapour Content in the Troposphere as Derived from Interferometric Observations of Celestial Radio Sources," *Journal of Atmospheric and Terrestrial Physics*, **34**, No. 7, 1171–1186 (July 1972).

[59] A. R. Jacobson and R. Sramek, "A Method for Improved Microwave-Interferometer Remote Sensing of Convective Boundary Layer Turbulence Using Water Vapor as a Passive Tracer," *Radio Science*, **32**, No. 5, 1851–1860 (September–October 1997).

[60] W. J. Altenhoff, J. W. M. Baars, D. Downes and J. E. Wink, "Observations of Anomalous Refraction at Radio Wavelengths," *Astronomy and Astrophysics*, **184**, Nos. 1–2, 381–385 (October 1987).

[61] U. Frisch, P. L. Sulem and M. Nelkin, "A Simple Dynamical Model of Intermittent Fully Developed Turbulence," *Journal of Fluid Mechanics*, **87**, Part 4, 719–736 (29 August 1978).

[62] V. I. Tatarskii and V. U. Zavorotnyi, "Wave Propagation in Random Media with Fluctuating Turbulent Parameters," *Journal of the Optical Society of America A*, **2**, No. 12, 2069–2076 (December 1985).

[63] V. I. Tatarskii, "Some New Aspects in the Problem of Waves and Turbulence," *Radio Science*, **22**, No. 6, 859–865 (November 1987).

[64] E. E. Gossard and K. C. Yeh, Forward, Special Issue on "Radar Investigations of the Clear Air," *Radio Science*, **15**, No. 2, 147–150 (March–April 1980).

[65] M. C. Thompson, F. E. Marler and K. C. Allen, "Measurement of the Microwave Structure Constant Profile," *IEEE Transactions on Antennas and Propagation*, **AP-28**, No. 2, 278–280 (March 1980).

[66] E. E. Gossard and R. G. Strauch, *Radar Observations of Clear Air and Clouds* (Elsevier, Amsterdam, 1983), 62–65.

[67] D. Atlas, K. R. Hardy and T. G. Konrad, "Radar Detection of the Tropopause and Clear Air Turbulence," *12th Weather Radar Conference, American Meteorological Society* (Boston, MA, 1966), 279–284.

[68] E. E. Gossard, R. B. Chadwick, T. R. Detman and J. Gaynor, "Capability of Surface-Based Clear-Air Doppler Radar for Monitoring Meteorological

Structure of Elevated Layers," *Journal of Climate and Applied Meteorology*, **23**, No. 3, 474–485 (March 1984).

[69] E. E. Gossard, R. B. Chadwick, K. P. Moran, R. G. Strauch, G. E. Morrison and W. C. Campbell, "Observations of Winds in the Clear Air Using an FM-CW Doppler Radar," *Radio Science*, **13**, No. 2, 285–289 (March–April 1978).

[70] R. A. Kropfli, I. Katz, T. G. Konrad and E. B. Dobson, "Simultaneous Radar Reflectivity Measurements and Refractive Index Spectra in the Clear Atmosphere," *Radio Science,* **3** (New Series), No. 10, 991–994 (October 1968).

[71] P. K. James, "A Review of Radar Observations of the Troposphere in Clear Air Conditions," *Radio Science*, **15**, No. 2, 151–175 (March– April 1980).

[72] R. K. Crane, "A Review of Radar Observations of Turbulence in the Lower Stratosphere," *Radio Science*, **15**, No. 2, 177–193 (March–April 1980).

[73] T. E. VanZandt, J. L. Green, K. S. Gage and W. L. Clark, "Vertical Profiles of Refractivity Structure Constant: Comparison of Observations by the Sunset Radar with a New Theoretical Model," *Radio Science*, **13**, No. 5, 819–829 (September–October 1978).

[74] R. F. Woodman and A. Guillen, "Radar Observations of Winds and Turbulence in the Stratosphere and Mesosphere," *Journal of the Atmospheric Sciences*, **31**, No. 2, 493–505 (March 1974).

[75] R. F. Woodman, "High-Altitude-Resolution Stratospheric Measurements with the Arecibo 430-MHz Radar," *Radio Science*, **15**, No. 2, 417–422 (March–April 1980).

[76] R. F. Woodman, "High-Altitude-Resolution Stratospheric Measurements with the Arecibo 2380-MHz Radar," *Radio Science*, **15**, No. 2, 423–430 (March–April 1980).

[77] R. E. Good, B. J. Watkins, A. F. Quesada, J. H. Brown and G. B. Loriot, "Radar and Optical Measurements of $C_n^2$," *Applied Optics*, **21**, No. 18, 3373–3376 (15 September 1982).

[78] B. J. Watkins and R. H. Wand, "Observations of Clear Air Turbulence and Winds with the Millstone Hill Radar," *Journal of Geophysical Research*, **86**, No. C10, 9605–9614 (20 October 1981).

[79] G. D. Nastrom, K. S. Gage and W. L. Ecklund, "Variability of Turbulence, 4–20 km, in Colorado and Alaska from MST Radar Observations," *Journal of Geophysical Research*, **91**, No. D6, 6722–6734 (20 May 1986).

[80] F. D. Eaton, W. A. Peterson, J. R. Hines, K. R. Peterman, R. E. Good, R. R. Beland and J. H. Brown, "Comparisons of VHF Radar, Optical, and Temperature Fluctuation Measurements of $C_n^2$, $r_0$ and $\theta_0$," *Theoretical and Applied Climatology*, **39**, No. 1, 17–29 (1988).

[81] J. Röttger, "Reflection and Scattering of VHF Radar Signals from Atmospheric Refractivity Structures," *Radio Science*, **15**, No. 2, 259–276 (March–April 1980).

[82] J. Röttger and M. F. Larsen, "UHF/VHF Radar Techniques for Atmospheric Research and Wind Profiler Applications," in itRadar in Meteorology, edited by D. Atlas (American Meteorological Society, Boston, MA, 1990), 235–281.

[83] R. F. Woodman and Yen-Hsyang Chu, "Aspect Sensitivity Measurements of VHF Backscatter Made with the Chung-Li Radar: Plausible Mechanisms," *Radio Science*, **24**, No. 2, 113–125 (March–April 1989).

[84] E. E. Gossard, D. C. Welsh and R. G. Strauch, "Radar-Measured Height Profiles of $C_n^2$ and Turbulence Dissipation Rate Compared with Radiosonde Data during October 1989 at Denver," NOAA Technical Report ERL 442-WPL 63, U.S. Department of Commerce, NOAA, ERL (August 1990).

[85] K. S. Gage, "Radar Observations of the Free Atmosphere: Structure and Dynamics," in *Radar in Meteorology*, edited by D. Atlas (American Meteorological Society, Boston, MA, 1990), 534–565.

[86] A. D. Wheelon, "Backscattering by Turbulent Irregularities: A New Analytical Description," *Proceedings of the IEEE*, **60**, No. 3, 252–265 (March 1972).

[87] R. J. Doviak and D. S. Zrinic, "Reflection and Scatter Formula for Anisotropically Turbulent Air," *Radio Science*, **19**, No. 1, 325–336 (January–February 1984).

[88] A. D. Wheelon, "Corona: The First Reconnaissance Satellites," *Physics Today*, **50**, No. 2, 24–30 (February 1997).

[89] C. W. Fairall and A. S. Frisch, "Diurnal and Annual Variations in Mean Profiles of $C_n^2$," NOAA Technical Memorandum ERL WPL-195, U.S. Department of Commerce, NOAA, ERL (March 1991).

[90] L. R. Tsvang, "Some Characteristics of the Spectra of Temperature Pulsations in the Boundary Layer of the Atmosphere," *Izvestiya Akademii Nauk SSSR, Seriya Geofizicheskaya (Bulletin of the Academy of Sciences of the USSR, Geophysical Series)*, No. 10, 961–965 (October 1963).

[91] L. R. Tsvang, "Measurements of the Spectrum of Temperature Fluctuations in the Free Atmosphere," *Izvestiya Akademii Nauk SSSR, Seriya Geofizicheskaya (Bulletin of the Academy of Sciences of the USSR, Geophysical Series)*, No. 1, 1117–1120 (January 1960).

[92] J. L. Bufton, P. O. Minott and M. W. Fitzmaurice, "Measurements of Turbulence Profiles in the Troposphere," *Journal of the Optical Society of America*, **62**, No. 9, 1068–1070 (September 1972).

[93] C. E. Coulman, "Vertical Profiles of Small-Scale Temperature Structure in the Atmosphere," *Boundary-Layer Meteorology*, **4**, 169–177 (1973).

[94] R. Barletti, G. Ceppatelli, L. Paterno, A. Righini and N. Speroni, "Mean Vertical Profile of Atmospheric Turbulence Relevant for Astronomical Seeing," *Journal of the Optical Society of America*, **66**, No. 12, 1380–1383 (December 1976).

[95] G. R. Ochs, Ting-i Wang, R. S. Lawrence and S. F. Clifford, "Refractive-Turbulence Profiles Measured by One-Dimensional Spatial Filtering of Scintillations," *Applied Optics*, **15**, No. 10, 2504–2510 (October 1976).

[96] G. C. Loos and C. B. Hogge, "Turbulence of the Upper Atmosphere and Isoplanatism," *Applied Optics*, **18**, No. 15, 2654–2661 (1 August 1979).

[97] J. Vernin, "Atmospheric Turbulence Profiles," in *Wave Propagation in Random Media (Scintillation)*, edited by V. I. Tatarskii, A. Ishimaru and V. U. Zavorotny (SPIE and Institute of Physics Publishing, Bristol, 1993), 248–260.

[98] J. C. Kaimal and J. J. Finnigan, *Atmospheric Boundary Layer Flows, their Structure and Measurement* (Oxford University Press, Oxford, 1994).

[99] A. Consortini, F. Cochetti, J. H. Churnside and R. J. Hill, "Inner-Scale Effect on Irradiance Variance Measured for Weak-to-Strong Atmospheric Scintillation," *Journal of the Optical Society of America A*, **10**, No. 11, 2354–2362 (November 1993).

[100] F. D. Eaton and G. D. Nastrom, "Preliminary Estimates of the Vertical Profiles of Inner and Outer Scales from White Sands Missile Range, New Mexico, VHF Radar Observations," *Radio Science*, **33**, No. 4, 895–903 (July–August 1998).

[101] See [2], pages 77–78.

[102] R. S. Cole, K. L. Ho and N. D. Mavrokoukoulakis, "The Effect of the Outer Scale of Turbulence and Wavelength on Scintillation Fading at Millimeter Wavelengths," *IEEE Transactions on Antennas and Propagation*, **AP-26**, No. 5, 712–715 (September 1978).

[103] G. D. Nastrom and K. S. Gage, "A Climatology of Atmospheric Wavenumber Spectra of Wind and Temperature Observed by Commercial Aircraft," *Journal of the Atmospheric Sciences*, **42**, No. 9, 950–960 (1 May 1985).

[104] See [2], pages 73–74.

[105] K. Davies, *Ionospheric Radio Propagation* (National Bureau of Standards Monograph 80, U.S. Government Printing Office, Washington, November 1965).

[106] R. C. Sagalyn and M. Smiddy, "Electrical Processes in the Nighttime Exosphere," *Journal of Geophysical Research*, **69**, No. 9, 1809–1823 (1 May 1964).

[107] J. S. Hey, S. J. Parsons and J. W. Phillips, "Fluctuations in Cosmic Radiation at Radio-Frequencies," *Nature*, **158**, 234 (17 August 1946).

[108] C. G. Little and A. C. B. Lovell, "Origin of the Fluctuations in the Intensity of Radio Waves from Galactic Sources: Jodrell Bank Observations," *Nature*, **165**, No. 4194, 423–424 (18 March 1950).

[109] F. G. Smith, "Origin of the Fluctuations in the Intensity of Radio Waves from Galactic Sources: Cambridge Observations," *Nature*, **165**, No. 4194, 422–423 (18 March 1950).

[110] M. Spencer, "The Shape of Irregularities in the Upper Ionosphere," *The Proceedings of the Physical Society*, Section B, **68**, 493–503 (August 1955).

[111] D. G. Singleton, "Dependence of Satellite Scintillations on Zenith Angle and Azimuth," *Journal of Atmospheric and Terrestrial Physics*, **32**, No. 5, 789–803 (May 1970).

[112] C. L. Rufenach, "Power-Law Wavenumber Spectrum Deduced from Ionospheric Scintillation Observations," *Journal of Geophysical Research*, **77**, No. 25, 4761–4772 (1 September 1972).

[113] D. G. Singleton, "Power Spectra of Ionospheric Scintillations," *Journal of Atmospheric and Terrestrial Physics*, **36**, 113–133 (January 1974).

[114] P. L. Dyson, J. P. McClure and W. B. Hanson, "*In situ* Measurements of the Spectral Characteristics of F Region Ionospheric Irregularities," *Journal of Geophysical Research*, **79**, No. 10, 1497–1502 (1 April 1974).

[115] R. C. Livingston, C. L. Rino, J. P. McClure and W. B. Hanson, "Spectral Characteristics of Medium-Scale Equatorial F Region Irregularities," *Journal of Geophysical Research*, **86**, No. A4, 2421–2428 (1 April 1981).

[116] C. L. Rino, R. Tsunoda, J. Petriceks, R. C. Livingston, M. C. Kelley and K. D. Baker, "Simultaneous Rocket-Borne Beacon and *in situ* Measurements of Equatorial Spread F – Intermediate Wavelength Results," *Journal of Geophysical Research*, **86**, No. A4, 2411–2420 (1 April 1981).

[117] H. G. Booker, "The Role of Acoustic Gravity Waves in the Generation of Spread-F and Ionospheric Scintillation," *Journal of Atmospheric and Terrestrial Physics*, **41**, No. 5, 501–515 (May 1979).

[118] S. Basu and S. Basu, "Ionospheric Structures and Scintillation Spectra," in *Wave Propagation in Random Media (Scintillation)*, edited by V. I. Tatarskii, A. Ishimaru and V. U. Zavorotny (SPIE and Institute of Physics Publishing, Bristol, 1993), 139–155.

[119] I. P. Shkarofsky, "Generalized Turbulence Space-Correlation and Wave-number Spectrum-Function Pairs," *Canadian Journal of Physics*, **46**, No. 19, 2133–2153 (October 1968).

[120] R. Umeki, C. H. Liu and K. C. Yeh, "Multifrequency Studies of Ionospheric Scintillations," *Radio Science*, **12**, No. 2, 311–317 (March–April 1977).

[121] A. Bhattacharyya, R. G. Rastogi and K. C. Yeh, "Signal Frequency Dependence of Ionospheric Amplitude Scintillations," *Radio Science*, **25**, No. 4, 289–297 (July–August 1990).

[122] C. L. Rino, "A Power Law Phase Screen Model for Ionospheric Scintillation, 1. Weak Scatter," *Radio Science*, **14**, No. 6, 1135–1145 (November–December 1979).

[123] I. L. Jones, "Further Observations of Radio Stellar Scintillation," *Journal of Atmospheric and Terrestrial Physics*, **19**, No. 1, 26–36 (1960).

[124] R. K. Crane, "Spectra of Ionospheric Scintillation," *Journal of Geophysical Research*, **81**, No. 13, 2041–2050 (1 May 1976).

[125] R. K. Crane, "Ionospheric Scintillation," *Proceedings of the IEEE*, **65**, No. 2, 180–199 (February 1977).

[126] See [2], page 13.

[127] R. J. Sasiela, *Electromagnetic Wave Propagation in Turbulence* (Springer-Verlag, Berlin, 1994), 24.

[128] J. W. Strohbehn, "Line-of-Sight Wave Propagation Through the Turbulent Atmosphere," *Proceedings of the IEEE*, **56**, No. 8, 1301–1318 (August 1968).

# 3
# Geometrical Optics Expressions

Geometrical optics provides an approximate solution for the electromagnetic field equations. It is valid when the wavelength is very small relative to other dimensions of the problem. Effects of diffraction are completely ignored in this approach. One can start with Maxwell's vector field equations and derive the geometrical optics expressions for the phase and amplitude of a signal from them [1]. We shall enter the process at a later stage and begin with the scalar wave equation for the electric field which ignores the very small changes in polarization induced by random media.[1] For a dielectric constant that varies with position

$$\nabla^2 E + k^2 \varepsilon(\mathbf{r}) E = \text{source} \tag{3.1}$$

where the source is described by a function that vanishes everywhere except at the transmitter. The solution would be a plane wave if $\varepsilon(\mathbf{r})$ were constant and the source at a great distance:

$$E = E_0 \exp(ikz)$$

This suggests that we try a solution that is the product of an amplitude term and an exponential phase term:

$$E(\mathbf{r}) = A(\mathbf{r}) \exp[ik\Psi(\mathbf{r})] \tag{3.2}$$

The function $\Psi(\mathbf{r})$ is called the *eikonal* and defines the phase of the signal when it is multiplied by the electromagnetic wavenumber $k$.

The basic assumption of geometrical optics is that neither the dielectric constant nor the boundary conditions change significantly over one wavelength. This suggests that the amplitude and phase terms are slowly

---

[1] The justification for this simplification is discussed in Chapter 11 of the second volume.

varying functions of position. We expand the amplitude in inverse powers of
the wavenumber to exploit that assumption:

$$A(\mathbf{r}) = A_1(\mathbf{r}) + \frac{A_2(\mathbf{r})}{ik} + \frac{A_3(\mathbf{r})}{(ik)^2} \cdots \tag{3.3}$$

The trial solution (3.2) with this expression for the amplitude is substituted
into the wave equation and the terms are grouped according to descending
powers of the wavenumber:

$$
\begin{aligned}
[\nabla^2 + k^2 \varepsilon(\mathbf{r})]E \;=\; & k^2[\varepsilon(\mathbf{r}) - (\nabla\Psi)^2] + ik[\,2\,\boldsymbol{\nabla}\Psi \cdot \boldsymbol{\nabla}A_1 + A_1\,\nabla^2\Psi] \\
& + [\,2\,\boldsymbol{\nabla}\Psi \cdot \boldsymbol{\nabla}A_2 + A_2\,\nabla^2\Psi + \nabla^2 A_1] \\
& + \frac{i}{k}[\,2\,\boldsymbol{\nabla}\Psi \cdot \boldsymbol{\nabla}A_3 + A_3\,\nabla^2\Psi + \nabla^2 A_2] = \text{source}
\end{aligned}
$$

$$\tag{3.4}$$

The coefficients of different powers of $k$ must vanish individually since this
equation must hold for all values of the frequency. The wavenumber is large
for the problems which lend themselves to geometrical optics and we keep
only the first two terms. This provides two differential equations that
together determine $A_1$ and $\Psi$.

The *eikonal equation* is the central result of geometrical optics and results
from the requirement that the coefficient of $k^2$ vanish:

$$(\boldsymbol{\nabla}\Psi)^2 = \varepsilon(\mathbf{r}) \tag{3.5}$$

The eikonal is determined uniquely by the dielectric constant and by the
type of wave transmitted. With it one can describe the ray path followed by
the signal from the transmitter to the receiver. The surfaces described by
$\Psi(r) = $ constant are called *geometrical wave-fronts*.

The amplitude of the signal is determined from the second term in
(3.4), which is proportional to $k$. Its necessary vanishing generates the
*transport equation*:

$$2\,\boldsymbol{\nabla}\Psi \cdot \boldsymbol{\nabla}A_1 + A_1\,\nabla^2\Psi = 0 \tag{3.6}$$

When the eikonal is known it can be substituted into this equation to
determine the amplitude. This equation does not contain the wavelength
and its solution cannot describe the influence of diffraction on the amplitude of the signal.

Once the eikonal and amplitude have been determined, the challenge is
that of finding out how they are perturbed when random fluctuations of the
dielectric constant are added to the normal atmospheric profile. The goal of

this chapter is to establish general expressions for fluctuations of the phase, amplitude and angle of arrival of the signal.

## 3.1 Solutions of the Eikonal Equation

The first task is that of determining the trajectory along which the signal travels. This is called the *ray path* and is illustrated in Figure 3.1. Geometrical optics provides an analytical description of this path when the profile of the dielectric constant is specified. That description depends on the solution of the eikonal equation (3.5), which can be expressed as a line integral

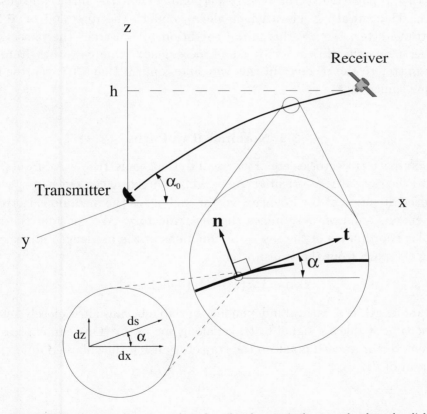

Figure 3.1: Rectangular coordinates used to describe the nominal ray path when the dielectric constant depends only on altitude. The transmitted signal is launched in the $(x, z)$ plane with an elevation angle $\alpha_0$. An expanded view of the path element identifies unit vectors normal to and at a tangent to the path. A further expanded view identifies the coordinate differentials.

taken along the nominal ray trajectory connecting the transmitter and the receiver:

$$\Psi(R) = \oint_T^R ds \sqrt{\varepsilon(s)} \tag{3.7}$$

This expression defines the *optical path length*. It satisfies the eikonal equation since, at any point along the trajectory,

$$\boldsymbol{\nabla}\Psi = \mathbf{t}\,\frac{\partial \Psi}{\partial R} = \mathbf{t}\sqrt{\varepsilon(R)} \tag{3.8}$$

where $\mathbf{t}$ is the unit vector at a tangent to the nominal ray path. This solution need not be unique because any unit vector multiplying $\sqrt{\varepsilon}$ would satisfy the eikonal equation. To show that it is a unique solution, we observe that the propagation vector must be perpendicular to the surface of constant phase. The initial ray elevation angle $\alpha_0$ defines the direction of $\boldsymbol{\nabla}\Psi$ at the transmitter. We use this initial condition to construct the phase front further along the trajectory by small increments. One can work from the transmitter to the receiver in this way and confirm that (3.7) represents a unique solution [2].

## 3.2 Nominal Ray Paths

The ray-path trajectories are orthogonal to the geometrical wave-fronts. We regard the ray paths as oriented curves whose direction coincides everywhere with the direction of the Poynting vector, so long as the medium is isotropic. The eikonal solution determines the ray trajectory. We consider the rectangular coordinates of the ray to be functions of the arc length $s$ measured along the path from the transmitter:

$$\mathbf{r}(s) = \mathbf{i}_x x(s) + \mathbf{i}_y y(s) + \mathbf{i}_x z(s)$$

The arc length is a natural independent variable because it is closely related to the time it takes a signal to travel along the path.[2] One can express the tangent vector at each point on the trajectory using the enlarged differential diagram of Figure 3.1:

$$\mathbf{t} = \frac{d\mathbf{r}}{ds} \tag{3.9}$$

---

[2] The actual travel time along the ray is given by

$$\frac{1}{c}\oint \frac{ds}{\sqrt{\varepsilon(s)}}.$$

Combining this with (3.8) gives

$$\sqrt{\varepsilon(\mathbf{r})}\,\frac{d\mathbf{r}}{ds} = \boldsymbol{\nabla}\Psi.$$

We differentiate this equation with respect to the arc length:

$$\frac{d}{ds}\left(\sqrt{\varepsilon(\mathbf{r})}\,\frac{d\mathbf{r}}{ds}\right) = \frac{d}{ds}(\boldsymbol{\nabla}\Psi)$$

$$= \left(\frac{d\mathbf{r}}{ds}\cdot\boldsymbol{\nabla}\right)(\boldsymbol{\nabla}\Psi)$$

$$= \frac{1}{\sqrt{\varepsilon(\mathbf{r})}}(\boldsymbol{\nabla}\Psi\cdot\boldsymbol{\nabla})(\boldsymbol{\nabla}\Psi)$$

In these steps we have again used (3.8) to replace the derivative with respect to the arc length by terms involving only $\Psi$ and $\varepsilon$. Using the vector identity

$$(\boldsymbol{\nabla}\Psi\cdot\boldsymbol{\nabla})(\boldsymbol{\nabla}\Psi) = \frac{1}{2}\,\boldsymbol{\nabla}(\boldsymbol{\nabla}\Psi)^2$$

and the eikonal equation (3.5) with the substitution

$$\boldsymbol{\nabla}(\sqrt{\varepsilon}) = \frac{1}{2}\frac{1}{\sqrt{\varepsilon}}\,\boldsymbol{\nabla}\varepsilon$$

the desired description of the ray-trajectory equations can be written

$$\frac{d}{ds}\left(\sqrt{\varepsilon(\mathbf{r})}\,\frac{d\mathbf{r}}{ds}\right) = \boldsymbol{\nabla}\left[\sqrt{\varepsilon(\mathbf{r})}\right]. \tag{3.10}$$

This vector equation defines the ray path once the profile of $\varepsilon(\mathbf{r})$ is known.

The first task is that of establishing the nominal ray trajectories. They are determined by the average dielectric constant which depends only on altitude in the troposphere and ionosphere. The rays are plane curves when the gradient of $\varepsilon$ has a constant direction. We use a flat-earth approximation and rectangular coordinates to write the vector equation (3.10) as three ordinary differential equations:

$$\frac{d}{ds}\left(\frac{dx}{ds}\,\sqrt{\varepsilon(z)}\right) = 0 \tag{3.11}$$

$$\frac{d}{ds}\left(\frac{dy}{ds}\,\sqrt{\varepsilon(z)}\right) = 0 \tag{3.12}$$

$$\frac{d}{ds}\left(\frac{dz}{ds}\,\sqrt{\varepsilon(z)}\right) = \frac{\partial}{\partial z}\sqrt{\varepsilon(z)} \tag{3.13}$$

The first ray equation is easily solved and the constant of integration set by the initial elevation angle of the ray:

$$\frac{dx}{ds}\sqrt{\varepsilon(z)} = \cos\alpha_0\sqrt{\varepsilon(0)} \tag{3.14}$$

This equation has a familiar meaning. From the inset to Figure 3.1 we see that the inclination angle of the tangent vector is defined by

$$\frac{dx}{ds} = \cos\alpha.$$

This holds at every point along the trajectory and, in combination with (3.14), it establishes *Snell's Law*:

$$\sqrt{\varepsilon(z)}\cos\alpha = \sqrt{\varepsilon(0)}\cos\alpha_0 \tag{3.15}$$

The progressive bending of the ray is thus related to the local dielectric constant. This relationship was first discovered experimentally and provides confirmation for a description based on geometrical optics. If the signals are launched in the $(x, z)$ plane, the second equation shows that the rays do not leave that plane:

$$\frac{dy}{ds} = 0 \tag{3.16}$$

With these results in hand, one can complete the description of the ray path. We rewrite the third trajectory equation (3.13) as follows:

$$\frac{\partial}{\partial z}\sqrt{\varepsilon(z)} = \frac{d}{ds}\left(\frac{dz}{ds}\sqrt{\varepsilon}\right) = \frac{dx}{ds}\frac{d}{dx}\left(\frac{dx}{ds}\frac{dz}{dx}\sqrt{\varepsilon}\right)$$

A differential equation that involves only $x$ and $z$ emerges if we again use (3.14) to express $dx/ds$:

$$\frac{d^2z}{dx^2} = \frac{\sec^2\alpha_0}{2\varepsilon(0)}\frac{\partial}{\partial z}\varepsilon(z)$$

One can verify that the solution of this equation is

$$x = \int_0^{z(x)} du\,\frac{\cos\alpha_0\sqrt{\varepsilon(0)}}{[\varepsilon(u) - \varepsilon(0)\cos^2\alpha_0]^{\frac{1}{2}}} \tag{3.17}$$

which tells one how $z$ and $x$ are related for an initial elevation angle $\alpha_0$ and a specified profile $\varepsilon(z)$. It is often used to describe ray bending both in

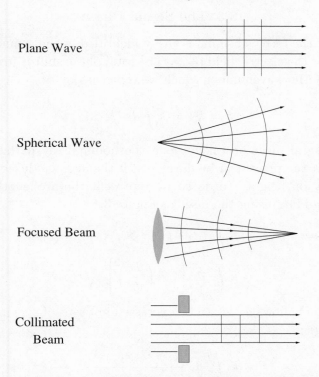

Figure 3.2: Geometrical optics approximations for four types of transmitted waves commonly used in propagation experiments. Plane waves are represented by an infinite array of parallel rays. Spherical waves are represented by diverging rays and focused beams by converging rays. Collimated beams are characterized by a spatially limited bundle of parallel rays.

the lower atmosphere and in the ionosphere. Similar trajectories describe propagation of acoustic signals in the ocean.

The ray trajectory reduces to a straight line when $\varepsilon(z)$ does not change appreciably over the path. This is usually the case both for astronomical observations and for terrestrial propagation experiments. These signals fall into the four classes illustrated by Figure 3.2. A plane wave represents the signals coming from distant stellar and satellite sources. In the geometrical optics approximation it is described by an infinite family of parallel rays if refractive bending in the atmosphere is ignored. A spherical wave is generated by a point source and is represented by a family of diverging rays that originate at the transmitter. Focused beams are described by invoking converging rays. A collimated beam is represented by a family of parallel rays bounded by the beam width.

## 3.3 The Signal Phase

The phase of the received signal is easily identified in the geometrical optics expression for the electric field (3.2). The total phase shift is just the eikonal multiplied by the wavenumber which we represent by $S$:

$$S = k\Psi = k \oint ds \sqrt{\varepsilon(s)} \tag{3.18}$$

The line integral must be taken along the nominal ray path. An explicit description of the phase can be developed if the average dielectric constant depends only on height. To do so we represent the arc-length element in terms of $dx$ and $dz$ using the inset to Figure 3.1:

$$S = k \int_0^h (dx^2 + dz^2)^{\frac{1}{2}} \sqrt{\varepsilon(z)}$$

$$= k \int_0^h dz \left[1 + \left(\frac{dx}{dz}\right)^2\right]^{\frac{1}{2}} \sqrt{\varepsilon(z)}$$

Equation (3.17) allows $dx/dz$ to be expressed in terms of $\varepsilon(z)$ and the initial elevation angle:

$$S = k \int_0^h dz \frac{\varepsilon(z)}{[\varepsilon(z) - \varepsilon(0)\cos^2\alpha_0]^{\frac{1}{2}}} \tag{3.19}$$

The phase of the signal can be determined from this expression when the dielectric profile is known. For a constant profile it becomes

$$\varepsilon(z) = \text{constant:} \qquad S = kh \text{ cosec } \alpha_0$$

since the surface refractivity is very close to unity. This represents a very large number of radians for atmospheric paths.

We are primarily interested in phase fluctuations that are induced by small changes in the dielectric constant:

$$\varepsilon(\mathbf{r}, t) = \varepsilon(z) + \Delta\varepsilon(x, y, z; t)$$

In this case the phase becomes

$$S(t) = k \oint ds [\varepsilon(z) + \Delta\varepsilon(x, y, z; t)]^{\frac{1}{2}}.$$

The square root can be expanded in powers of $\Delta\varepsilon$ so long as $\varepsilon(z)$ is large relative to the fluctuation:

$$S(t) = k \oint ds \left(\sqrt{\varepsilon(z)} + \frac{\Delta\varepsilon(x, y, z; t)}{2\sqrt{\varepsilon(z)}}\right)$$

We separate the phase into a nominal value and a random component:

$$S(t) = S_0 + \varphi(t) \tag{3.20}$$

The term $S_0$ represents the signal phase shift which would be measured in the absence of dielectric variations and can be estimated from (3.19). Phase fluctuations are identified with the second term $\varphi(t)$, which will be the main focus of this first volume:

$$\varphi(t) = \frac{1}{2} k \oint ds \, \frac{1}{\sqrt{\varepsilon(z)}} \, \Delta\varepsilon(x, y, z; t) \tag{3.21}$$

The phase fluctuation is a stochastic function of time because $\Delta\varepsilon$ is a random function of space and time. The denominator term $\sqrt{\varepsilon}$ weights the contributions of irregularities along the path. It is greatest where the nominal dielectric constant is smallest $d$ as it is for short-wave propagation near a reflection point in the ionosphere. It is important to remember that the arc length $ds$ and integration path also depend on $\varepsilon(z)$.

For a refracted ray that passes completely through the atmosphere, one can express the line element as we did before:

$$\varphi(t) = \frac{1}{2} k \int_0^h dz \, \frac{\Delta\varepsilon(z; t)}{[\varepsilon(z) - \varepsilon(0) \cos^2 \alpha_0]^{\frac{1}{2}}} \tag{3.22}$$

When the average dielectric constant does not change with altitude[3] and the source is a distant star or satellite:

$$\text{Star:} \quad \varphi(t) = \frac{k}{2} \operatorname{cosec} \alpha_0 \int_0^\infty dz \, \Delta\varepsilon(z; t) \tag{3.23}$$

This form is used to describe stellar phase fluctuations and transionospheric microwave propagation. For horizontal propagation $\varepsilon(z)$ is nearly constant and close to unity. In this case the ray path is a straight line connecting the transmitter and receiver,

$$\text{Link:} \quad \varphi(t) = \frac{k}{2} \int_0^R dx \, \Delta\varepsilon(x; t) \tag{3.24}$$

where $R$ is the path length to the receiver. This expression is widely used to describe phase fluctuations for microwave and optical links.

---

[3] We neglect the ground-level departure from unity since $\varepsilon(0) \approx 1 + 300 \times 10^{-6}$.

### 3.4 The Signal Amplitude

The transport equation describes the amplitude and its fluctuating component. We consider a plane wave that travels along a refracted ray path. Because the eikonal gradient is always aligned with the tangent vector we can rewrite the transport equation as follows:

$$A \, \nabla^2 \Psi = -2 \, \boldsymbol{\nabla} \Psi \cdot \boldsymbol{\nabla} A = -2 \left( \mathbf{t} \, \frac{\partial \Psi}{\partial s} \right) \cdot \boldsymbol{\nabla} A = -2 \sqrt{\varepsilon(s)} \, \frac{\partial A}{\partial s}$$

where we have dropped the subscript from $A_1$ for simplicity. This partial differential equation can be solved in terms of a line integral:

$$A(R) = A(T) \exp \left( -\frac{1}{2} \oint_T^R ds \, \frac{1}{\sqrt{\varepsilon(s)}} \, \nabla^2 \Psi \right)$$

If we substitute for $\Psi$ from (3.7) the amplitude measured at the receiver is expressed as a double line integral:

$$A(R) = A(T) \exp \left[ -\frac{1}{2} \oint_T^R \frac{ds}{\sqrt{\varepsilon(s)}} \, \nabla^2 \left( \oint_T^s ds' \, \sqrt{\varepsilon(s')} \right) \right] \tag{3.25}$$

This solution can be simplified by expressing the Laplacian in terms of derivatives taken along the ray path and normal to it:

$$\nabla^2 = \frac{\partial^2}{\partial s^2} + \boldsymbol{\nabla}_\perp \cdot \boldsymbol{\nabla}_\perp$$

The exponential of the sum is the product of the exponentials of individual terms:

$$A(R) = A(T) \exp \left( -\frac{1}{2} \oint_T^R \frac{ds}{\sqrt{\varepsilon(s)}} \, \frac{\partial^2}{\partial s^2} \oint_T^s ds' \, \sqrt{\varepsilon(s')} \right)$$

$$\times \exp \left( -\frac{1}{2} \oint_T^R \frac{ds}{\sqrt{\varepsilon(s)}} \, \boldsymbol{\nabla}_\perp \cdot \boldsymbol{\nabla}_\perp \oint_T^R ds' \, \sqrt{\varepsilon(s')} \right)$$

The second exponential can be written as follows:

$$\exp \left( -\frac{1}{2} \oint_T^R \frac{ds}{\sqrt{\varepsilon(s)}} \, \boldsymbol{\nabla}_\perp \cdot \left[ \boldsymbol{\nabla}_\perp \Psi(s) \right] \right)$$

which vanishes because the gradient of $\Psi$ always lies along the ray tangent. Derivatives with respect to $s$ operate along the nominal ray path so the first exponential can be evaluated as follows:

$$\exp\left(-\frac{1}{2}\oint_T^R ds\,\frac{1}{\sqrt{\varepsilon(s)}}\,\frac{\partial^2}{\partial s^2}\oint_T^s ds'\,\sqrt{\varepsilon(s')}\right)$$
$$= \exp\left[-\frac{1}{4}\ln\left(\frac{\varepsilon(R)}{\varepsilon(T)}\right)\right]$$

The evolution of the amplitude of the signal is therefore described by

$$A(R)[\varepsilon(R)]^{\frac{1}{4}} = A(T)[\varepsilon(T)]^{\frac{1}{4}}. \tag{3.26}$$

This relationship has an important consequence. The transport equation can be written in the form

$$\boldsymbol{\nabla}\left(A^2\,\boldsymbol{\nabla}\varphi\right) = 0$$

and, on replacing the eikonal gradient by $\sqrt{\varepsilon}\,\mathbf{t}$, we have

$$\boldsymbol{\nabla}\left(I\sqrt{\varepsilon}\,\mathbf{t}\right) = 0.$$

On integrating this equation over the volume of a ray tube with cross sectional area $\mathsf{A}$ and applying Gauss' theorem [1], we find that

$$I(R)\sqrt{\varepsilon(R)}\,\mathsf{A} = \text{constant}. \tag{3.27}$$

This relation describes conservation of energy in the context of geometrical optics.

A fluctuating amplitude component is generated when the dielectric constant has a random component:

$$A(R) + \delta A = A(T)\exp\left[-\frac{1}{2}\oint_T^R\frac{ds}{\sqrt{\varepsilon(s)}}\,\nabla^2\oint_T^s ds'\left(\sqrt{\varepsilon(s')}+\frac{1}{2}\frac{\Delta\varepsilon(s',t)}{\sqrt{\varepsilon(s')}}\right)\right]$$

The first term in the exponent leads to variation of the nominal signal amplitude:

$$A(R) + \delta A = A(R)\exp\left(-\frac{1}{4}\oint_T^R\frac{ds}{\sqrt{\varepsilon(s)}}\,\nabla^2\oint_T^s ds'\,\frac{\Delta\varepsilon(s',t)}{\sqrt{\varepsilon(s')}}\right)$$

This is usually expressed in terms of the *logarithmic amplitude fluctuation* $\chi$ and is small when geometrical optics is applicable:

$$\log\left(\frac{A(R)+\delta A}{A(R)}\right) = \log\left(1+\frac{\delta A}{A}\right) = \ln(1+\chi) \simeq \chi \qquad (3.28)$$

Combining these expressions yields

$$\chi(t) = -\frac{1}{4}\oint_T^R \frac{ds}{\sqrt{\varepsilon(s)}}\, \nabla^2 \oint_T^s ds'\, \frac{\Delta\varepsilon(s',t)}{\sqrt{\varepsilon(s')}}.$$

For a terrestrial path near the surface one can assume that the average dielectric constant does not change very much. In this case the path direction coincides with the horizontal coordinate $x$ identified in Figure 3.1:

$$\chi(t) = -\frac{1}{4\varepsilon_0}\int_0^R dx \left(\frac{\partial^2}{\partial x^2} + \frac{\partial^2}{\partial y^2} + \frac{\partial^2}{\partial z^2}\right) \int_0^z dx'\Delta\varepsilon(x',y,z;t)$$

The second derivative with respect to $x$ can be integrated twice to give

$$-\frac{1}{4\varepsilon_0}\left[\Delta\varepsilon(x,y,R;t) - \Delta\varepsilon(x,y,0;t)\right].$$

The difference between $\Delta\varepsilon$ at the end points is very small in comparison with the other terms which describe the logarithmic amplitude:

$$\chi(t) = -\frac{1}{4\varepsilon_0}\int_0^R dx \int_0^x dx' \left(\frac{\partial^2}{\partial y^2} + \frac{\partial^2}{\partial z^2}\right) \Delta\varepsilon(x',y,z;t)$$

The first integration can be carried out:

$$\text{Plane:}\quad \chi(t) = -\frac{1}{4\varepsilon_0}\int_0^R dx\,(R-x)\left(\frac{\partial^2}{\partial y^2} + \frac{\partial^2}{\partial z^2}\right)\Delta\varepsilon(x,y,z;t) \quad (3.29)$$

The path weighting $(R-x)$ indicates that eddies near the receiver have little influence on the scintillation. There is an important difference between this result and the corresponding expression for spherical waves [3]:

$$\text{Spherical:}\quad \chi(t) = -\frac{1}{4}\int_0^R dr\, r\left(1-\frac{r}{R}\right)\nabla_\perp^2\,\Delta\varepsilon(r,\theta,\phi;t) \qquad (3.30)$$

The additional weighting term $r$ means that eddies have little effect on the level of scintillation if they are near either the transmitter or the receiver.

These expressions were widely used in early studies but were found *not* to agree with experimental data. The problem is that measured scintillations depend on the wavelength of the radiation, which is entirely absent from these expressions. Later in this chapter we examine the conditions which must be met before one can use geometrical optics. These conditions relate primarily to amplitude fluctuations and are seldom satisfied in practice. We shall not pause here to develop expressions for the variance of $\chi$ from the expressions given above. Such results are primarily of historical interest and readers who wish to explore scintillation in the context of geometrical optics are referred to earlier books [4][5].

## 3.5 The Angle of Arrival

Fluctuations of the dielectric constant along the nominal trajectory also cause the ray path to wander as suggested in Figure 3.3. The integrated effect of many small refractive bendings produces an angular error in the tangent vector relative to the nominal ray. This error varies with time and accounts for the dancing of stellar images on photographic plates. It also induces tracking errors in precision radars. If the receiver is large enough to capture the wandering ray, one can describe the angular error with equations (3.11) through (3.13).

Let us begin by considering a signal traveling along a refracted path. The displacement of the ray *normal to the ray path* is described by the second ray equation and is illustrated in the top view of the propagation path provided in Figure 3.3. If we replace the dielectric constant by its perturbed value

$$\frac{d}{ds}\left(\frac{dy}{ds}\sqrt{\varepsilon + \Delta\varepsilon}\right) = \frac{\partial}{\partial y}\sqrt{\varepsilon + \Delta\varepsilon} = \frac{1}{2}\frac{\partial\Delta\varepsilon/\partial y}{\sqrt{\varepsilon + \Delta\varepsilon}}.$$

It is appropriate to neglect $\Delta\varepsilon$ in comparison with $\varepsilon$. One can then integrate both sides of this equation along the nominal ray path to establish the instantaneous cross-plane angular error:

$$\delta\theta_\perp = \frac{dy}{ds} = \frac{1}{2\sqrt{\varepsilon(R)}}\int_T^R ds\,\frac{1}{\sqrt{\varepsilon(s)}}\frac{\partial}{\partial y}\Delta\varepsilon(s,t) \qquad (3.31)$$

This expression tells one how the gradient components are weighted by the average dielectric constant at each point.

Random deflections *in the ray plane* can be analyzed using the side view of the ray trajectory illustrated in Figure 3.3. We are interested in a small variation in the normal vector of the ray normal. This is related to fluctuations in the coordinates that describe the trajectory:

$$\delta\mathbf{n} = \mathbf{i}_x(-\delta x\,\sin\alpha) + \mathbf{i}_z(\delta z\,\cos\alpha) \qquad (3.32)$$

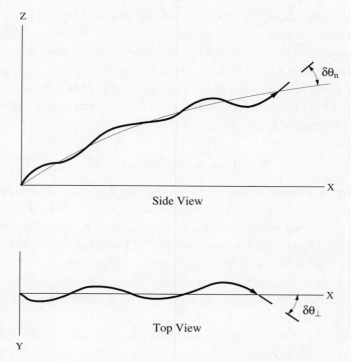

Figure 3.3: Two views of the ray path indicating the wandering induced by random variations of the dielectric constant and identifying the orthogonal components of the angle-of-arrival error.

The fluctuating components $\delta x$ and $\delta z$ can be found from the first and third ray equations:

$$\frac{d}{ds}\left(\frac{dx}{ds}\sqrt{\varepsilon}\right) = \frac{1}{2\sqrt{\varepsilon}}\,\frac{\partial}{\partial x}\Delta\varepsilon(x,y,z;t)$$

$$\frac{d}{ds}\left(\frac{dz}{ds}\sqrt{\varepsilon}\right) = \frac{1}{2\sqrt{\varepsilon}}\,\frac{\partial}{\partial z}[\varepsilon(z) + \Delta\varepsilon(x,y,z;t)]$$

We are interested in coordinate fluctuations about the nominal trajectory:

$$x(s) = x_0(s) + \delta x$$

$$z(s) = z_0(s) + \delta z$$

Substituting these expressions into the ray equations yields

$$\frac{d}{ds}(\delta x) = \frac{1}{2\sqrt{\varepsilon}} \int_T^R ds \; \frac{1}{\sqrt{\varepsilon}} \; \frac{\partial}{\partial x} \Delta\varepsilon(x, y, z; t)$$

$$\frac{d}{ds}(\delta z) = \frac{1}{2\sqrt{\varepsilon}} \int_T^R ds \; \frac{1}{\sqrt{\varepsilon}} \; \frac{\partial}{\partial z} \Delta\varepsilon(x, y, z; t)$$

The normal derivative is related to the local ray inclination $\alpha$ by the following expression:

$$\frac{\partial}{\partial n} = \cos\alpha \; \frac{\partial}{\partial z} - \sin\alpha \; \frac{\partial}{\partial x}$$

The final expression for the in-plane angular error becomes

$$\delta\theta_n = \frac{1}{2\sqrt{\varepsilon(R)}} \int_T^R ds \; \frac{1}{\sqrt{\varepsilon(s)}} \; \frac{\partial}{\partial n} \Delta\varepsilon(x, y, z; t) \qquad (3.33)$$

which is similar to the out-of-plane error (3.31).

From the two expressions, one can relate the angular-error *vector* to the normal component of the gradient of the dielectric variations integrated along the ray path:

$$\delta\boldsymbol{\theta} = \frac{1}{2\sqrt{\varepsilon(R)}} \int_T^R ds \; \frac{1}{\sqrt{\varepsilon(s)}} \; \boldsymbol{\nabla}_\perp \Delta\varepsilon(x, y, z; t) \qquad (3.34)$$

Using this form it is a simple matter to describe the angular error for a diverging wave. One casts the gradient operator in spherical coordinates and proceeds to calculate variances and correlations with it.

Recalling the definition for phase fluctuations (3.21), the angular error can be written in a form first established by Chandrasekhar [6]:

$$\delta\boldsymbol{\theta} = \frac{1}{k\sqrt{\varepsilon}} \; \boldsymbol{\nabla}_\perp \varphi \qquad (3.35)$$

It is often more convenient to use this version since we will already have established phase correlations for the various wave types of the transmitted signals. We shall develop this subject more fully in Chapter 7 and show how the result is influenced by the type of instrument used to measure the bearing angle.

## 3.6  Validity Conditions

It is important to know when the preceding formulas can be used to describe propagation in random media. Different conditions are imposed on the amplitude and phase of the signal. The signal phase is completely determined by values of the dielectric constant that lie along a single ray in geometrical optics. This places almost no restriction on phase fluctuations generated by random media. The angle of arrival is proportional to the derivative of phase and can also be used over a wide range of situations. By contrast, expressions for the amplitude and intensity of the signal are severely restricted. Three conditions limit one's ability to use geometrical optics.

### 3.6.1 The Smooth-medium Condition

The first restriction relates to the random medium itself. It is derived from the general conditions that allow one to replace the wave equation by the eikonal and transport equations. This becomes clear if one substitutes the trial solution (3.2) into the wave equation without resorting to the expansion for $A(\mathbf{r})$ given by (3.3). On separating the result into real and imaginary parts one finds that $A(\mathbf{r})$ and $\Psi(\mathbf{r})$ must satisfy two partial differential equations that together determine them exactly:

$$(\boldsymbol{\nabla}\Psi)^2 - \frac{1}{k^2}\frac{\nabla^2 A}{A} = \varepsilon(r) \tag{3.36}$$

$$2\,\boldsymbol{\nabla} A \cdot \boldsymbol{\nabla}\Psi + A\,\nabla^2\Psi = 0 \tag{3.37}$$

The second result is identical to the transport equation (3.6). The first equation reduces to the eikonal equation if one can neglect

$$\frac{1}{k^2}\frac{\nabla^2 A}{A}$$

The smallness of this term in comparison with other terms in (3.36) is central to the validity of geometrical optics. Freehafer showed that it can be ignored if two conditions are satisfied [2].

The first requirement is that the dielectric constant must not change appreciably over one wavelength:

$$\lambda\frac{|\boldsymbol{\nabla}\varepsilon|}{\varepsilon} \ll 1 \tag{3.38}$$

This is often called the smooth-medium condition and is the basis for the expansion (3.3). Our task is to relate this condition to the random fluctuations that occur in the dielectric constant. In the lower atmosphere we can approximate

$$\varepsilon = 1 + \Delta\varepsilon(\mathbf{r}, t)$$

and the smooth-medium condition becomes

$$\lambda \left| \boldsymbol{\nabla}[\Delta\varepsilon(\mathbf{r},t)] \right| \ll 1$$

Since $\Delta\varepsilon$ is a random function we must compute the variance of the left-hand side:

$$\lambda^2 \langle \boldsymbol{\nabla}_r[\Delta\varepsilon(\mathbf{r},t)] \cdot \boldsymbol{\nabla}_{r'}[\Delta\varepsilon(\mathbf{r}',t)] \rangle_{r=r'} \ll 1$$

The Fourier decomposition of the spatial correlation introduced in (2.44) allows this to be written as a moment of the turbulence spectrum:

$$\lambda^2 \int d^3\kappa\, \Phi_\varepsilon(\boldsymbol{\kappa})\, \boldsymbol{\nabla}_r \cdot \boldsymbol{\nabla}_{r'} \exp[i\boldsymbol{\kappa} \cdot (\mathbf{r} - \mathbf{r}')]\big|_{r=r'} \ll 1$$

or

$$\lambda^2 \int d^3\kappa\, \kappa^2 \Phi_\varepsilon(\boldsymbol{\kappa}) \ll 1$$

It is sufficient for this estimate to consider only isotropic irregularities:

$$4\pi\lambda^2 \int_0^\infty d\kappa\, \kappa^4 \Phi_\varepsilon(\kappa) \ll 1$$

The fourth moment emphasizes small wavenumbers and one can use the Kolmogorov model (2.62) to describe the lower atmosphere if an inner-scale cutoff is retained:

$$\Phi_\varepsilon(\kappa) = 4\Phi_n(\kappa) = 0.132 C_n^2 \kappa^{-\frac{11}{3}} \qquad 0 < \kappa < \kappa_{\mathrm{m}} \tag{3.39}$$

The requirement that the medium change by very little over a wavelength is therefore equivalent to the following statement:

$$C_n^2 \lambda^2 (\kappa_{\mathrm{m}})^{\frac{4}{3}} \ll 1 \tag{3.40}$$

This can also be expressed in terms of the structure function for refractive-index variations described by (2.43):

$$(C_n^2 \lambda^{\frac{2}{3}})(\kappa_{\mathrm{m}}\lambda)^{\frac{4}{3}} \ll 1 \tag{3.41}$$

With measured values for $C_n^2$ and $\kappa_{\mathrm{m}}$ one finds that this quantity is very small both for optical and for microwave transmissions. The smooth-medium condition therefore places no practical limit on geometrical optics when one is describing electromagnetic propagation through random media. We must look elsewhere to establish appropriate limits on its use.

Previous examination of the smooth-medium condition had approximated the gradient by

$$\frac{\nabla \varepsilon}{\varepsilon} \approx \frac{1}{\ell_0} \tag{3.42}$$

where $\ell_0$ is the smallest eddy size in the medium. This suggests that ray theory is valid only if the wavelength is much less than the inner scale length [7]:

$$\lambda \ll \ell_0 \tag{3.43}$$

It is now clear that the approximation (3.42) is too crude[4] and leads to a condition for geometrical optics that is unnecessarily restrictive. We shall learn later that the relationship (3.43) is required for modern propagation theories that are based on small-angle forward scattering and the parabolic-equation method. It is significant that geometrical optics is not constrained by (3.43). This explains why its predictions are valid and are confirmed by experiments over a surprisingly wide range of propagation conditions.

### 3.6.2  The Caustic Condition

A single ray path by itself cannot define the level of amplitude scintillation. One must examine the behavior of ray bundles to do so. In the geometrical optics approximation, amplitude and intensity are defined in terms of the closeness of one ray to another. The fractional change of the spacing between rays must be small over a distance of one wavelength. This means that one cannot use ray theory where the rays are converging or diverging rapidly, especially in regions where they cross to create caustics or come to a focus. This conclusion evidently depends on the type of signal that is launched into the medium.

One can describe this situation in a qualitative manner. The flow of energy is proportional to the intensity of the signal times the area of the ray bundle. This must be constant to satisfy conservation of energy. Expression (3.27) tells one how the local intensity must vary for a bundle of rays. In terms of the logarithmic amplitude fluctuation:

$$\sqrt{\varepsilon}\, I_0 (1 + \chi)^2 = \frac{\text{constant}}{A}$$

---

[4] Feinberg noted this problem in his treatment of radio-wave propagation around the earth [8] and set

$$\nabla \varepsilon = \frac{\varepsilon - \varepsilon_0}{\ell}$$

where $\ell$ is a general scale length in the spectrum of irregularities and need not be the inner scale length.

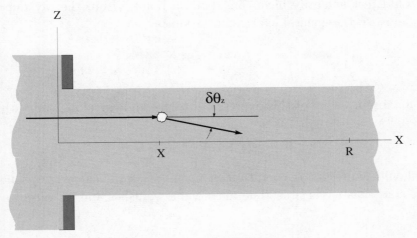

Figure 3.4: Random deflection of a ray component in a collimated beam.

The cross sectional area A becomes very small where rays converge and $\chi$ must be large at such places. Conversely the area cannot go to zero if $\chi \ll 1$ and a caustic is thereby avoided.

One can place this criterion on a quantitative basis. We consider a point $(x, z)$ within the ray bundle and examine the effect of ray bending as shown in Figure 3.4. Phase fluctuations imposed by the random medium along the route leading to this point cause random deflections of the ray. To avoid creating a caustic at a point further along the beam, we insist that

$$\delta\theta_z(R - x) \ll z$$

The angular error is given by (3.33) for a plane wave propagating along a horizontal path:

$$\frac{1}{2}(R - x)\int_0^x dx' \frac{\partial}{\partial z}\Delta\varepsilon(x', y, z; t) \ll z$$

One can write the same relation for a nearby point at the same depth:

$$\frac{1}{2}(R - x)\int_0^x dx' \frac{\partial}{\partial z}\Delta\varepsilon(x', y, z + \delta z; t) \ll z + \delta z$$

Taking the difference and dividing by the small separation:

$$\frac{1}{2}(R - x)\int_0^x dx' \frac{\partial^2}{\partial z^2}\Delta\varepsilon(x', y, z; t) \ll 1$$

We could just as easily have considered a point within the ray bundle on the $y$ axis, so the general expression becomes

$$\frac{1}{2}(R-x)\int_0^x dx' \left(\frac{\partial^2}{\partial y^2} + \frac{\partial^2}{\partial z^2}\right)\Delta\varepsilon(x', y, z; t) \ll 1$$

The maximum value of this expression occurs midway along the path:

$$\frac{1}{4}R\int_0^{\frac{R}{2}} dx\, \nabla_\perp^2 \Delta\varepsilon(x, y, z; t) \ll 1 \tag{3.44}$$

We should express this condition in terms of statistical averages because $\Delta\varepsilon$ is a stochastic variable. The mean value vanishes and its variance must satisfy

$$\left\langle \left( \frac{R}{4}\int_0^{\frac{R}{2}} dx\, \nabla_\perp^2 \Delta\varepsilon(x, y, z; t) \right)^2 \right\rangle \ll 1$$

The quantity inside the large parentheses is similar to the expression for logarithmic amplitude fluctuations given by (3.29). Using the Kolmogorov model, one can show that

$$\left\langle \left( \frac{R}{4}\int_0^{\frac{R}{2}} dx\, \nabla_\perp^2 \Delta\varepsilon(x, y, z; t) \right)^2 \right\rangle = \frac{3}{2}\langle\chi^2\rangle$$

This means that we can write the caustic condition in terms of a measured quantity [9]:

$$\langle\chi^2\rangle \ll 1 \tag{3.45}$$

This agrees with the qualitative argument given above and restricts the ability of geometrical optics to predict levels of scintillation in many situations.

Our characterization of the caustic condition thus far relates only to the logarithmic amplitude. We also need to know how the phase and angle of arrival are bounded in the situations which can be described by geometrical optics. The generic evolution of a plane wave in the strong-scattering region is suggested in Figure 3.5. As the signal penetrates into the random medium, the parallel rays gradually converge and diverge causing caustics where they combine. The signal phase changes by $\pi/2$ radians at these junctions. On the other hand, that change is relatively small compared to the large random phase shift which has accumulated before the signal reaches the region where caustics form [10]. This suggests that the measured phase variance can be accurately described by geometrical optics even in the region of strong scattering.

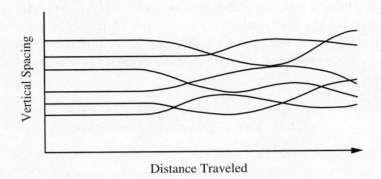

Figure 3.5: Evolution of the paths of parallel rays which represent a plane wave in the geometrical optics approximation. This diagram indicates the formation of caustics in the region of strong scattering and is adapted from Barabanenkov, Kravtsov, Rytov and Tatarskii [10].

This conclusion is confirmed in part by angle-of-arrival measurements made with laser signals on long horizontal paths [11]. The data agrees with the predictions of geometrical optics that will be developed in Chapter 7. This agreement can be appreciated by using the following qualitative argument. We know from (3.35) that the angular error is proportional to the lateral gradient of phase fluctuations. While the phase itself might change significantly at the caustics, Figure 3.5 provides no indication that this lateral change is large. That conclusion is confirmed by careful analysis [12]. These calculations and experiments confirm that angular errors are accurately described by geometrical optics for values as large as

$$\delta\theta_{\text{rms}} \simeq 10 \text{ radians} \tag{3.46}$$

This description does not guarantee that the phase variations themselves can be large. On the other hand, there is now general agreement [10][13] that geometrical optics accurately describes $\langle \varphi^2 \rangle$ for situations that correspond to values of the logarithmic amplitude as large as $\langle \chi^2 \rangle = 40$. This means that one can use this relatively simple method to describe phase fluctuations, even for optical signals on terrestrial paths as discussed in Chapter 4. Since the variance of the logarithmic amplitude is constrained to small values by (3.45), we expect that the relationship

$$\langle \chi^2 \rangle \ll \langle \varphi^2 \rangle \tag{3.47}$$

is valid in those situations which can be described by geometrical optics. This means that the field strength is determined primarily by its phase when geometrical optics is valid. That will provide an important simplification when the average field strength and mutual coherence function are calculated in Chapter 9.

### 3.6.3 The Diffraction Condition

A third condition results because the electromagnetic wavelength appears nowhere in the two basic equations of geometrical optics. The argument breaks down when diffraction becomes important. Consider the bundle of rays formed by a circular opening of diameter $\ell$ that is illustrated in Figure 3.6. At a distance $R$ from this opening, experiments show that interference fringes occur with a lateral dimension given by the *Fresnel length* which is defined by $\sqrt{\lambda R}$. This spreading must be substantially less than the diameter of the ray bundle if diffraction is to be ignored:

$$\sqrt{\lambda R} \ll \ell$$

The same thing happens when parallel rays intercept a transparent object with a dielectric constant different than unity $d$ such as a lens or raindrop. Experiments show that an object of size $\ell$ scatters almost all its energy forward in a cone of angular width

$$\delta\theta = \frac{\lambda}{\ell}$$

and energy is spread over a lateral distance

$$\delta x = R\frac{\lambda}{\ell}$$

This spreading must be considerably less than the size of the object if diffraction is to be ignored:

$$R\frac{\lambda}{\ell} \ll \ell$$

The scattering is necessarily three-dimensional when the cone of significant scattering is larger than the object and geometrical optics cannot properly describe the amplitude fluctuations. We identify the scattering object in Figure 3.6 with a turbulent eddy of size $\ell$. The condition for geometrical optics to be valid is therefore

$$\sqrt{\lambda R} \ll \ell \tag{3.48}$$

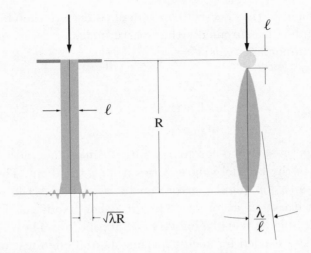

Figure 3.6: Diffraction effects measured on the left-hand side by a collimated beam formed by a circular opening of diameter $\ell$ and on the right-hand side by scattering by a transparent object of the same size.

There is a wide range of eddy sizes in the atmosphere. The critical question is this: *Which eddy sizes are important for the application of geometrical optics?* The answer depends on the type of measurement one is making. We noted in the introduction that the phase variance is determined almost entirely by the largest eddies. This suggests that the most influential eddy size for phase measurements is the outer scale length:

$$\text{Phase:} \quad \sqrt{\lambda R} \ll L_0 \tag{3.49}$$

This condition is satisfied for optical and microwave propagation in many cases of practical interest and one can use geometrical optics to estimate phase fluctuations with high confidence.

The phase structure function describes phase-difference measurements taken between adjacent receivers and is the figure of merit for interferometers. It is determined both by large eddies and by the inertial range. The diffraction condition for these experiments depends primarily on the separation between receivers:

$$\text{Structure function:} \quad \sqrt{\lambda R} \ll \rho \tag{3.50}$$

This condition is usually satisfied for microwave and optical interferometers because long baselines are favored for greater accuracy.

Angle-of-arrival measurements also depend on the phase difference but are usually made with smaller receivers. The angular variance is proportional to the third moment of the spectrum and small eddies are therefore influential. For most angular measurements the aperture size $a_r$ replaces the smallest eddies as the important scale:

$$\text{Angle of arrival:} \quad \sqrt{\lambda R} \ll a_r \qquad (3.51)$$

This inequality is satisfied for many optical links and most astronomical telescopes since the Fresnel length varies from 1 to 10 cm. The geometrical optics expressions for angular error can be used to describe many optical applications with good accuracy. On the other hand, the Fresnel length for microwave links is usually between 10 and 50 m. These values exceed the antenna sizes needed for such experiments and one must use diffraction theory to make good estimates.

The situation is entirely different for amplitude fluctuations. The variance of the logarithmic amplitude is proportional to the fifth moment of the wavenumber spectrum. This means that the smallest eddies are uniquely important for amplitude and intensity measurements. The inner scale length sets the standard for a point receiver:

$$\text{Amplitude:} \quad \sqrt{\lambda R} \ll \ell_0 \qquad (3.52)$$

The inner scale length is several millimeters in the lower atmosphere and the maximum transmission distance for optical signals is *80 m*. This condition is never satisfied for microwave signals. One must use diffraction theory to describe amplitude scintillation for almost every case encountered in practice.

There are two exceptions to this rule and both relate to systems that employ large apertures. In these cases the aperture radius replaces the inner scale length in (3.52). Astronomical telescopes are a common example. One can show that the variations of amplitude scintillation with the zenith angle and aperture size are correctly predicted by geometrical optics. Scintillometers that use large transmitting and receiving apertures to measure $C_n^2$ also operate in the regime for which this approximation is valid. We shall address both problems with diffraction theory in the second volume and show how its predictions reduce to those of geometrical optics when the apertures are large.

## 3.7 Problems

### Problem 1

Fermat's Principle requires that the ray follow a trajectory in space that renders the phase integral stationary:

$$\oint ds \sqrt{\varepsilon(s)} = \text{extremum}$$

This usually means a minimum. Use the flat-earth approximation and the rectangular coordinates of Figure 3.1 to express this condition. If the dielectric constant depends only on altitude, show that this is equivalent to the following variational principle:

$$\delta \left\{ \oint_T^R ds \sqrt{\varepsilon(z)} \left[ \left( \frac{dx}{ds} \right)^2 + \left( \frac{dy}{ds} \right)^2 + \left( \frac{dz}{ds} \right)^2 \right]^{\frac{1}{2}} \right\} = 0$$

Use the calculus of variations to find the three Euler–Lagrange equations which must be satisfied [14]. Show that these equations are equivalent to the ray equations (3.11) through (3.13). Fermat's Principle is sometimes used as the starting point in developing the theory of geometrical optics [15].

### Problem 2

Use the polar coordinates illustrated in Figure 3.7 to analyze refractive bending for long-distance propagation. Assume that the dielectric constant depends only on the radial distance from the center of the earth and derive Bourger's Law:

$$r \cos \alpha \sqrt{\varepsilon(r)} = \text{constant}$$

which is the appropriate generalization of Snell's Law for spherical atmospheres. Show that the ray trajectory in this case is given by

$$\theta = \int_{r_E}^{r(\theta)} \frac{dr}{r} \left( \frac{r^2 \varepsilon(r) \sec^2 \alpha_0}{r_E^2 \varepsilon(0)} - 1 \right)^{-\frac{1}{2}}.$$

Establish explicit expressions for the phase and amplitude.

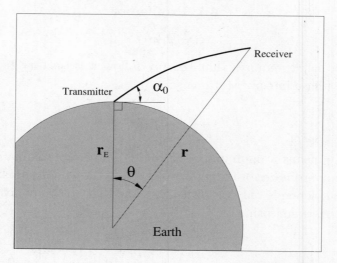

Figure 3.7: Refractive bending in a spherically symmetrical atmosphere that describes long-distance propagation.

## Problem 3

Use expression (3.29) to show that the variance of the logarithmic amplitude for a plane wave traveling along a horizontal path is related to the wavenumber spectrum of dielectric variations by

$$\langle \chi^2 \rangle = \frac{\pi^2}{12} R^3 \int_0^\infty d\kappa \, \kappa^5 \Phi_\varepsilon(\kappa).$$

Use the Kolmogorov model (3.39) to show that

$$\langle \chi^2 \rangle = 0.044 R^3 C_n^2 (\kappa_{\mathrm{m}})^{\frac{7}{3}}.$$

Can you guess how this result changes if the receiver is larger than the inner scale length?

## References

[1] M. Born and E. Wolf, *Principles of Optics*, 6th Ed. (Pergamon Press, New York, 1980), 109 *et seq.*

[2] J. E. Freehafer, "Geometrical Optics," in *Propagation of Short Radio Waves*, vol. 13, edited by D. E. Kerr (MIT Radiation Laboratory Series, McGraw-Hill, New York, 1951), 41 *et seq.*

[3] V. I. Tatarskii, *The Effects of the Turbulent Atmosphere on Wave Propagation* (Translated from the Russian and issued by the National Technical

Information Office, U.S. Department of Commerce, Springfield, VA 22161, 1971), 197–199.

[4] S. M. Rytov, Yu. A. Kravtsov and V. I. Tatarskii, *Principles of Statistical Radiophysics 4, Wave Propagation Through Random Media* (Springer-Verlag, Berlin, 1989), 21–32.

[5] See [3], pages 177–208.

[6] S. Chandrasekhar, "A Statistical Basis for the Theory of Stellar Scintillation", *Monthly Notices of the Royal Astronomical Society,* **112**, No. 5, 475–483 (1952).

[7] See [4], pages 2–3.

[8] E. L. Feinberg, *The Propagation of Radio Waves along the Surface of the Earth* (Translated from the Russian and issued by Foreign Technology Division of the US Air Force Systems Command, 1967).

[9] See [3], pages 212–214.

[10] Yu. N. Barabanenkov, Yu. A. Kravtsov, S. M. Rytov and V. I. Tatarskii, "Status of the Theory of Propagation of Waves in a Randomly Inhomogeneous Medium," *Uspekhi Fizicheskikh Nauk (Soviet Physics Uspekhi),* **13**, No. 5, 551–580 (March–April 1971).

[11] A. S. Gurvich and M. A. Kallistratova, "Experimental Study of the Fluctuations in Angle of Incidence of a Light Beam under Conditions of Strong Intensity Fluctuations," *Izvestiya Vysshikh Uchebnykh Zavedenii, Radiofizika (Soviet Radiophysics),* **11**, No. 1, 37–40 (January 1968).

[12] V. I. Klyatskin, "Variance of the Angle of Arrival of a Plane Light Wave Propagating in a Medium with Weak Random Irregularities," *Izvestiya Vysshikh Uchebnykh Zavedenii, Radiofizika (Soviet Radiophysics),* **12**, No. 5, 578–580 (May 1969).

[13] See [4], page 24.

[14] P. M. Morse and H. Feshbach, *Methods of Theoretical Physics, Part I* (McGraw-Hill, New York, 1953), 275 *et seq.*

[15] See [1], page 128 *et seq.*

# 4

# The Single-path Phase Variance

Geometrical optics provides an accurate description of electromagnetic phase fluctuations under a wide range of conditions. The phase variance computed in this way is a benchmark parameter for describing propagation in random media. One can calculate this quantity for most situations of practical interest. We shall find that it is proportional to the first moment of the spectrum of irregularities and is therefore sensitive to the small-wavenumber portion of the spectrum. This is the region where energy is fed into the turbulent cascade process. We have no universal physical model for the spectrum in this wavenumber range and phase measurements provide an important way of exploring that region.

In analyzing these situations, we must recognize the anisotropic nature of irregularities in the troposphere and ionosphere. Large structures are highly elongated in both regions and exert a strong influence on phase. These measurements are also sensitive to trends in the data that are caused by nonstationary processes in the atmosphere. Sample length, filtering and other data-processing procedures thus have an important influence on the measured quantities. By contrast, aperture smoothing has a negligible effect.

Single-path phase measurements have been made primarily at microwave frequencies because phase-stable transmitters and receivers were available in these bands. Early experiments were performed on horizontal paths using signals in the frequency range 1–10 GHz. At least one experiment has measured the single-path phase variance at optical wavelengths. Phase-stable signals from navigation satellites and other spacecraft are beginning to provide information about the upper atmosphere. Refractive bending of the rays can be neglected in all these cases and the nominal ray path is a straight

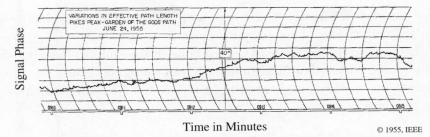

Figure 4.1: Single-path phase fluctuations at 1046 MHz measured on a 16-km terrestrial path by Herbstreit and Thompson [1]. A 5-min data sample from this pioneering experiment is reproduced here.

line connecting the transmitter and receiver. The line integral taken along the ray path in (3.21) becomes a simple integration on the slant range:[1]

$$\varphi(t) = \tfrac{1}{2}k \int_T^R ds \, \Delta\varepsilon(s,t) \tag{4.1}$$

The phase fluctuation is a *stochastic* function of time because it is a linear function of the stochastic variable $\Delta\varepsilon$. This random behavior is apparent in the experimental phase history reproduced in Figure 4.1.

Since the phase of the signal is a random function of time, we must describe its behavior in terms of moments and correlation functions. These are computed by taking the ensemble average over all possible configurations of the random medium. As a practical matter, we usually approximate ensemble averages by taking time averages with finite sample lengths. The average phase vanishes since $\langle\Delta\varepsilon\rangle = 0$. The phase variance is obtained by squaring the expression above and averaging over all possible atmospheric configurations:

$$\langle\varphi^2\rangle = \tfrac{1}{4}\,k^2 \int_T^R ds \int_T^R ds' \langle\Delta\varepsilon(s,t)\Delta\varepsilon(s',t)\rangle$$

$$= \tfrac{1}{4}\,k^2 \int_T^R ds \int_T^R ds' \langle\Delta\varepsilon^2(s,s')\rangle C(\mathbf{s}-\mathbf{s}')$$

The second expression separates the effects of local dielectric variations and inhomogeneous changes in the level of turbulent activity. Models for the

---

[1] We can write the phase more accurately in vector notation:

$$\varphi(t) = \tfrac{1}{2}k \int_{\mathbf{T}}^{\mathbf{R}} d\mathbf{s} \, \Delta\varepsilon(\mathbf{s},t)$$

and one should keep this larger meaning in mind when using the simple expression given above.

spatial correlation function $C(R)$ were used to evaluate the double integral in early descriptions of line-of-sight propagation [2][3][4].

A vastly superior strategy is to use the Fourier wavenumber decomposition for the spatial correlation function [5][6][7]. In the general case, we consider the spectrum to be a function both of the wavenumber and of the position along the path:

$$\langle \Delta\varepsilon(\mathbf{r}, t)\, \Delta\varepsilon(\mathbf{r} + \mathbf{R}, t)\rangle = \int d^3\kappa\, \Phi_\varepsilon(\boldsymbol{\kappa},\, \mathbf{r}) \exp(i\boldsymbol{\kappa} \cdot \mathbf{R}) \qquad (4.2)$$

With this representation the phase variance becomes

$$\langle \varphi^2 \rangle = \tfrac{1}{4}k^2 \int_0^R ds \int_0^R ds' \int d^3\kappa\, \Phi_\varepsilon[\boldsymbol{\kappa}, \tfrac{1}{2}(\mathbf{s} + \mathbf{s}')] \exp[i\boldsymbol{\kappa} \cdot (\mathbf{s} - \mathbf{s}')]. \qquad (4.3)$$

The vector product in the exponent can be expressed in terms of the slant-range difference and the wavenumber component in the line-of-sight direction:

$$\boldsymbol{\kappa} \cdot (\mathbf{s} - \mathbf{s}') = \kappa_{\mathrm{s}}(s - s')$$

Variation of the spectrum with position is usually described in terms of the average distance along the rays, which translates the expression into

$$\langle \varphi^2 \rangle = \tfrac{1}{4}k^2 \int_0^R ds \int_0^R ds' \int d^3\kappa\, \Phi_\varepsilon[\boldsymbol{\kappa}, \tfrac{1}{2}(s + s')] \exp[i\kappa_{\mathrm{s}}(s - s')]. \qquad (4.4)$$

The remaining integrations are handled differently, depending on whether one is describing horizontal transmissions near the surface or elevated paths between earth-orbiting satellites and their ground stations.

## 4.1 Terrestrial Links

Most of the available measurements of phase variance have been made with horizontal microwave links near the surface as illustrated in Figure 4.2. Such experiments directly simulate terrestrial communication and geodetic ranging applications. They were also used to predict the accuracy of radio guidance systems for ballistic missiles, earth-orbiting satellites and planetary spacecraft before such vehicles became available.

### 4.1.1 Microwave Phase Measurements

Single-path phase variations are equivalent to changes in the electrical path length. One is looking for changes that are a small fraction of a wavelength over a path that is typically several million wavelengths long. This explains

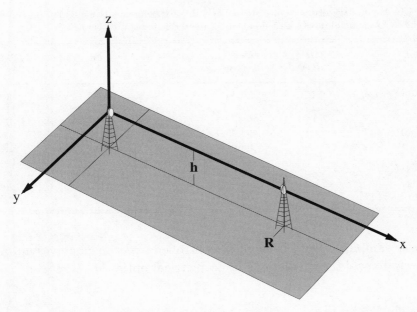

Figure 4.2: A typical microwave link used to measure phase variance on a terrestrial link parallel to the surface. Coordinates used to describe these experiments are shown. The line-of-sight path coincides with the $x$ axis and is at a height h above the surface.

why the transmitter and receiver must be securely mounted at both ends of the link. More fundamentally, it means that very stable microwave sources are required. The development of frequency standards accurate to several parts in $10^9$ for several hours made such measurements possible [1][8]. Identical microwave transceivers are employed at both ends of the link. The frequencies of the signals transmitted from the two ends are offset from one another by an audio frequency. The arriving signal and the local frequency standard are mixed together in a multiplier. The mixer output contains the audio reference and the phase shift imposed by the medium. This combination is modulated on a low-frequency signal (typically 100 kHz) and transmitted back to the originating site. The return signal is not influenced by the random medium because its phase shift is proportional to the wavenumber $k$. The phase fluctuation emerges from a comparison of this modulation with the reference audio signal.

The same technique has been used to generate a large body of phase data for other frequencies, path lengths and geographic locations. These experiments are summarized in Table 4.1. The measured phase fluctuations are typically less than 30° rms. The index of refraction was often measured at the transmitter and receiver sites – and sometimes along the path, using

Table 4.1: A summary of microwave single-path phase measurements made on horizontal paths near the surface; simultaneous measurements of refractive index are also noted

| Location | $R$ (km) | Frequency (GHz) | $\langle \varphi^2 \rangle$ | $\delta n$ | Ref. |
|---|---|---|---|---|---|
| Colorado Springs | 16 | 1.046 | x | x | 1 |
| Maui | 25 | 9.4 | x | x | 9, 10, 11, 12, 13 |
| Boulder | 15 | 9.4 | x | x | 14, 15, 16 |
| Maui to Hawaii | 64 | 9.6 and 34.5 | x |  | 17 |
|  |  | 9.6 | x | x | 18 |
|  |  | 9.6, 19, 22, 25 and 33 | x |  | 19 |
| Bahamas | 47 | 9.2 and 9.4 | x | x | 20 |
| Maui to Hawaii | 150 | 9.6 | x | x | 21 |

airborne sampling. Data from these controlled experiments gives an important way to test the predictions of geometrical optics.

### 4.1.2 Optical Phase Measurements

Phase fluctuations imposed on optical signals are much greater than those experienced at microwave frequencies. These phase shifts regularly exceed $2\pi$ radians and present a difficult measurement problem. Lukin and his colleagues in the USSR solved this problem by using an optical heterodyning technique [22][23] with the basic Michelson-interferometer arrangement illustrated in Figure 4.3.

A helium–neon laser signal is divided into two components by a beam splitter. One component is collimated, travels through the random medium and is then reflected back to the transmitter along substantially the same path. The second component from the beam splitter acts as a reference signal. It is guided to a mirror that is vibrated at 5 kHz by a piezoelectric crystal. The reflected reference signal is thus modulated with an audio-frequency carrier. The propagating component and modulated reference signal are then multiplied together. The output from the mixer contains the low-frequency audio signal and the phase fluctuations imposed by the random medium.[2] The atmospheric phase shift never exceeds the wavelength of the low-frequency carrier. One can follow the large phase excursions, although with declining accuracy as they increase. The rapid response of

---

[2] The output from the mixer also contains a constant phase term corresponding to $k$ times the path differential for the two optical signals but this does not influence the measurement of time-varying phase fluctuations.

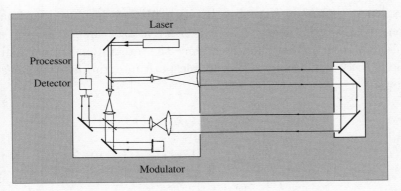

Figure 4.3: The experimental arrangement for measuring phase variance, power spectra and auto-correlation at optical wavelengths employed by Lukin, Pokasov and Khmelevtsov [22]. The laser signal is divided into two beams. One beam samples the random medium on a reflected path. The other acts as a reference signal that is harmonically modulated. The two optical signals are recombined and their phase difference detected.

the phase detector employed made it possible to track fluctuations up to 500 Hz. This technique was used to measure the phase variance, autocorrelation of phase and phase power spectrum for path lengths ranging from 15 to 200 m.

### 4.1.3 A Basic Expression for the Phase Variance

The optical measurements summarized above used propagation paths that were nearly horizontal, which means that the level of turbulent activity was relatively constant. Several of the microwave experiments also satisfied this condition. As a first approximation we ignore variations of the spectrum with position along the path. Because these terrestrial links travel in the lower atmosphere we shift from the spectrum of dielectric variations to the spectrum for refractive-index fluctuations using $\Phi_\varepsilon = 4\Phi_n$. The rectangular coordinates illustrated in Figure 4.2 describe the propagation. The distance along the path $s$ is identical to the horizontal coordinate $x$:

$$\langle \varphi^2 \rangle = k^2 \int_0^R dx \int_0^R dx' \int d^3\kappa \, \Phi_n(\boldsymbol{\kappa}) \exp\left[i\kappa_x (x - x')\right] \qquad (4.5)$$

The path integrations can now operate on the kernel without making an assumption for the turbulent spectrum. In this way we can postpone the choice of a model until the propagation calculations are almost completed. We focus first on the integrations along the ray paths:

$$\mathcal{J} = \int_0^R dx \int_0^R dx' \exp\left[i\kappa_x (x - x')\right]$$

The exponential can be replaced by its real part because the imaginary component is as often negative as it is positive over the integration square. The integration is further simplified since the integral is symmetrical about the line $x = x'$, which allows the double path integrations to be evaluated as follows:

$$
\begin{aligned}
\mathcal{J} &= 2 \int_0^R dx \int_0^x dx' \cos[\kappa_x(x - x')] \\
&= 2 \int_0^R dx\, x \int_0^1 du \cos[\kappa_x x(1 - u)] \\
&= 2 \int_0^R dx\, x \int_0^1 dw \cos(\kappa_x x w) \\
&= 2 \left( \frac{1 - \cos(\kappa_x R)}{\kappa_x^2} \right)
\end{aligned}
$$

and the phase-variance expression becomes

$$
\langle \varphi^2 \rangle = 2k^2 \int d^3\kappa\, \Phi_n(\boldsymbol{\kappa}) \left( \frac{1 - \cos(\kappa_x R)}{\kappa_x^2} \right). \tag{4.6}
$$

To make further progress one must describe the turbulence spectrum. Our first approximation is to assume that the spectrum of irregularities is isotropic, which provides a fair description for propagation paths that are close to the surface. The spectrum depends only on the magnitude of the wavenumber vector when the random medium is isotropic:

$$
\Phi_n(\boldsymbol{\kappa}) = \Phi_n \left( \sqrt{\kappa_x^2 + \kappa_y^2 + \kappa_z^2} \right) \tag{4.7}
$$

It is convenient to introduce spherical wavenumber coordinates, which use the line-of-sight direction as the polar axis:

$$
\begin{aligned}
\kappa_x &= \kappa \cos \psi \\
\kappa_y &= \kappa \sin \psi \sin \omega \\
\kappa_z &= \kappa \sin \psi \cos \omega
\end{aligned} \tag{4.8}
$$

With this transformation the phase-variance expression (4.6) becomes

$$
\langle \varphi^2 \rangle = 2k^2 \int_0^\infty d\kappa\, \Phi_n(\kappa) \int_0^\pi d\psi \sin \psi \int_0^{2\pi} d\omega \left( \frac{1 - \cos(\kappa R \cos \psi)}{\cos^2 \psi} \right).
$$

The angular integrations can be done analytically and the phase variance written as a weighted average of the turbulence spectrum:

$$
\langle \varphi^2 \rangle = 4\pi^2 R k^2 \int_0^\infty d\kappa\, \kappa \Phi_n(\kappa) F(\kappa R) \tag{4.9}
$$

The *finite-path spectral weighting function* is defined by[3]

$$F(x) = \frac{2}{\pi}\left(\text{Si}(x) - \frac{1 - \cos(x)}{x}\right). \tag{4.10}$$

This function rises slowly from zero to unity and suppresses wavenumbers that are comparable to the inverse path length [5][6]. The outer scale lengths for eddies *close to the surface* are substantially smaller than the path lengths employed both at microwave and at optical frequencies. In these situations one can replace $F(\kappa R)$ by its asymptotic value:

$$\kappa_0 R \gg 1: \qquad \langle \varphi^2 \rangle = 4\pi^2 R k^2 \int_0^\infty d\kappa\, \kappa \Phi_n(\kappa) \tag{4.11}$$

### 4.1.4 Frequency Scaling

The frequency does not influence the wavenumber integration in the phase-variance expression in the above approximation. This implies that the rms phase shift should be linearly proportional to the carrier frequency:

$$\varphi_{\text{rms}} = \text{constant} \times f$$

This scaling law was tested by comparing simultaneous phase histories of microwave signals for different frequencies that traveled along the same path. The first experiment used 9.2- and 9.4-GHz signals transmitted over a 47-km path in the Bahamas [20]. The phase fluctuations were highly correlated to one another – and to the surface refractive index measured at the transmitter and receiver.

A second experiment was performed using 9.6- and 34.52-GHz signals on a 64-km path between Maui and Hawaii [17]. Careful examination of the phase histories revealed two features. The variances were highly correlated when the phase data was averaged over a few minutes to approximate the ensemble average implied by (4.11). A comparison of the long-term behaviors of the rms phase fluctuations at these frequencies is reproduced in Figure 4.4 and confirms the linearity of the scaling with frequency. Very rapid high-frequency phase fluctuations were often quite different for the two signals. To resolve the apparent conflict between short-term and time-averaged behavior one must examine the frequency spectrum of the phase variations.

---

[3] The sine integral function is defined by

$$\text{Si}(x) = \int_0^x \frac{du}{u}\sin(u).$$

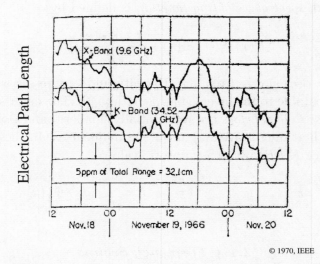

Figure 4.4: Phase-variance histories at 9.6 and 34.52 GHz measured on a 64-km path by Janes, Thompson, Smith and Kirkpatrick [17].

### 4.1.5 Phase-variance Estimates

To proceed further we must characterize the spectrum of refractive-index fluctuations. The phase-variance expression depends on the first moment of the spectrum and is therefore determined primarily by the largest eddies. We cannot ignore the small-wavenumber portion of the spectrum, even though we have no valid physical description for it. In Section 2.2.6 we discussed several analytical models that extend the Kolmogorov spectrum into this region. We will use the von Karman model which attempts to represent large irregularities and the inertial range:

$$\Phi_n(\kappa) = \frac{0.033 C_n^2}{\left(\kappa^2 + \kappa_0^2\right)^{\frac{11}{6}}} \qquad 0 < \kappa < \kappa_s$$

On substituting this model into (4.11) we find that

$$\langle \varphi^2 \rangle = 4\pi^2 R k^2 \int_0^{\kappa_s} d\kappa\, \kappa \frac{0.033 C_n^2}{\left(\kappa^2 + \kappa_0^2\right)^{\frac{11}{6}}}$$

$$= 0.782 R k^2 C_n^2 \left( \frac{1}{(\kappa_0)^{\frac{5}{3}}} - \frac{1}{\left(\kappa_s^2 + \kappa_0^2\right)^{\frac{11}{6}}} \right)$$

The inner-scale wavenumber is always much greater than $\kappa_0$ in the lower atmosphere. The second term is thus unimportant and the dissipation region should not influence phase-variance measurements:

$$\langle\varphi^2\rangle = 0.782Rk^2C_n^2\kappa_0^{-\frac{5}{3}} \tag{4.12}$$

It is important to know whether this result gives estimates that agree with the measured values. This was done using the optical heterodyne technique [22][23], as shown in Figure 4.3. For terrestrial experiments one knows $R$ and $k$ with considerable precision. The value of $C_n^2$ was established from temperature measurements made at the site. The outer scale length was assumed equal to the height of the path above the terrain. The measured and calculated values for the phase variance agreed remarkably well. This process was then reversed to establish a distribution of outer scale lengths using the measured values of $R, k, C_n^2$ and $\langle\varphi^2\rangle$. The resulting values for $L_0$ were clustered about the path height ($\mathsf{h} = 1.5$ m), as they should be near the surface.

### 4.1.5.1 Optical Link Estimates

With this validation in hand, one can use the basic expression to estimate the phase variance for a variety of frequencies and path lengths. For visible wavelengths the data reproduced in Figure 2.14 suggests that $C_n^2 = 3 \times 10^{-14} \mathrm{m}^{-\frac{2}{3}}$ is a typical value near the surface. Optical links are usually close to the ground and one can assume that the outer scale length is approximately 1 m. For visible light and a path length of 1000 m we would expect

$$\varphi_{\mathrm{rms}} = \left[0.782 \times 10^3 \times \left(\frac{2\pi}{5\times10^{-7}}\right)^2 \times 3 \times 10^{-14} \times \left(\frac{1}{2\pi}\right)^{\frac{5}{3}}\right]^{\frac{1}{2}}$$

$$\simeq 13 \text{ radians.}$$

This is more than two complete phase revolutions and indicates the care that is required in making such measurements.

### 4.1.5.2 Microwave Link Estimates

The basic expression (4.12) also describes microwave propagation if we are careful to use the appropriate values for $C_n^2$. The programs identified in Table 4.1 usually measured surface refractivity directly rather than $C_n^2$. They did so using microwave refractometers that measure fluctuations about the vacuum value, which are expressed in terms of parts per million or "$N$ units"

$$n = 1 + \delta n = 1 + \Delta N\, 10^{-6}$$

so that

$$\langle \delta n^2 \rangle = \langle \Delta N^2 \rangle 10^{-12}.$$

We found in (2.70) that the refractive-index variance is related to $C_n^2$ and $\kappa_0$ by the following expression:

$$\langle \delta n^2 \rangle = 0.622 C_n^2 \kappa_0^{-\frac{2}{3}}$$

The phase variance can now be connected to the microwave refractometer measurements by expressing $C_n^2$ in terms of $\langle \Delta N^2 \rangle$ and the outer scale length:

$$\langle \varphi^2 \rangle = 11.1 \frac{R\, L_0}{\lambda^2} \langle \Delta N^2 \rangle 10^{-12} \tag{4.13}$$

The same result can be established without resorting to specific models for the spectrum. Using (2.49) one can define an effective outer scale length as the ratio of the first two moments [24]:

$$\frac{L_0}{2\pi} = \frac{\displaystyle\int_0^\infty d\kappa\, \kappa \Phi_n(\kappa)}{\displaystyle\int_0^\infty d\kappa\, \kappa^2 \Phi_n(\kappa)}$$

The denominator is proportional to $\langle \delta n^2 \rangle$ as defined by (2.48) and hence to $\langle \Delta N^2 \rangle$, which leads to (4.13). The constant is slightly different and depends on the specific definition of the outer scale length.

An early series of phase measurements used 1046-MHz signals on a 16-km path in Colorado [1]. The refractive index was measured at both ends of the path with refractometers. A segment of the time histories for simultaneous single-path phase and surface refractivity is reproduced in Figure 4.5. This data indicates that the refractivity variations were approximately two $N$ units both at the transmitter and at the receiver. The horizontal outer scale length was estimated from phase-difference measurements at the receivers to be about 100 m:

$$\varphi_{\text{rms}} \simeq \left( 9.39 \times \frac{16\,000 \times 100}{(0.3)^2} \times \left( 2 \times 10^{-6} \right)^2 \right)^{\frac{1}{2}} \simeq 3^\circ$$

This estimate agrees with the measured values if one subtracts the phase trend.

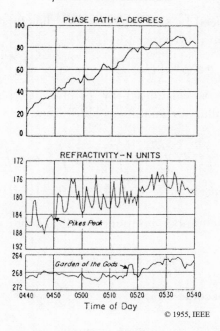

© 1955, IEEE

Figure 4.5: Simultaneous histories of the single-path phase at 1046 MHz and surface refractivity measured by Herbstreit and Thompson [1].

One should question the use of end-point values for $\Delta n$ and $L_0$ since much of this steep path was well above the boundary layer. It is likely that the outer scale length was greater and the refractivity fluctuations smaller along the elevated portions of this path. We return to these questions in Problem 2 after expressions for propagation along inclined paths in the troposphere have been established.

### 4.1.6 Anisotropic Media and Distance Scaling

The previous descriptions have assumed that the irregularities in refractive index are isotropic. This is a reasonable approximation when the propagation path is close to the surface. Now we want to consider horizontal paths higher up in the troposphere. This situation is encountered with air-to-air links between aircraft flying at nearly the same altitude. It also describes communications links between relay stations on mountains for which the line-of-sight is well above the boundary layer.

Irregularities in the troposphere are elongated at the surface and become more so as one goes up in altitude, as illustrated in Figure 2.18. For elevated horizontal propagation paths this means that we cannot safely assume that the path length is greater than the outer scale length. To describe the phase fluctuations experienced by such links we use the ellipsoidal description for the turbulent spectrum introduced in (2.87):

$$abc\Phi_n\left(\sqrt{a^2\kappa_x^2 + b^2\kappa_y^2 + c^2\kappa_z^2}\right) \qquad (4.14)$$

The phase variance is calculated by combining this model with (4.5) expressed in rectangular wavenumber coordinates:

$$\langle\varphi^2\rangle = k^2 \int_0^R dx \int_0^R dx' \int_{-\infty}^{\infty} d\kappa_x \int_{-\infty}^{\infty} d\kappa_y \int_{-\infty}^{\infty} d\kappa_z$$
$$\times abc\Phi_n\left(\sqrt{a^2\kappa_x^2 + b^2\kappa_y^2 + c^2\kappa_z^2}\right) \exp[i\kappa_x(x - x')]$$

If we rescale the wavenumber components [25] using

$$\kappa_x = \frac{q_x}{a} \qquad \kappa_y = \frac{q_y}{b} \qquad \kappa_z = \frac{q_z}{c} \qquad (4.15)$$

the phase variance becomes

$$\langle\varphi^2\rangle = k^2 \int_0^R dx \int_0^R dx' \int_{-\infty}^{\infty} dq_x \int_{-\infty}^{\infty} dq_y \int_{-\infty}^{\infty} dq_z$$
$$\times \Phi_n\left(\sqrt{q_x^2 + q_y^2 + q_z^2}\right) \exp\left(i\frac{q_x}{a}(x - x')\right).$$

The spectrum of turbulent irregularities is now isotropic in the stretched wavenumber coordinates:

$$\langle\varphi^2\rangle = k^2 \int_0^R dx \int_0^R dx' \int d^3q\, \Phi_n(q) \exp\left(i\frac{q_x}{a}(x - x')\right)$$

This result is identical to the isotropic expression (4.5) except that the argument of the trigonometric term is $q_x/a$ rather than $\kappa_x$. Keeping this exception in mind, one can use (4.10) to describe the finite-path spectral weighting function for signals that travel through anisotropic media:

$$\langle\varphi^2\rangle = 4\pi^2 R k^2 a \int_0^{\infty} dq\, q\Phi_n(q)\left[\frac{2}{\pi}\left(\text{Si}(qR/a) - \frac{1 - \cos(qR/a)}{qR/(ax)}\right)\right]$$

$$(4.16)$$

It is tempting to assume that the path length $R$ is much larger than the important eddies. If this were true one could replace the function in square brackets by unity, as we did in describing isotropic media. This is *not* a safe strategy here because the horizontal scale length in the free atmosphere is often comparable to the transmission distance, as noted in Section 2.3.5.

To complete the analysis we must describe the turbulence spectrum. We choose the von Karman model because it represents a gradual transition to the spectral region of large eddies and because it can be treated analytically:

$$\langle \varphi^2 \rangle = 0.264 \pi R k^2 C_n^2 a \int_0^\infty \frac{dq\, q}{(q^2 + \kappa_0^2)^{\frac{11}{6}}} \left( \mathrm{Si}(qR/a) - \frac{1 - \cos(qR/a)}{qR/a} \right)$$

Integrating by parts and setting $u = qR/a$ yields

$$\langle \varphi^2 \rangle = 0.158 \pi R k^2 C_n^2 \left( \frac{R}{a} \right)^{\frac{5}{3}} a \int_0^\infty \frac{du}{(u^2 + \eta^2)^{\frac{5}{6}}} \left( \frac{1 - \cos u}{u^2} \right). \tag{4.17}$$

The remaining integral depends on the following dimensionless parameter:

$$\eta = \frac{\kappa_0 R}{a} = 2\pi \frac{R}{(aL_0)} \tag{4.18}$$

We learned in Section 2.2.7 that $aL_0$ is the correlation length in the direction of propagation. For an elevated path it should be identified with the very large horizontal scale lengths that are measured in the free atmosphere. It is then quite possible that $\eta$ is small and we must evaluate the $u$ integral in (4.17) with care in order to not let the integrand blow up. The denominator is raised to exponential form by treating it as a Laplace transform:

$$\frac{1}{(u^2 + \eta^2)^{\frac{5}{6}}} = \frac{1}{\Gamma\left(\frac{5}{6}\right)} \int_0^\infty d\zeta\, \zeta^{-\frac{1}{6}} \exp[-\zeta(u^2 + \eta^2)]$$

so that

$$\langle \varphi^2 \rangle = \frac{0.158 \pi R k^2 C_n^2}{\Gamma\left(\frac{5}{6}\right)} \left( \frac{R}{a} \right)^{\frac{5}{3}} a \int_0^\infty d\zeta\, \zeta^{-\frac{1}{6}} e^{-\eta^2 \zeta} \int_0^\infty du \left( \frac{1 - \cos u}{u^2} \right) e^{-u^2 \zeta}$$

The $u$ integral is found in Appendix C and depends on the error function:

$$\langle \varphi^2 \rangle = \frac{0.158 \pi R k^2 C_n^2}{\Gamma\left(\frac{5}{6}\right)} \left( \frac{R}{a} \right)^{\frac{5}{3}} a \int_0^\infty d\zeta\, \zeta^{-\frac{1}{6}} e^{-\eta^2 \zeta}$$

$$\times \left\{ \frac{\pi}{2} \, \mathrm{erf}\left( \frac{1}{2\sqrt{\zeta}} \right) - \sqrt{\pi \zeta} \left[ 1 - \exp\left( -\frac{1}{4\zeta} \right) \right] \right\}$$

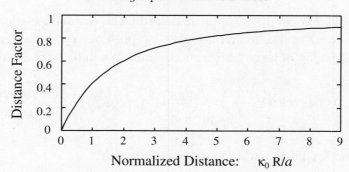

Figure 4.6: The distance factor which multiplies the basic expression for the phase variance and describes the influence of anisotropy and large horizontal scale lengths.

The remaining integration can be expressed in terms of MacDonald functions as noted in Appendix D. The result is written as a factor multiplying the basic expression (4.12):

$$\langle \varphi^2 \rangle = (0.782 R k^2 C_n^2 a \kappa_0^{-\frac{5}{3}}) \mathcal{L}(\kappa_0 R/a) \tag{4.19}$$

The distance factor is defined by the following expression and is plotted in Figure 4.6:

$$\mathcal{L}(\eta) = \frac{2}{\sqrt{\pi}\,\Gamma\left(\frac{5}{6}\right)} \left[ \left(\frac{\eta}{2}\right)^{\frac{1}{3}} K_{\frac{4}{3}}(\eta) - \frac{\Gamma\left(\frac{4}{3}\right)}{\eta} + \frac{1}{2^{\frac{1}{3}}} \int_0^{\eta} dw\, w^{\frac{1}{3}} K_{\frac{1}{3}}(w) \right] \tag{4.20}$$

Let us examine how the result (4.19) depends on distance. When the path length is much greater than the horizontal scale, the key parameter $\kappa_0 R/a$ is large and the distance factor is unity. This situation corresponds to optical and microwave links near the surface, where the scale lengths are small. The phase variance scales linearly with the path length in this case:

$$\kappa_0 R/a \gg 1: \qquad \langle \varphi^2 \rangle = \text{constant} \times R$$

There is a simple explanation for this relationship. When the transmission distance is large relative to the outer scale length one can break the path into $N = R/(aL_0)$ segments, which are for the most part uncorrelated. Each segment causes a random speeding up or slowing down of the signal. These events are largely independent and the rms phase shift should be proportional to the square root of $N$ by the Central Limit theorem.

A different situation occurs for horizontal propagation in the free atmosphere. The horizontal scale length $aL_0$ can be tens or even hundreds of kilometers in this region. It can be larger than the path length, in which case the key parameter $\eta$ is then less than unity. Using small-argument expansions for the MacDonald functions from Appendix D, one calculates

$$\lim_{\eta \to 0} \mathcal{L}(\eta) = 0.6695\eta$$

which shows that the phase variance now scales as the square of the distance:

$$\kappa_0 R/a < 1: \qquad \langle \varphi^2 \rangle = \text{constant} \times R^2$$

One can understand this result by returning to (4.1). If the signal sees only a portion of the large eddies, the path integration is approximated by

$$\varphi(t) \simeq \tfrac{1}{2} Rk \, \Delta\varepsilon(t)$$

and the variance should be proportional to $R^2$.

This suggests that the distance scaling could be verified by comparing simultaneous phase measurements at several distances. This was tried at 1046 MHz using path lengths of 5.6, 16 and 97 km in Colorado [1]. The data showed that longer distances generally did produce larger phase variations. However, those experiments cannot confirm our result because the three line-of-sight routes necessarily passed through different portions of the atmosphere. One could confirm these predictions by comparing the phase variance of signals that make multiple round trips between two transceivers since the random medium would then be common.

### 4.1.7 Phase Trends and the Sample-length Effect

Trends are often encountered in phase measurements and are apparent in the data reproduced in Figures 4.1 and 4.5. The phase of a signal is influenced both by random variations of the refractive index and by gradual changes in the ambient value:

$$n(\mathbf{r}, t) = n_0(t) + \delta n(\mathbf{r}, t) \tag{4.21}$$

The mean value $n_0$ changes during the day and with the seasons. It can also change suddenly when a weather front or intermittent structure moves through the path. These changes need not be turbulent but they are reflected in phase shifts through the basic relation (4.1). The simulated phase history of Figure 4.7 provides a basis for discussing such problems. It represents a random process superimposed on a trend that is increasing linearly with time.

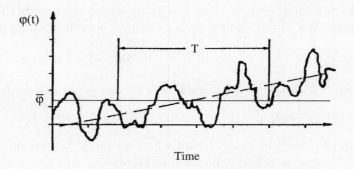

Figure 4.7: A simulated phase history showing random fluctuations that are superimposed on a linear trend caused by changing ambient conditions. The average value measured with a finite sample length is suggested by $\overline{\varphi}$.

A good deal of research has gone into developing *detrending* procedures for handling phase data. These involve fitting the running average phase with straight lines or polynomials. The fitted curve is then subtracted from the raw data and the remaining fluctuations are treated as a random process. There is always an element of uncertainty in doing so, since one can never be sure whether one is eliminating a trend or erasing the low-frequency contributions of large eddies.

The phase history is detrended in most experiments and the residual phase is treated as a stationary process. In this case, phase drift should be regarded as segments of low-frequency components in a Fourier time analysis of $\varphi(t)$. Trends would presumably bend over and develop into low-frequency harmonics if we could examine an infinitely long record. The basic problem is that one cannot distinguish between the effects of large eddies and secular trends when one is using a finite data sample.

This concern brings one back to the ensemble-averaging process implied by the brackets in our phase expressions. In an ideal world, this averaging should extend over all possible configurations of the turbulent atmosphere. One must approximate the ensemble average by a time average in practical situations:

$$\langle \varphi^2 \rangle \simeq \frac{1}{T} \int_0^T dt\, \varphi^2(t) \tag{4.22}$$

We believe that the two approaches will agree if the sample is infinitely long. On the other hand, we usually cannot wait long enough to realize that ideal and it is important to identify the sample size employed. It is likely that the measured phase variance will be quite different for sample lengths of one minute, one hour, one day and one month. A finite data sample necessarily discriminates against low frequencies in the temporal phase spectrum and therefore suppresses the contributions of large eddies which would otherwise play a prominent role.

The simulated phase record in Figure 4.7 illustrates random variations superimposed on a gradual change of the electric path length. If this trend is ignored and the fluctuations are measured relative to an estimated average value, the phase excursions appear progressively larger as one goes to either end of the data sample. These large positive and negative values exaggerate the estimate of the phase variance.

We can analyze this effect if the detrended phase record is a stationary process [5][6][26]. The first step is to establish a mean value from which the phase fluctuations can be measured,

$$\overline{\varphi(T)} = \frac{1}{T} \int_0^T dt \, \varphi(t) \tag{4.23}$$

which should approach zero as the sample length increases. However, the measured average need not be zero when a finite sample length is used. In that case the mean square phase and the phase variance are different. The variance for a single data sample becomes:

$$\overline{|\varphi(T) - \overline{\varphi(T)}|^2} = \frac{1}{T} \int_0^T dt \left| \varphi(t) - \frac{1}{T} \int_0^T dt \, \varphi(t) \right|^2$$

$$= \frac{1}{T} \int_0^T dt \, \varphi^2(t) - \frac{1}{T^2} \int_0^T dt \int_0^T dt' \, \varphi(t)\varphi(t')$$

One usually averages over a family of samples. The result should approach the ensemble average if there are enough samples to establish a statistically significant estimate:

$$\langle |\varphi(T) - \overline{\varphi(T)}|^2 \rangle = \frac{1}{T} \int_0^T dt \, \langle \varphi^2(t) \rangle - \frac{1}{T^2} \int_0^T dt \int_0^T dt' \, \langle \varphi(t)\varphi(t') \rangle$$

To complete this calculation we need the autocorrelation of phase. The usual assumption is that a constant wind bears a frozen turbulent structure across the line of sight at a uniform speed $v$. When this is a valid description we shall find in Chapter 6 that

$$\langle \varphi(t)\varphi(t')\rangle = 4\pi^2 Rk^2 \int_0^\infty d\kappa\, \kappa \Phi_n(\kappa) J_0(\kappa v |t - t'|).$$

Combining these expressions and carrying out the time averages gives

$$\overline{\langle |\varphi(T) - \overline{\varphi(T)}|^2\rangle} = 4\pi^2 Rk^2 \int_0^\infty d\kappa\, \kappa \Phi_n(\kappa) S(\kappa v T) \qquad (4.24)$$

where the *finite-sample-length* weighting function

$$S(x) = 1 + 2\frac{J_1(x)}{x} - \frac{2}{x}\int_0^x du\, J_0(u) \qquad (4.25)$$

describes how the contributions of different wavenumbers are modified. This function rises gradually from zero and approaches unity for large values of $x$.

It is evident that a finite data-sample length tends to suppress the contributions of large eddies which are ordinarily the primary source of phase fluctuations. The sample length $T$ and outer scale length $L_0$ therefore play competing roles in determining measured values of the phase variance. To investigate their relative influence we introduce the von Karman spectrum into (4.24) and express the result as a factor applied to the phase-variance expression (4.12) which corresponds to an infinitely long averaging time:

$$\overline{\langle |\varphi(T) - \overline{\varphi(T)}|^2\rangle} = (0.782 Rk^2 C_n^2 \kappa_0^{-\frac{5}{3}})\mathcal{S}(\kappa_0 v T) \qquad (4.26)$$

where

$$\mathcal{S}(\eta) = \frac{5}{3}\eta^{\frac{5}{3}} \int_0^\infty \frac{du\, u}{(u^2 + \eta^2)^{\frac{11}{6}}}\left(1 + 2\frac{J_1(u)}{u} - \frac{2}{u}\int_0^u dw\, J_0(w)\right). \qquad (4.27)$$

The function $\mathcal{S}(\eta)$ is plotted in Figure 4.8. It begins at zero and rises slowly to an asymptotic value of unity.

The dependence of the phase variance on the sample length was measured at 9414 MHz on Maui [9] using the path profile reproduced in Figure 4.9. The transmitter was placed at the summit of the dormant volcano Haleakala to simulate space-to-ground propagation conditions. Receivers were deployed at sea level near Puunene and were thus in the atmospheric boundary layer. It is evident from Figure 4.9 that the signal traveled mostly in the free atmosphere. This region is characterized by very large horizontal scale lengths and relatively small vertical scales, as we learned in Section 2.3.

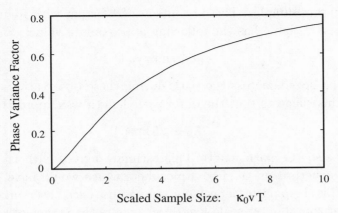

Figure 4.8: The phase-variance factor $\mathcal{S}(\kappa_0 v T)$ which describes the influence of the sample size $T$. Here $v$ is the wind speed and $\kappa_0$ is the outer-scale wavenumber.

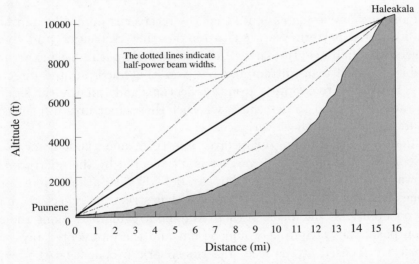

Figure 4.9: The propagation path profile for the phase measurements made by Norton, Herbstreit, Janes, Hornberg, Peterson, Barghausen, Johnson, Wells, Thompson, Vetter and Kirkpatrick on Maui using 9414-MHz signals [9].

Using this experimental arrangement, single-path phase variations for various sample sizes were measured and compared with a standard 15-min result. The largest increase in variance occurred between samples of durations 15 and 30 min, whereas very little increase was observed between

60- and 120-min samples. These results are consistent with Figure 4.8 if the dimensionless variable has the following approximate value:

$$\kappa_0 v \, 15 \, \text{min} = 1$$

Typical wind speeds across the path during these observations were about 5 m s$^{-1}$. This suggests that the outer scale length was approximately

$$L_0 \simeq 25 \, \text{km}$$

which was also the path length. This estimate agrees with the very large horizontal structures in the free atmosphere which would have been traveling across this steep path. Conversely, one can use such observations to verify that the horizontal outer scale length on such paths is tens of kilometers.

### *4.1.8 Receiver-aperture Averaging*

The foregoing discussion assumes that the receiver is point-like and has an omnidirectional pattern. We need to remove that restriction and consider receivers with finite capture areas. All of the single-path phase experiments in Table 4.1 used transmitters and receivers that incorporated directional gain. They did so to suppress ground reflections and enhance the signal-to-noise ratio. We need to determine whether those apertures influenced the measurements of the phase variance.

To describe the coupling of receiver apertures and random media, one must consider the family of rays that is intercepted by the aperture. Most microwave receivers employ circular paraboloids with a focal detector like the one shown in Figure 4.10. All signals that strike the reflector are combined coherently at the focus. If there were no irregularities along the propagation path, the wave-front would be flat and the intercepted rays would be perfectly focused at the feed. In atmospheric propagation the arriving wave-front is crinkled and/or tilted because the phase shift varies from ray to ray across the aperture. Since the radius of the reflector was substantially larger than the wavelength in these experiments, the field strength induced in the feed can be represented by the aperture average of the incident field taken over the area of the reflector:

$$\overline{E} = \frac{1}{\mathsf{A}} \int \int_{\mathsf{A}} d^2\sigma \, E(\sigma) \tag{4.28}$$

There is therefore a one-to-one relationship between the field at the surface element of the reflector and the field at the entry plane of the receiver. The

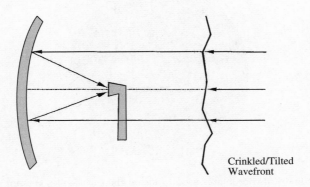

Figure 4.10: The incidence of a randomly crinkled and/or tilted wave on a parabolic microwave antenna with a feed at its focus.

distance $\Delta R$ traveled from the entry plane to the focus is the same for each reflected ray if the antenna is boresighted on the signal source:

$$E(\sigma) = E_r(\sigma)\exp(ik\,\Delta R)$$

The arriving field contains both phase and amplitude fluctuations:

$$E_r(\sigma) = [A_0 + \delta A(\sigma)]\exp[i\varphi(\sigma)]$$

Combining these expressions gives the following general description for the aperture-averaged field strength:

$$\overline{E} = [A_0\exp(ik\,\Delta R)]\frac{1}{\mathsf{A}}\int\int_{\mathsf{A}}d^2\sigma\left(1 + \frac{\delta A(\sigma)}{A_0}\right)\exp[i\varphi(\sigma)]$$

The combination in parentheses outside the surface integral is the field strength $E_0$ that would be measured if the medium were uniform.

We found in (3.45) that the logarithmic amplitude fluctuation is negligible when geometrical optics is valid and this simplifies the integrand. Separating the measured field into its phase and amplitude:

$$\overline{E} = \overline{A}\exp(i\,\overline{\varphi}) = E_0\frac{1}{\mathsf{A}}\int\int_{\mathsf{A}}d^2\sigma\exp[i\varphi(\sigma)] \tag{4.29}$$

One cannot assume that the aperture-averaged amplitude is constant, even though we have ignored amplitude fluctuations in the integrand. Smoothing produced by the reflector–feed combination reduces the measured amplitude

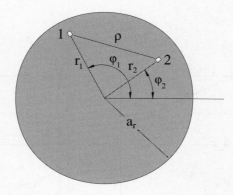

Figure 4.11: Cylindrical coordinates used to locate the surface elements on a circular parabolic reflector of radius $a_\mathrm{r}$.

through the exponential phase term. This is called *gain loss* and will be addressed in the second volume.

For the present purpose we must treat the aperture-averaged phase and amplitude as random variables. The phase is extracted by taking the ratio of the imaginary and real parts of (4.29):

$$\tan \overline{\varphi} = \frac{\iint_\mathsf{A} d^2\sigma \sin \varphi(\sigma)}{\iint_\mathsf{A} d^2\sigma \cos \varphi(\sigma)} \tag{4.30}$$

The rms phase fluctuation is measured as 3° to 30° in the frequency bands employed in the experiments identified in Table 4.1 and one can therefore safely expand the trigonometric terms:

$$\overline{\varphi} = \frac{1}{\mathsf{A}} \iint_\mathsf{A} d^2\sigma \, \varphi(\sigma) \tag{4.31}$$

This result is the starting point for most studies of phase smoothing. It is important to remember that it is valid only for microwave signals that suffer small phase shifts. It seldom applies to millimeter-wave links and *cannot be used* for optical propagation.

The phase variance is naturally expressed with the cylindrical coordinates identified in Figure 4.11 which locate the surface elements on the reflector:

$$\overline{\langle \varphi^2 \rangle} = \frac{1}{\pi^2 a_\mathrm{r}^4} \int_0^{a_\mathrm{r}} dr_1 \, r_1 \int_0^{2\pi} d\phi_1 \int_0^{a_\mathrm{r}} dr_2 \, r_2 \int_0^{2\pi} d\phi_2 \langle \varphi(r_1, \phi_1) \varphi(r_2, \phi_2) \rangle$$

This depends on the phase covariance between parallel rays that are intercepted by the receiver. In Chapter 5 we show that it can be represented as

$$\langle \varphi(\mathbf{r})\varphi(\mathbf{r}+\boldsymbol{\rho})\rangle = 4\pi^2 R\, k^2 \int_0^\infty d\kappa\, \kappa \Phi_n(\kappa) J_0(\kappa|\boldsymbol{\rho}|).$$

The scalar distance between surface elements is described by the cylindrical coordinates introduced above:

$$\overline{\langle \varphi^2 \rangle} = 4\, k^2 R\, \frac{1}{a_{\mathrm{r}}^4} \int_0^{a_{\mathrm{r}}} dr_1\, r_1 \int_0^{2\pi} d\phi_1 \int_0^{a_{\mathrm{r}}} dr_2\, r_2 \int_0^{2\pi} d\phi_2$$
$$\times \int_0^\infty d\kappa\, \kappa \Phi_n(\kappa) J_0\!\left(\kappa\sqrt{r_1^2 + r_2^2 - 2r_1 r_2 \cos(\phi_1 - \phi_2)}\right)$$

The addition theorem for Bessel functions separates the integrations:

$$J_0\!\left(\kappa\sqrt{r_1^2 + r_2^2 - 2r_1 r_2 \cos(\phi_1 - \phi_2)}\right)$$
$$= \sum_0^\infty \varepsilon_n J_n(\kappa r_1) J_n(\kappa r_2) \cos[n\,(\phi_1 - \phi_2)] \qquad (4.32)$$

The angular integrations eliminate each term in the series except the first. The remaining radial integrations can be done analytically [5][6] using a result from Appendix D:

$$\overline{\langle \varphi^2 \rangle} = 4\pi^2 R k^2 \int_0^\infty d\kappa\, \kappa \Phi_n(\kappa)\left(\frac{2J_1(\kappa a_{\mathrm{r}})}{\kappa a_{\mathrm{r}}}\right)^2 \qquad (4.33)$$

The *aperture-averaging spectral weighting function* in parentheses indicates how the contributions of different wavenumbers are weighted. This correction is also called the *piston* phase-variance filter function [27]. It declines rapidly with increasing $\kappa a_{\mathrm{r}}$. This means that irregularities smaller than the size of the antenna influence the phase-variance measurement very little. To be more specific, we evaluate (4.33) with the von Karman spectrum:

$$\overline{\langle \varphi^2 \rangle} = 4\pi^2 R k^2 \int_0^\infty d\kappa\, \kappa \frac{0.033 C_n^2}{(\kappa^2 + \kappa_0^2)^{\frac{11}{6}}}\left(\frac{2J_1(\kappa a_{\mathrm{r}})}{\kappa a_{\mathrm{r}}}\right)^2$$

The result is expressed as a factor that multiplies the point-detector result established in (4.12):

$$\overline{\langle \varphi^2 \rangle} = \langle \varphi^2 \rangle\left(\frac{20}{3}(\kappa_0 a_{\mathrm{r}})^{\frac{5}{3}} \int_0^\infty dx\, \frac{J_1^2(x)}{x[x^2 + (\kappa_0 a_{\mathrm{r}})^2]^{\frac{11}{6}}}\right) \qquad (4.34)$$

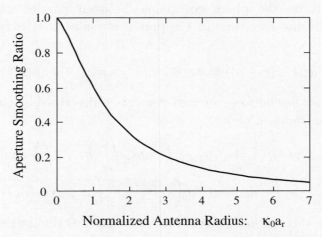

Figure 4.12: The aperture smoothing factor which describes the modification of the phase variance by a circular antenna of radius $a_r$.

The function in the large parentheses was evaluated numerically and is plotted as a function of $\kappa_0 a_r$ in Figure 4.12.

It is evident that aperture averaging of phase is not important unless the radius of the receiver is comparable to the outer scale length measured in the directions normal to the line of sight. At microwave frequencies the outer scale length is 40–100 m near the surface and is much greater in the free atmosphere. This suggests that there should be negligible aperture smoothing for terrestrial links. That prediction was confirmed by an experiment at 9.4 GHz in which the phase spectra measured with receivers having diameters of 46 and 300 cm were compared [13]. This means that one cannot easily suppress phase noise on line-of-sight links by increasing the size of the antenna.

## 4.2 Satellite Signals

All of the preceding discussion has dealt with signals that propagate along horizontal paths near the surface. A second important class of experiments uses microwave signals from distant sources that pass through the entire atmosphere before reaching a receiver at ground level. In doing so, the signals encounter a wide variety of turbulent conditions along typical slanted paths. Anisotropy and the variations of $C_n^2$ and $\kappa_0$ with altitude play important roles for these experiments.

Signals from several types of source are measured. Radio astronomers routinely monitor microwave signals from distant galaxies. Although these signals are valuable for many purposes, they cannot be used to measure the phase variance because the sources are not phase stable. Fortunately satellite transmissions can be stabilized to the needed accuracy.

The obvious candidates are communications satellites. There are over a hundred microwave relay stations in geosynchronous orbit. Their signals are influenced differently by the troposphere and the ionosphere. They operate in the frequency bands between 4 and 30 GHz and are influenced primarily by tropospheric irregularities. Mobile communication services also use a narrow band of frequencies near 1.6 GHz, and are influenced by the ionosphere. Satellite operators are motivated to increase transmitter power and bandwidth to enhance service but have little incentive to make the signals phase stable. Attempts to measure phase variations with communications satellites have been frustrated thus far because of these limitations [28][29]. One must look elsewhere for microwave sources with which to measure the phase variance.

There is a class of earth satellites that radiate microwave signals whose frequency and phase are tightly controlled. These are the spacecraft used for navigation and position location. The American Global Positioning System (GPS) operates twenty-four satellites in six orbital planes at an altitude of 20 000 km. Each satellite radiates precisely controlled signals at 1.228 and 1.575 GHz. The Russian Glonass system is quite similar. A user typically receives simultaneous transmissions from four to eight spacecraft. By comparing the arrival times of the pseudo-random modulation, one can establish his position to an accuracy of within approximately 1 m anywhere on the earth's surface with a hand- held device. Using differential GPS measurements, one can establish one's position relative to a known point with an inaccuracy of less than 1 cm [30].

The GPS signals are strongly influenced by the ionosphere but this effect can be eliminated since the effect depends on the microwave frequency. One can remove the influence of the ionosphere by scaling and subtracting the times of arrival for the two L-band signals. The same technique cannot eliminate range errors introduced by the troposphere because the travel time is independent of frequency. Turbulent irregularities in the lower atmosphere therefore set an irreducible threshold for the accuracy of location achievable with the GPS. The corollary is that ranging errors measured with this system may some day provide an important insight into the physics of the lower atmosphere. Our first task is to develop a model with which we can interpret the GPS data.

### 4.2.1 GPS Range Errors

The phase variance for a satellite signal that has passed completely through the terrestrial atmosphere is given by the basic expression (4.3). Since the orbits of earth satellites are far above the troposphere, we can take the path integrations to infinity:

$$\langle \varphi^2 \rangle = k^2 \int_0^\infty ds \int_0^\infty ds' \int d^3\kappa \, \Phi_n[\boldsymbol{\kappa}, \tfrac{1}{2}(s+s')] \exp[i\kappa_{\mathrm{s}}(s-s')] \quad (4.35)$$

It is left to the height-dependent spectrum to select the important regions of the atmosphere. GPS signals are seldom used when the angle of elevation is small. This means that one can neglect refractive ray bending and assume that the signal travels along a straight-line path connecting the satellite and the ground station. We use the *flat-earth* approximation illustrated in Figure 4.13 to describe these satellite signals.[4]

One cannot assume that the spectrum of irregularities is isotropic since much of the propagation path lies in the free atmosphere. Horizontal scales in this region are much larger than those in the vertical direction. We use the ellipsoidal spectrum (2.87) to describe this situation. Following the discussion of Section 2.2.8, we split the spectrum into a term that describes the local properties of the random medium and a second term that characterizes the height variation of turbulent activity:

$$\Phi_n\left(\boldsymbol{\kappa}, \frac{s+s'}{2}\right) = \Omega(\boldsymbol{\kappa})C_n^2\left(\frac{s+s'}{2}\right)$$

With this separation the phase variance (4.35) becomes

$$\langle \varphi^2 \rangle = k^2 \int_0^\infty ds \int_0^\infty ds' \int_{-\infty}^\infty d\kappa_x \int_{-\infty}^\infty d\kappa_y \int_{-\infty}^\infty d\kappa_z \exp[i\kappa_{\mathrm{s}}(s-s')]$$

$$\times abc\Omega\left(\sqrt{a^2\kappa_x^2 + b^2\kappa_y^2 + c^2\kappa_z^2}\right)C_n^2\left(\frac{s+s'}{2}\right)$$

---

[4] GPS signals are also measured by satellites in low earth orbit to establish meteorological profiles [31][32]. These limb-sounding signals employ very small angles of elevation and the spherical nature of the atmosphere plays an important role in their interpretation.

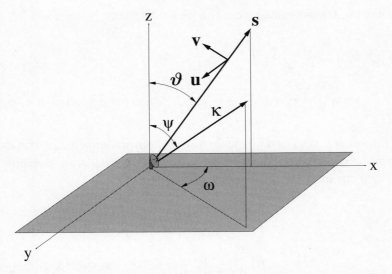

Figure 4.13: Rectangular coordinates used for analyzing phase fluctuations imposed on GPS signals by the troposphere. The line-of-sight vector lies in the (x, z) plane. The wavenumber vector which describes turbulent irregularities is given in spherical coordinates that are referenced to the local vertical.

We now transform the path integrations to sum and difference coordinates,

$$u = s - s' \quad \text{and} \quad r = \tfrac{1}{2}(s + s') \tag{4.36}$$

where the sum coordinate is the average slant range to the irregularities:

$$\int_0^\infty ds \int_0^\infty ds' \, C_n^2\left(\frac{s+s'}{2}\right) \exp[i\kappa_{\mathrm{s}}(s-s')] = \int_0^\infty dr \, C_n^2(r) \int_{-r}^r du \, e^{i\kappa_{\mathrm{s}} u}$$

We express the radial wavenumber component $\kappa_{\mathrm{s}}$ in terms of the rectangular components of $\boldsymbol{\kappa}$ by observing that

$$\mathbf{s} = s(\mathbf{i}_x \sin\vartheta + \mathbf{i}_z \cos\vartheta)$$

thus obtaining

$$\kappa_s = \kappa_x \sin\vartheta + \kappa_z \cos\vartheta.$$

On combining these expressions, the phase variance becomes

$$\langle \varphi^2 \rangle = k^2 \int_0^\infty dr \, C_n^2(r) \int_{-\infty}^\infty d\kappa_x \int_{-\infty}^\infty d\kappa_y \int_{-\infty}^\infty d\kappa_z \, abc$$

$$\times \, \Omega\Big(\sqrt{a^2\kappa_x^2 + b^2\kappa_y^2 + c^2\kappa_z^2}\Big) \int_{-r}^r du \exp[iu(\kappa_x \sin\vartheta + \kappa_z \cos\vartheta)].$$

The important step is to rescale the $\boldsymbol{\kappa}$ components using the stretched wavenumber coordinates defined by (4.15). This yields an isotropic turbulence spectrum in the new coordinates:

$$\langle \varphi^2 \rangle = k^2 \int_0^\infty dr \, C_n^2(r) \int d^3q \, \Omega(|\mathbf{q}|)$$

$$\times \int_{-r}^r du \exp\Big[iu\Big(\frac{q_x \sin\vartheta}{a} + \frac{q_z \cos\vartheta}{c}\Big)\Big]$$

It is now convenient to introduce spherical wavenumber coordinates:

$$\mathbf{q} = q(\mathbf{i}_x \sin\psi \cos\omega + \mathbf{i}_y \sin\psi \sin\omega + \mathbf{i}_z \cos\psi)$$

so that

$$\langle \varphi^2 \rangle = k^2 \int_0^\infty dr \, C_n^2(r) \int_{-r}^r du \int_0^\infty dq \, q^2 \Omega(q) \int_0^\pi d\psi \sin\psi \int_0^{2\pi} d\omega$$

$$\times \exp\Big[iqu\Big(\frac{\sin\vartheta \sin\psi \cos\omega}{a} + \frac{\cos\vartheta \cos\psi}{c}\Big)\Big].$$

The azimuth integration gives the zeroth-order Bessel function and the resulting polar integration is found in Appendix B:

$$\int_0^\pi d\psi \sin\psi \int_0^{2\pi} d\omega \exp\Big[iqu\Big(\frac{\sin\vartheta \sin\psi \cos\omega}{a} + \frac{\cos\vartheta \cos\psi}{c}\Big)\Big]$$

$$= 2\pi \int_0^\pi d\psi \sin\psi \, J_0\Big(\frac{qu}{a} \sin\vartheta \sin\psi\Big) \exp\Big(i\frac{qu}{c} \cos\vartheta \cos\psi\Big)$$

$$= 4\pi \frac{\sin(qu\mathsf{p})}{qu\mathsf{p}}$$

The parameter $\mathsf{p}$ depends on the zenith angle and the scaling parameters which describe the medium's elongation:

$$\mathsf{p} = \sqrt{\frac{\sin^2 \vartheta}{a^2} + \frac{\cos^2 \vartheta}{c^2}} \tag{4.37}$$

Letting $\zeta = uq\mathsf{p}$, the phase variance for a satellite signal can be expressed as

$$\langle \varphi^2 \rangle = 4\pi^2 k^2 \int_0^\infty dr\, C_n^2(r) \frac{1}{\mathsf{p}} \int_0^\infty dq\, q\Omega(q) \left( \frac{2}{\pi} \int_0^{rq\mathsf{p}} \frac{\sin \zeta}{\zeta} d\zeta \right). \tag{4.38}$$

The term in large parentheses can be replaced by unity if the slant range is substantially greater than the size of the stretched irregularities,

$$rq\mathsf{p} \gg 1: \qquad \langle \varphi^2 \rangle = 4\pi^2 k^2 \int_0^\infty dr\, C_n^2(r) \frac{1}{\mathsf{p}} \int_0^\infty dq\, q\Omega(q) \tag{4.39}$$

and one obtains the result established by Zavorotny if the scaling parameters do not vary along the path [25]:

$$\langle \varphi^2 \rangle_{\text{anisotropic}} = \langle \varphi^2 \rangle_{\text{isotropic}} \left( \frac{\sin^2 \vartheta}{a^2} + \frac{\cos^2 \vartheta}{c^2} \right)^{-\frac{1}{2}} \tag{4.40}$$

There is a potential problem in this line of reasoning that relates to the actual values for the upper limit in the $\zeta$ integration of (4.38). We know that the slant range is quite large for most points along the ray path of a satellite signal. On the other hand, it is not necessarily larger than the horizontal outer scale lengths encountered in the free atmosphere. The relative roles of vertical and horizontal scale lengths are reflected in the parameter $\mathsf{p}$ which also depends on the zenith angle. To relate the slant range to the two scale sizes, we introduce the von Karman spectrum for the isotropic spectrum in stretched coordinates:

$$\langle \varphi^2 \rangle = 8\pi \times 0.033 k^2 \int_0^\infty dr\, C_n^2(r)$$
$$\times \int_0^r du \left( \frac{1}{u\mathsf{p}} \int_0^\infty \frac{dq\, q}{(q^2 + \kappa_0^2)^{\frac{11}{6}}} \sin(qu\mathsf{p}) \right)$$

Integrating on $q$ by parts and using a result from Appendix B, one finds that the phase variance for a satellite signal is given by

$$\langle \varphi^2 \rangle = 0.782 k^2 \int_0^\infty dr\, C_n^2(r) \frac{1}{\mathsf{p}} \kappa_0^{-\frac{5}{3}} \mathcal{H}(r\kappa_0 \mathsf{p}) \tag{4.41}$$

where the function

$$\mathcal{H}(x) = \frac{2^{\frac{2}{3}}}{\sqrt{\pi}\,\Gamma\left(\frac{5}{6}\right)} \int_0^x d\zeta\, \zeta^{\frac{1}{3}} K_{\frac{1}{3}}(\zeta) \tag{4.42}$$

indicates how irregularities along the inclined path are weighted. It begins linearly and gradually approaches an asymptotic value of unity. This focuses attention on the following combination:

$$r\kappa_0\mathsf{p} = 2\pi r \sqrt{\frac{\sin^2\vartheta}{(L_0 a)^2} + \frac{\cos^2\vartheta}{(L_0 c)^2}} \tag{4.43}$$

The horizontal scale length $aL_0$ is much greater than the vertical scale $cL_0$. The first term in the square root is therefore considerably smaller than the second unless the zenith angle is very close to $90°$. Neither GPS nor communications satellite signals are commonly used when the angle of elevation is small and (4.43) is approximated as

$$r\kappa_0\mathsf{p} \simeq \frac{2\pi z}{cL_0}$$

The outer scale length in the vertical direction is much smaller than the height of the ray along most of the path and we can replace the function $\mathcal{H}(x)$ in (4.41) by unity. The slant range is related to the height at each point on the ray by $r = z\sec\vartheta$. The result is a relatively simple result for the phase variance that depends on the height profile of $C_n^2$ and $\kappa_0$:

$$\langle \varphi^2 \rangle = 0.782\,k^2\sec^2\vartheta \int_0^\infty dz\, C_n^2(z) c\kappa_0^{-\frac{5}{3}} \tag{4.44}$$

This agrees with (4.40) for $a \gg c$ and spacecraft transmitters that are not too close to the horizon.[5]

The final result is somewhat surprising in two respects. The first is that it depends on the vertical outer scale length $cL_0$ rather than the horizontal outer scale length $aL_0$. We began our investigation of propagation in anisotropic media because irregularities in the free atmosphere are highly

---

[5] The derived result (4.44) can also be expressed in terms of refractive-index variations measured by microwave refractometers. To do so we regard the expression for $\langle \delta n^2 \rangle$ stated above (4.13) as a local relationship that connects values of $C_n^2$, $\kappa_0$ and $\langle \delta n^2 \rangle$ at each point along the ray path:

$$\langle \varphi^2 \rangle = 0.238 \times 10^{-12} k^2 \sec^2\vartheta \int_0^\infty dz \langle \delta n^2 \rangle c\, L_0$$

This version makes clear the unique role played by the vertical outer scale length, which is the component measured by profiling radars.

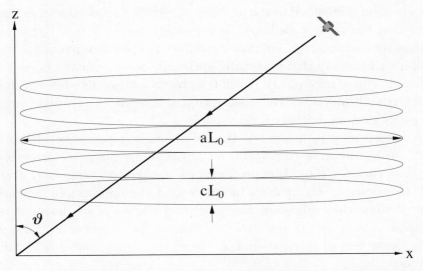

Figure 4.14: A satellite signal passing through elongated irregularities in the free atmosphere.

elongated in the horizontal direction. This does not seem to matter – at least for sources not too close to the horizon. We need to understand this conclusion on physical grounds rather than accept it as an analytical result. To do so, we refer to Figure 4.14, which shows a satellite signal passing through a number of *elongated irregularities*. Recall that phase fluctuations are induced by changes in the refractive index along the ray path. A satellite signal sees considerable change in the vertical direction as it travels through a succession of *pancakes*. By contrast, each transit of the irregularity encounters very little change in the horizontal direction because the refractive index is highly correlated over large distances. Because it is a sensitive indicator of a change in refractive index, the phase is influenced heavily by the vertical scale length and very little by the horizontal scale length. This is also what the analytical result (4.44) tells us.

There is an important exception to this description. This occurs for truly horizontal propagation between an elevated transmitter and an elevated receiver. In this case the ray might see only a single irregularity and will be sensitive primarily to the horizontal outer scale length $aL_0$.

This conclusion emerges also from the analytical description (4.41) for a source close to the horizon:

$$\vartheta \simeq \frac{\pi}{2}: \qquad r\kappa_0 \mathsf{p} \simeq \frac{2\pi z}{aL_0}$$

This situation is identical to the problem of terrestrial links discussed earlier, except that the outer scale length is very much larger.

Notice that atmospheric curvature modifies the physical picture presented in Figure 4.14. A ray that is initially horizontal will eventually emerge from a single elongated irregularity. It will then begin cutting distant irregularities in the vertical direction. This means that the result (4.44) might describe long paths that are initially horizontal.

The second surprise occurs in the zenith-angle dependence of the basic result. An estimate of $\langle \varphi^2 \rangle$ based on isotropic irregularities suggests that the phase variance should be proportional to the slant range and hence to $\sec \vartheta$. By contrast, our analysis indicates that the appropriate dependence is $\sec^2 \vartheta$ when the eddies are elongated and the angle of elevation is not too small. This result is apparent from the scaling relation of (4.40) when one neglects $1/a$ in relation to $1/c$. Here anisotropy makes a significant difference.

The *zenith-angle scaling* of the phase variance was confirmed by Naudet's study of single-path GPS range measurements [33]. Range residuals were analyzed for satellite positions above $20°$ at four widely separated stations. Experimental data taken between 12 and 24 June 1994 are reproduced in Figure 4.15 and were fitted to the following scaling law:

$$\langle \delta R^2 \rangle = a + b \operatorname{cosec}^2 E \qquad (4.45)$$

This conclusion suggests that two error sources limit accuracy with the GPS. The same data set exhibited significant variation of the coefficients $a$ and $b$ between dry and wet sites. The second term was initially identified with tropospheric path-length fluctuations [33]. ts scaling with the angle of elevation confirms the $\sec^2 \vartheta$ prediction of (4.44) since $E = \pi/2 - \vartheta$ and $\varphi = k \, \delta R$. This agreement further strengthens the identification of the second term with tropospheric irregularities.

We can take this comparison a step further by considering the absolute value of the range residuals. We identify the second term in (4.45) with the mean-square range error which is proportional to the phase variance. The zenith-angle dependence is common to both expressions and their equivalence implies that

$$b = 7 \times 10^{-5} \, \mathrm{m}^2 = 0.8 \int_0^\infty dz \, C_n^2(z) \frac{cL_0}{2\pi} \left( \frac{L_0}{2\pi} \right)^{\frac{2}{3}}$$

In writing this we have separated the two terms that depend on $L_0$ in order to identify the roles played by the vertical and horizontal outer scale lengths.

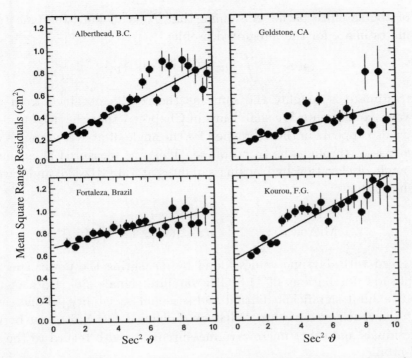

Figure 4.15: GPS range residuals measured as a function of the zenith angle at four widely separated sites. The range fluctuations taken at these sites were analyzed and presented in this form by Naudet [33].

We first summarize the available data on vertical outer scales that was presented in Section 2.3.2. This component was carefully measured over a five-year period by VHF radar-sounding experiments in New Mexico [34]. The profile values ranged from 10 to 60 m for the height range 4–18 km. By contrast, the value of $C_n^2$ often changes by two orders of magnitude in the first 5 km of the troposphere. Over the same height range, the variations of $cL_0$ are therefore quite small compared with those of $C_n^2$. This means that one can hold $cL_0$ constant in the vertical integration above:

$$10^{-6}\,\mathrm{m} = \int_0^\infty dz\, C_n^2(z)\left(\frac{L_0}{2\pi}\right)^{\frac{2}{3}}$$

The same argument can be applied to the term containing only $L_0$. We associate $L_0$ with the large horizontal outer length scales experienced over most of the satellite path. Assuming that $L_0 = 6$ km, we come to the following estimate for the integrated profile:

$$\text{GPS:} \qquad \int_0^\infty dz\, C_n^2(z) = 10^{-8}\, \text{m}^{\frac{1}{3}} \qquad\qquad (4.46)$$

This estimate is roughly the same as the results obtained from other microwave experiments. We shall learn in Chapter 7 that the height integral of $C_n^2(z)$ also appears in the expression for the angle-of-arrival variance when a signal passes through the troposphere. This was measured as a function of the angle of elevation by tracking satellites at 7.3 GHz [35] and gave the following estimate:

$$\text{Microwave:} \qquad \int_0^\infty dz\, C_n^2(z) \simeq 10^{-8}\, \text{m}^{\frac{1}{3}} \qquad\qquad (4.47)$$

This agreed with daytime values found by measuring the power spectrum of amplitude fluctuations of 11.8-GHz satellite signals [36]. Both of these methods exhibit significant diurnal and seasonal variability, so one cannot claim to have achieved closure. What we do have is fair agreement between GPS residuals and other microwave measurements with regard to the integrated profile.

### 4.2.2 The Ionospheric Influence

Satellite frequencies below the civilian band are strongly influenced by plasma irregularities because the ionized medium is dispersive. Satellite signals have been used to measure phase variations imposed by the ionosphere. The concept is to use a narrow-band, high-frequency satellite signal as a phase reference. Signals transmitted at lower frequencies are phase locked to the reference signals and are compared with it at a ground station after they have all passed through the ionosphere. The troposphere does not influence the low-frequency signals in view of the wavelength scaling indicated above.

This technique was first used with navigation satellites whose 150- and 400-MHz transmissions were made phase coherent for other reasons [37]. A scientific experiment using a spacecraft specifically designed for studying ionospheric scintillations was then planned. The DNA Wideband Satellite was launched into a polar circular orbit at 1000 km in 1976. It radiated ten mutually coherent signals at frequencies ranging from 138 to 2891 MHz. The phase and amplitude of the complex signals were measured at each frequency by comparing them with the 2891-MHz reference signal [38]. To interpret such data we need an expression for the phase of a signal that has passed through ionized irregularities. The microwave frequencies used in these experiments are well above the plasma frequency of the ionosphere. A simple relation therefore connects variations of the dielectric constant with fluctuations of the electron density:

$$\Delta \varepsilon = r_e \lambda^2 \, \delta N$$

Here $r_e = 2.8 \times 10^{-13}$ cm is the classical electron radius. When this is combined with (4.1), an expression for the signal phase in ionized media emerges:

$$\varphi(t) = \pi r_e \lambda \int_0^\infty ds \, \delta N(s, t) \tag{4.48}$$

The phase variance evidently depends on the spatial correlation of the electron-density fluctuations:

$$\langle \varphi^2 \rangle = \pi^2 r_e^2 \lambda^2 \int_0^\infty ds \int_0^\infty ds' \langle \delta N(s, t) \, \delta N(s', t) \rangle$$

We express the spatial correlation in terms of a wavenumber spectrum introduced in Section 2.4 for the plasma irregularities:

$$\langle \delta N(\mathbf{r}) \, \delta N(\mathbf{r}') \rangle = \int d^3 \kappa \, \Psi_N[\boldsymbol{\kappa}, \tfrac{1}{2}(\mathbf{r} + \mathbf{r}')] \exp[i\boldsymbol{\kappa} \cdot (\mathbf{r} - \mathbf{r}')] \tag{4.49}$$

One must not confuse the plasma spectrum $\Psi_N$ with the spectrum $\Phi_n$ which describes temperature and humidity fluctuations in the troposphere. A power-law model with an outer-scale cutoff was suggested in (2.135) to describe elongated ionospheric irregularities. The estimation of the phase variance is not sensitive to the choice of the exponent $\nu$ in that model so long as it is greater than 3. We take $\nu = 4$ for analytical convenience and write the spectrum as follows:

$$\Psi_N[\boldsymbol{\kappa}, \tfrac{1}{2}(\mathbf{r} + \mathbf{r}')] = \frac{\langle \delta N^2 \rangle Q_4 \kappa_0}{2\pi\kappa^4 [\sin^2 \Theta + \mathcal{A}^2 \cos^2 \Theta]^2} \qquad \kappa_0 < \kappa < \infty \tag{4.50}$$

where $\Theta$ is the angle between $\boldsymbol{\kappa}$ and the terrestrial magnetic field $\mathbf{H}$. The dimensionless axial ratio $\mathcal{A}$ describes the elongation of plasma irregularities. The normalization constant $Q_4$ ensures that the relationship (4.49) is satisfied when $\mathbf{r} = \mathbf{r}'$:

$$Q_4 = \left( \frac{1}{\sqrt{\mathcal{A}^2 - 1}} \tan^{-1} \left( \sqrt{\mathcal{A}^2 - 1} \right) + \frac{1}{\mathcal{A}^2} \right)^{-1}$$

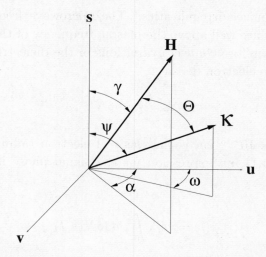

Figure 4.16: Coordinates used for analyzing phase fluctuations imposed by field-aligned electron-density irregularities in the ionosphere. The magnetic-field and wavenumber vectors are described by spherical coordinates that use the line of sight as their polar axis.

Since the axial ratio $\mathcal{A}$ varies between 10 and 30 in the ionosphere,

$$Q_4 \simeq 2\mathcal{A}/\pi$$

and the plasma spectrum becomes

$$\mathcal{A} \gg 1: \quad \Psi_N[\boldsymbol{\kappa}, \tfrac{1}{2}(\mathbf{r} + \mathbf{r}')] = \frac{\langle \delta N^2 \rangle \mathcal{A} \kappa_0}{\pi^2 \kappa^4 [1 + (\mathcal{A}^2 - 1) \cos^2 \Theta]^2} \quad \kappa_0 < \kappa < \infty.$$

(4.51)

The spectrum changes with position because the electron-density variance $\langle \delta N^2 \rangle$ and the outer-scale wavenumber $\kappa_0$ depend on altitude.

The coordinates identified in Figure 4.16 describe the path integrations and the parameters in the spectral model. The terrestrial magnetic field and the wavenumber vector are both referenced to the line-of-sight vector:

$$\boldsymbol{\kappa} = \kappa(\mathbf{i}_s \cos \psi + \mathbf{i}_u \sin \psi \cos \omega + \mathbf{i}_v \sin \psi \sin \omega)$$

$$\mathbf{H} = H \left( \mathbf{i}_s \cos \gamma + \mathbf{i}_u \sin \gamma \cos \alpha + \mathbf{i}_v \sin \gamma \sin \alpha \right)$$

The angle $\Theta$ between these vectors is defined by their scalar product:

$$\cos \Theta = \frac{\boldsymbol{\kappa} \cdot \mathbf{H}}{\kappa H} = \cos \psi \cos \gamma + \sin \psi \sin \gamma \cos(\omega - \alpha) \qquad (4.52)$$

We use the sum and difference path coordinates defined by (4.36) to evaluate the phase variance:

$$\langle\varphi^2\rangle = \mathcal{A}r_e^2\lambda^2 \int_0^\infty dr \langle\delta N^2(r)\rangle\kappa_0(r) \int_{\kappa_0(r)}^\infty \frac{d\kappa}{\kappa^3} \int_0^\pi d\psi \sin\psi \int_0^{2\pi} d\omega$$

$$\times \frac{\displaystyle\int_{-r}^r du \exp(i\kappa u \cos\psi)}{\{1 + (\mathcal{A}^2 - 1)[\cos\psi\cos\gamma + \sin\psi\sin\gamma\cos(\omega - \alpha)]^2\}^2}$$

The slant range to ionospheric layers is more than 300 km. This is substantially greater than the largest eddies and the integration on $u$ can be replaced by the delta function described in Appendix F:

$$\int_{-r}^r du \exp(i\kappa u \cos\psi) = \frac{2\sin(\kappa r \cos\psi)}{\kappa\cos\psi} = 2\pi\delta(\kappa\cos\psi)$$

This collapses the polar integration and simplifies the denominator:

$$\langle\varphi^2\rangle = \pi\mathcal{A}r_e^2\lambda^2 \int_0^\infty dr \frac{\langle\delta N^2(r)\rangle}{\kappa_0(r)}$$

$$\int_0^{2\pi} d\omega \frac{1}{[1 + (\mathcal{A}^2 - 1)\sin\gamma^2\cos^2(\omega - \alpha)]^2}$$

The azimuth integration is found in Appendix B and we convert from slant range to height above the earth:

$$\langle\varphi^2\rangle = \frac{\pi}{2}r_e^2\lambda^2 \frac{\mathcal{A}[2 + (\mathcal{A}^2 - 1)\sin^2\gamma]}{[1 + (\mathcal{A}^2 - 1)\sin\gamma^2]^{\frac{3}{2}}} \sec\vartheta \int_0^\infty dz \langle\delta N^2(z)\rangle L_0(z) \quad (4.53)$$

The angle $\gamma$ is not small because the terrestrial magnetic field and line of sight are seldom aligned. Since the axial ratio is large, the phase variance is nearly independent of $\mathcal{A}$ and the influence of anisotropy occurs only through a factor cosec $\gamma$:

Ionosphere: $\quad \langle\varphi^2\rangle \simeq \frac{\pi}{2}r_e^2\lambda^2 \operatorname{cosec}\gamma \sec\vartheta \int_0^\infty dz \langle\delta N^2(z)\rangle L_0(z) \quad (4.54)$

This expression shows how $\langle\varphi^2\rangle$ depends on the frequency, the zenith angle and the orientation of the local magnetic field. Notice that the wavelength scaling is the inverse of that for tropospheric phase variations.

Our knowledge of the profiles for $L_0$ and $\langle \delta N^2 \rangle$ is not good enough to estimate the phase variance with confidence. One can appeal to experimental data from the DNA Wideband Satellite to establish the phase variance for various microwave frequencies [38]. Probability density functions for phase fluctuations were measured at 138 MHz and gave $\varphi_{\rm rms} = 180°$. The rms values were approximately 90° at 380 MHz and 20° at 1239 MHz. This data tracks the linear wavelength scaling for the rms phase suggested by our analysis.

A sounding rocket was used to measure phase fluctuations over the Kwajalein Atoll in the central Pacific [39]. The rocket rose to an altitude of 600 km and carried coherent transmitters at 145, 291 and 437 MHz that were locked to a reference signal at 874 MHz. Phase fluctuations imposed by the ionosphere were measured as a function of altitude, both as the rocket rose and as it fell. The rms phase scaled linearly with wavelength to high accuracy, which confirms the expressions above. This experiment was unique in that it correlated phase data to other measurements that were made simultaneously. The ion density was measured along the trajectory with a Langmuir probe carried on the same rocket. Backscattered signals from the 155.5-MHz Altair radar on Kwajalein provided another source of data for the ionosphere on that day and at that location.

Measurements of the phase variance for signals passing through the ionosphere can be influenced both by the outer scale length and by the length of the data sample [40]. The outer scale of plasma irregularities is best understood as the *transition* from horizontal traveling structures to field-aligned irregularities. The scale at which this happens is probably greater than 1 km and perhaps as great as 10 km. The plasma moves at approximately $100 \text{ m s}^{-1}$ so the scaled sample length identified in Figure 4.8 becomes

$$\kappa_0 vT \simeq 4T_{\rm min}.$$

In the DNA satellite experiments a sample size of 20 s was used so that $\kappa_0 vT \simeq 1$ and one must include the factor defined by (4.27). If the sample size is very small one can use the small-argument expansion

$$\lim_{\kappa_0 vT \to 0} \mathcal{S}(\kappa_0 vT) = 0.3812(\kappa_0 vT)^{\frac{5}{3}}$$

to write

$$\langle \varphi^2 \rangle = 12.81 r_{\rm e}^2 \lambda^2 (vT)^{\frac{5}{3}} \operatorname{cosec} \gamma \sec \vartheta \int_0^\infty dz \, \langle \delta N^2(z) \rangle [\mathrm{L}_0(z)]^{-\frac{2}{3}}. \qquad (4.55)$$

but one is again left with an expression that depends on unknown profiles.

## 4.3 Problems

### *Problem 1*

Consider a microwave signal that travels horizontally in a random medium that is anisotropic. Assume that the link is close to the surface and that the transmission distance is large relative to the eddies encountered. Confirm the following result which is independent of the spectral model [41]:

$$\langle \varphi^2 \rangle_{\text{anisotropic}} = a \langle \varphi^2 \rangle_{\text{isotropic}}$$

Use the alternative description of the isotropic phase variance given by (4.13) to establish the following expression:

$$\langle \varphi^2 \rangle_{\text{anisotropic}} = 11.1 \frac{R(aL_0)}{\lambda^2} \langle \Delta N^2 \rangle 10^{-12}$$

Here $aL_0$ is clearly identified with the *effective* outer scale length or the correlation length along the line of sight.

### *Problem 2*

Reconsider the estimation of the phase variance given in Section 4.1.5 for the microwave propagation conditions that correspond to the data reproduced in Figure 4.5. The transmission path was 16 km long with the transmitter on Pike's Peak (14 000 ft) and the receivers deployed at Garden-of-the-Gods (6500 ft). Combine the expression for satellite signals passing through anisotropic media (4.41) with the simple expression for the phase variance (4.13). Show that the result agrees with the earlier estimate if the vertical outer scale length is 50 m. Why does this situation depend on the vertical scale length, in contrast to the solution of Problem 1 which is expressed in terms of the horizontal outer scale length?

## References

[1] J. W. Herbstreit and M. C. Thompson, "Measurements of the Phase of Radio Waves Received over Transmission Paths with Electrical Lengths Varying as a Result of Atmospheric Turbulence," *Proceedings of the IRE*, **43**, No. 10, 1391–1401 (October 1955).

[2] P. G. Bergmann, "Propagation of Radiation in a Medium with Random Inhomogeneities," *Physical Review*, **70**, Nos. 7 and 8, 486–492 (1 and 15 October 1946).

[3] V. A. Krasil'nikov, "On Fluctuations of the Angle-of-Arrival in the Phenomenon of Twinkling of Stars," *Doklady Akademii Nauk SSSR, Seriya Geofizicheskaya*, **65**, No. 3, 291–294 (1949) and "On Phase Fluctuations of

Ultrasonic Waves Propagating in the Layer of the Atmosphere Near the Earth," *Doklady Akademii Nauk SSSR, Seriya Geofizicheskaya*, **88**, No. 4, 657–660 (1953). (These references are in Russian and no translations are currently available.)

[4] R. B. Muchmore and A. D. Wheelon, "Line-of-Sight Propagation Phenomenon – I. Ray Treatment," *Proceedings of the IRE*, **43**, No. 10, 1437–1449 (October 1955).

[5] A. D. Wheelon, "Relation of Radio Measurements to the Spectrum of Tropospheric Dielectric Fluctuations," *Journal of Applied Physics*, **28**, No. 6, 684–693 (June 1957).

[6] A. D. Wheelon, "Radiowave Scattering by Tropospheric Irregularities," *Journal of Research of the NBS – D. Radio Propagation*, **63D**, No. 2, 205–233 (September–October 1959).

[7] V. I. Tatarskii, *The Effects of the Turbulent Atmosphere on Wave Propagation* (Translated from the Russian and issued by the National Technical Information Office, U.S. Department of Commerce, Springfield, VA 22161, l971), 181–208.

[8] M. C. Thompson and M. J. Vetter, "Single Path Phase Measuring System for Three-Centimeter Radio Waves," *Review of Scientific Instruments*, **29**, No. 2, 148–150 (February 1958).

[9] K. A. Norton, J. W. Herbstreit, H. B. Janes, K. O. Hornberg, C. F. Peterson, A. F. Barghausen, W. E. Johnson, P. I. Wells, M. C. Thompson, M. J. Vetter and A. W. Kirkpatrick, *An Experimental Study of Phase Variations in Line-of-Sight Microwave Transmissions* (National Bureau of Standards Monograph 33, U.S. Government Printing Office, Washington, 1 November 1961).

[10] M. C. Thompson, H. B. Janes and A. W. Kirkpatrick, "An Analysis of Time Variations in Tropospheric Refractive Index and Apparent Radio Path Length," *Journal of Geophysical Research*, **65**, No. 1, 193–201 (January 1960).

[11] M. C. Thompson, H. B. Janes and F. E. Freethey, "Atmospheric Limitations on Electronic Distance-Measuring Equipment," *Journal of Geophysical Research*, **65**, No. 2, 389–393 (February 1960).

[12] H. B. Janes and M. C. Thompson, "Comparison of Observed and Predicted Phase-Front Distortion in Line-of-Sight Microwave Signals," *IEEE Transactions on Antennas and Propagation*, **AP-21**, No. 2, 263–266 (March 1973).

[13] M. C. Thompson and H. B. Janes, "Antenna Aperture Size Effect on Tropospheric Phase Noise," *IEEE Transactions on Antennas and Propagation*, **AP-14**, No. 6, 800–802 (November 1966).

[14] M. C. Thompson and H. B. Janes, "Preliminary Measurements of Phase Stability over Low-Level Tropospheric Paths," National Bureau of Standards Report No. 6010 (29 September 1958).

[15] M. C. Thompson and H. B. Janes, "Measurements of Phase Stability over a Low-Level Tropospheric Path," *Journal of Research of the NBS – D. Radio Propagation*, **63D**, No. 1, 45–51 (July–August 1959).

[16] H. B. Janes and M. C. Thompson, "Errors Induced by the Atmosphere in Microwave Range Measurements," *Radio Science, Journal of Research NBS/USNC-URSI*, **68D**, No. 11, 1229–1235 (November 1964).

[17] H. B. Janes, M. C. Thompson, D. Smith and A. W. Kirkpatrick, "Comparison of Simultaneous Line-of-Sight Signals at 9.6 and 34.5 GHz," *IEEE*

*Transactions on Antennas and Propagation,* **AP-18**, No. 4, 447–451 (July 1970).

[18] H. B. Janes, M. C. Thompson and D. Smith, "Tropospheric Noise in Microwave Range-Difference Measurements," *IEEE Transactions on Antennas and Propagation,* **AP-21**, No. 2, 256–260 (March 1973).

[19] M. C. Thompson, L. E. Wood, H. B. Janes and D. Smith, "Phase and Amplitude Scintillations in the 10 to 40 GHz Band," *IEEE Transactions on Antennas and Propagation,* **AP-23**, No. 6, 792–797 (November 1975).

[20] H. B. Janes, "Correlation of the Phase of Microwave Signals on the Same Line-of-Sight Path at Different Frequencies," *IEEE Transactions on Antennas and Propagation,* **AP-11**, No. 6, 716–717 (November 1963).

[21] M. C. Thompson, H. B. Janes, L. E. Wood and D. Smith, "Phase and Amplitude Scintillations at 9.6 GHz on an Elevated Path," *IEEE Transactions on Antennas and Propagation,* **AP-23**, No. 6, 850–854 (November 1975).

[22] V. P. Lukin, V. V. Pokasov and S. S. Khmelevtsov, "Investigation of the Time Characteristics of Fluctuations of the Phases of Optical Waves Propagating in the Bottom Layer of the Atmosphere," *Izvestiya Vysshikh Uchebnykh Zavedenii, Radiofizika (Soviet Radiophysics),* **15**, No. 12, 1426–1430 (December 1972). See also V. P. Lukin, V. L. Mironov, V. V. Pokasov and S. S. Khmelevtsov, "Phase Fluctuations of Optical Waves Propagating in a Turbulent Atmosphere," *Radiotekhnika i Elektronika (Radio Engineering and Electronic Physics),* **20**, No. 6, 28–34 (June 1975).

[23] V. P. Lukin, *Atmospheric Adaptive Optics* (SPIE Optical Engineering Press, Bellingham, Washington, 1995; originally published in Russian in 1986), 85–90.

[24] A. Ishimaru, *Wave Propagation and Scattering in Random Media,* vol. 2 (Academic Press, San Diego, CA, 1978), 362.

[25] V. U. Zavorotny, private communication on 30 July 1998.

[26] W. B. Davenport, Jr and W. L. Root, *An Introduction to the Theory of Random Signals and Noise* (IEEE Press, New York, 1987; originally published by McGraw Hill, 1958), 68–70.

[27] R. J. Sasiela, *Electromagnetic Wave Propagation in Turbulence* (Springer-Verlag, Berlin, 1994), 49.

[28] E. Vilar, J. Haddon, P. Lo and T. J. Moulsley, "Measurement and Modelling of Amplitude and Phase Scintillations in an Earth–Space Path," *Journal of the Institution of Electronic and Radio Engineers,* **55**, No. 3, 87–96 (March 1985).

[29] D. C. Cox, H. W. Arnold and R. P. Leck, "Phase and Amplitude Dispersion for Earth–Satellite Propagation in the 20- to 30-GHz Frequency Range," *IEEE Transactions on Antennas and Propagation,* **AP-28**, No. 3, 359–366 (May 1980).

[30] B. Hofmann-Wellenhof, H. Lichtenegger and J. Collins, *Global Positioning System: Theory and Practice,* Fourth Revised Edition (Springer-Verlag, Vienna, 1997).

[31] L. L. Yuan, R. A. Anthes, R. H. Ware, C. Rocken, W. D. Bonner, M. G. Bevis and S. Businger, "Sensing Climate Change Using the Global Positioning System," *Journal of Geophysical Research,* **98**, No. D8, 14.925–14.937 (20 August 1993).

[32] R. Ware, M. Exner, D. Feng, M. Gorbunov, K. Hardy, B. Herman, Y. Kuo, T. Meehan, W. Melbourne, C. Rocken, W. Schreiner, S. Sokolovskiy, F. Solheim, X. Zou, R. Anthes, S. Businger and K. Trenberth, "GPS

Sounding of the Atmosphere from Low Earth Orbit: Preliminary Results," *Bulletin of the American Meteorological Society*, **77**, No. 1, 19–40 (January 1996).

[33] C. J. Naudet, "Estimation of Tropospheric Fluctuations using GPS Data," TDA Progress Report 42–126, Jet Propulsion Laboratory, Pasadena, CA (15 April 1996).

[34] F. D. Eaton and G. D. Nastrom, "Preliminary Estimates of the Vertical Profiles of Inner and Outer Scales from White Sands Missile Range, New Mexico, VHF Radar Observations," *Radio Science*, **33**, No. 4, 895–903 (July–August 1998).

[35] R. K. Crane, "Low Elevation Angle Measurement Limitations Imposed by the Troposphere: An Analysis of Scintillation Observations Made at Haystack and Millstone," MIT Lincoln Laboratory Technical Report No. 518 (18 May 1976).

[36] E. Vilar and J. Haddon, "Measurement and Modeling of Scintillation Intensity to Estimate Turbulence Parameters in an Earth–Space Path," *IEEE Transactions on Antennas and Propagation*, **AP-32**, No. 4, 340–346 (April 1984).

[37] R. K. Crane, "Ionospheric Scintillation," *Proceedings of the IEEE*, **65**, No. 2, 180–199 (February 1977).

[38] E. J. Fremouw, R. C. Livingston and D. A. Miller, "On the Statistics of Scintillating Signals," *Journal of Atmospheric and Terrestrial Physics*, **42**, No. 8, 717–731 (August 1980).

[39] C. L. Rino, R. T. Tsunoda, J. Petriceks, R. C. Livingston, M. C. Kelley and K. D. Baker, "Simultaneous Rocket-Borne Beacon and *in situ* Measurements of Equatorial Spread F – Intermediate Wavelength Results," *Journal of Geophysical Research*, **86**, No. A4, 2411–2420 (1 April 1981).

[40] C. L. Rino, "A Power Law Phase Screen Model for Ionospheric Scintillation I. Weak Scatter," *Radio Science*, **14**, No. 6, 1135–1145 (November– December 1979).

[41] A. I. Kon, "Qualitative Theory of Amplitude and Phase Fluctuations in a Medium with Anisotropic Turbulent Irregularities," *Waves in Random Media*, **4**, No. 3, 297–306 (July 1994). (Notice that there is an error in equation (32) of this paper.)

# 5

# The Phase Structure Function

The phase difference measured between adjacent receivers is a fundamental quantity both for practical applications and for scientific research. It has several important advantages. Phase trends are often common to the two paths and are canceled out by taking the difference. This means that one can usually deal with stationary data streams. Large-scale inhomogeneities and intermittent structures often enclose both paths and their influence is suppressed when one takes the phase difference. The *phase structure function* describes these measurements:

$$\mathcal{D}_\varphi(\rho) = \langle (\varphi_1 - \varphi_2)^2 \rangle = 2[\langle \varphi^2 \rangle - \langle \varphi_1 \varphi_2 \rangle] \qquad (5.1)$$

The structure function is the natural performance measure for many instruments and systems. The phase difference is the basic measurement of optical and microwave interferometers used to study distant objects in the universe. Radio direction finding and microwave tracking of spacecraft rely on measurements of the phase difference between adjacent receivers. One needs similar information to design phase-tilt correctors for optical trackers. The lack of phase coherence across an aperture limits the effective resolution and light-gathering capacity of many large telescopes.

It is relatively easy to measure the phase difference at microwave frequencies and there is a wealth of experimental data with which to compare the predictions. Similar experiments at visible wavelengths are more difficult. Horizontal measurements are invariably made with receivers deployed perpendicular to the direction of propagation of the reference signal. This is the most sensitive arrangement for measuring the location of a source. One can generate data for many separations with a judicious choice of the locations of receivers. The situation is unavoidably different for interferometers used to observe stars and galaxies. The source location determines the orientation of a fixed receiver baseline relative to the line of sight for these

179

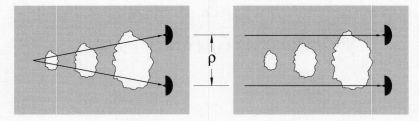

Figure 5.1: The influence of various eddy sizes on the signal phase measured at adjacent receivers for spherical and planar reference signals.

applications. The usual approach is to project the receiver locations onto a plane that is normal to the line of sight and thereby define an *effective baseline*.

Phase-difference measurements have used several types of wave to carry the reference signal. Most terrestrial experiments employ adjacent receivers to monitor the signal from a common transmitter, as shown on the left-hand side of Figure 5.1. The transmitted signal can usually be described as a spherical wave in this case. By contrast, astronomical observations and transmissions from distant spacecraft are described by plane waves, as suggested on the right-hand side of Figure 5.1. The phase correlations for plane and spherical waves are quite different, even if the medium is the same. The reason is apparent from Figure 5.1. Only those eddies which are larger than the separation between receivers can influence the signals traveling along parallel paths. By contrast, signals traveling along rotated paths are influenced by eddies of every size. Small eddies near the transmitter can modify both signals and the phase correlation for a spherical wave is significantly larger than that for a plane wave.

The phase fluctuations imposed on a single ray are described by (4.1) in the geometrical optics approximation. In this chapter we will deal only with microwave and optical signals, for which the ionosphere plays no meaningful role. In this case we can set $\Delta\varepsilon = 2\,\delta n$ and the covariance between the phases measured at separated receivers becomes

$$\langle \varphi_1 \varphi_2 \rangle = k^2 \int_0^R ds_1 \int_0^R ds_2 \,\langle \delta n(s_1)\,\delta n(s_2) \rangle \qquad (5.2)$$

which can be expressed using the wavenumber representation of the spatial correlation:

$$\langle \varphi_1 \varphi_2 \rangle = k^2 \int_0^R ds_1 \int_0^R ds_2 \int d^3\kappa \,\Phi_n[\boldsymbol{\kappa}, \tfrac{1}{2}(\mathbf{s}_1 + \mathbf{s}_2)] \exp[i\boldsymbol{\kappa} \cdot (\mathbf{s}_1 - \mathbf{s}_2)]$$
$$(5.3)$$

This is similar to the phase-variance expression (4.4) except that the two rays now travel along different paths; the sum and difference vectors depend on the separation between receivers. This general result will be used to describe optical and microwave signals that propagate through the atmosphere.

## 5.1 Terrestrial Links

Many experiments have measured phase differences between adjacent paths near the surface of the earth. Most of these have been performed using microwave and millimeter-wave signals. Good measurements of the phase structure function have also been made at optical wavelengths. In these situations one can assume that the spectrum does not change significantly along the path:

$$\langle \varphi_1 \varphi_2 \rangle = k^2 \int d^3\kappa \, \Phi_n(\boldsymbol{\kappa}) \int_0^R ds_1 \int_0^R ds_2 \exp[i\boldsymbol{\kappa} \cdot (\mathbf{s}_1 - \mathbf{s}_2)] \tag{5.4}$$

If the irregularities are isotropic one can simplify this expression still further using spherical wavenumber coordinates centered on the difference vector:

$$\langle \varphi_1 \varphi_2 \rangle = k^2 \int_0^\infty d\kappa \, \kappa^2 \Phi_n(\kappa) \int_0^R ds_1 \int_0^R ds_2$$

$$\times \int_0^\pi d\psi \sin \psi \int_0^{2\pi} d\omega \exp(i\kappa \cos \psi \, |\mathbf{s}_1 - \mathbf{s}_2|)$$

The angular integrations are elementary:

$$\langle \varphi_1 \varphi_2 \rangle = 4\pi k^2 \int_0^\infty d\kappa \, \kappa \Phi_n(\kappa) \int_0^R ds_1 \int_0^R ds_2 \, \frac{\sin(\kappa \, |\mathbf{s}_1 - \mathbf{s}_2|)}{|\mathbf{s}_1 - \mathbf{s}_2|} \tag{5.5}$$

This is the starting point for discussions of horizontal propagation and represents a fair approximation for paths near the surface. We will reexamine the assumption of isotropy later and learn how the results depend on irregularities that are different in the horizontal and vertical directions. To complete the description based on (5.5) one must identify the type of reference signal and express the scalar distance in terms of coordinates that describe the experimental situation.

### 5.1.1 *Spherical Waves on Horizontal Paths*

Numerous phase-difference measurements have been performed at microwave frequencies. In these cases the reference signal can be described by a spherical wave since the transmitters are not highly directional. Two or more

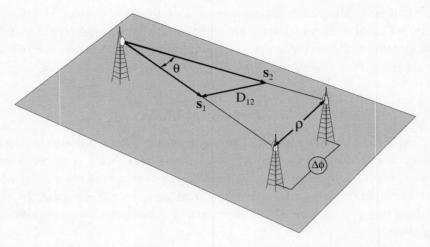

Figure 5.2: The geometry for analyzing the phase correlation at adjacent receivers for a spherical wave represented by rotated ray paths.

receivers are deployed along a baseline that is perpendicular to the normal line of sight, as illustrated in Figure 5.2. The receivers are connected by waveguide or coaxial cable. Signals received at the various locations are compared coherently and the phase difference is measured as a function of time.

### 5.1.1.1 A Basic Expression for the Phase Structure Function

To analyze this situation we assume that the profile of turbulent activity is constant along the route and that the spectrum of irregularities is isotropic. This allows us to use the simplified expression (5.5) for the phase covariance. The scalar distance between the two integration points is expressed in terms of the ray-path coordinates identified in Figure 5.2:

$$D_{12} = |\mathbf{s}_1 - \mathbf{s}_2| = \sqrt{s_1^2 + s_2^2 - 2s_1 s_2 \cos\theta}$$

The phase covariance becomes

$$\langle \varphi_1 \varphi_2 \rangle = 4\pi k^2 \int_0^\infty d\kappa\, \kappa \Phi_n(\kappa) \int_0^R ds_1$$

$$\times \int_0^R ds_2 \left( \frac{\sin\left( \kappa \sqrt{s_1^2 + s_2^2 - 2s_1 s_2 \cos\theta} \right)}{\sqrt{s_1^2 + s_2^2 - 2s_1 s_2 \cos\theta}} \right).$$

Since the ray-path integrations are symmetrical in $s_1$ and $s_2$, one can write

$$\mathcal{J} = 2 \int_0^R ds_1 \int_0^{s_1} ds_2 \left( \frac{\sin\left( \kappa \sqrt{s_1^2 + s_2^2 - 2s_1 s_2 \cos\theta} \right)}{\sqrt{s_1^2 + s_2^2 - 2s_1 s_2 \cos\theta}} \right).$$

Letting $s_2 = w s_1$ and dropping the subscript on $s_1$,

$$\mathcal{J} = 2 \int_0^R ds \int_0^1 dw \left( \frac{\sin\left( \kappa s \sqrt{1 + w^2 - 2w \cos\theta} \right)}{\sqrt{1 + w^2 - 2w \cos\theta}} \right).$$

With the substitution

$$\zeta \sin\theta = \sqrt{1 + w^2 - 2w \cos\theta}$$

one finds

$$\mathcal{J} = 2 \int_0^R ds \int_{\sec\left(\frac{\theta}{2}\right)}^{\mathrm{cosec}\,\theta} d\zeta \, \frac{\sin(\zeta \kappa s \sin\theta)}{\sqrt{\zeta^2 - 1}}.$$

The angle subtended by the receivers is related to their separation and the path length by $\rho = R\theta$. The spacing between receivers is much smaller than the path length in practical situations and the separation angle $\theta$ is quite small. To a good approximation

$$\mathcal{J} = 2 \int_0^R ds \int_1^{\infty} d\zeta \, \frac{\sin(\zeta \kappa s \theta)}{\sqrt{\zeta^2 - 1}} = \pi \int_0^R ds \, J_0(\kappa s \theta)$$

where we have used an integral representation for the Bessel function from Appendix D. On changing to a new variable $s = Ru$, the phase covariance is

Spherical: $\quad \langle \varphi_1 \varphi_2 \rangle = 4\pi^2 R k^2 \int_0^{\infty} d\kappa \, \kappa \Phi_n(\kappa) \left( \int_0^1 du \, J_0(\kappa \rho u) \right)$

$$(5.6)$$

The spectral weighting function in large parentheses indicates how different wavenumbers influence the spatial covariance. It begins at unity and falls to 0.15 at $\kappa\rho = 5$, indicating that large eddies have the greatest influence, as they did for the phase variance. The curve flattens out above $\kappa\rho = 5$ and eventually goes to zero as $1/(\kappa\rho)$. This means that eddies that are small relative to the spacing can also influence the covariance, as suggested by Figure 5.1. The spatial correlation is formed by normalizing the phase covariance with its mean-square value:

$$\mu(\rho) = \frac{\langle\varphi_1\varphi_2\rangle}{\langle\varphi^2\rangle} = \frac{\int_0^\infty d\kappa\,\kappa\Phi_n(\kappa)\left(\int_0^1 du\,J_0(\kappa\rho u)\right)}{\int_0^\infty d\kappa\,\kappa\Phi_n(\kappa)} \qquad (5.7)$$

Early studies of propagation in random media concentrated on estimating $\mu(\rho)$ for various spectrum models. That approach has been replaced by emphasizing a quantity that is more directly related to the system's performance.

The phase structure function for spherical waves traveling along horizontal paths was established by Fried [1][2]:

$$\text{Spherical:}\qquad \mathcal{D}_\varphi(\rho) = 8\pi^2 Rk^2 \int_0^\infty d\kappa\,\kappa\Phi_n(\kappa)\int_0^1 du\,[1 - J_0(\kappa\rho u)] \tag{5.8}$$

This expression is valid for homogeneous, isotropic irregularities. The same expression defines the *wave structure function* which will play a prominent role in subsequent developments.

### 5.1.1.2 Results for the Kolmogorov Spectrum

The structure function described by (5.8) has a spectral weighting function that vanishes for small values of $\kappa\rho$. This suggests that the inertial range of the spectrum is important for its evaluation and one can use the Kolmogorov model to describe it. In doing so one must retain the inner- and outer-scale wavenumber limits because they set the behavior for very small and very large separations between receivers:

$$\mathcal{D}_\varphi(\rho) = 8\pi^2 Rk^2 \int_{\kappa_0}^{\kappa_s} d\kappa\,\kappa\left(\frac{0.033C_n^2}{\kappa^{\frac{11}{3}}}\right)\int_0^1 du\,[1 - J_0(\kappa\rho u)]$$

Changing to $x = \kappa\rho$ makes the dependence on the spacing more evident:

$$\mathcal{D}_\varphi(\rho) = 2.606 Rk^2 C_n^2\rho^{\frac{5}{3}}\int_{\rho\kappa_0}^{\rho\kappa_s}\frac{dx}{x^{\frac{8}{3}}}\int_0^1 du\,[1 - J_0(xu)] \tag{5.9}$$

The structure function has different scaling laws in the three regions identified by Figure 5.3. The curve begins quadratically but quickly shifts to the similarity result, which is often approximated by a 5/3 relation. The function bends over as the spacing increases and approaches twice the phase variance. We need to examine these regimes separately with some care.

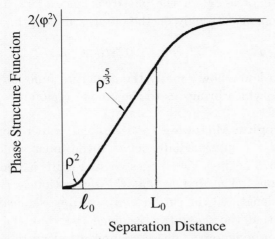

Figure 5.3: The approximate behavior of the phase structure function.

**Small separations** If the spacing is less than the inner scale length, the upper limit $\kappa_s\rho$ is small and one can use a small-argument expansion for the Bessel function. The lower limit goes to zero because $\kappa_s$ is much greater than $\kappa_0$ in the troposphere:

$$\kappa_s\rho < 1: \qquad \mathcal{D}_\varphi(\rho) = 2.606Rk^2C_n^2\rho^{\frac{5}{3}} \int_0^{\kappa_s\rho} \frac{dx}{x^{\frac{8}{3}}} \int_0^1 du\left(\frac{1}{4}x^2u^2\right)$$

The integrations are easily done and show that the structure function has a quadratic dependence on the separation:

$$\kappa_s\rho < 1: \qquad \mathcal{D}_\varphi(\rho) = \rho^2\left(0.651Rk^2C_n^2\kappa_s^{\frac{1}{3}}\right) \qquad (5.10)$$

This relationship provides one method for determining the inner scale length from phase measurements, although the size of the receiver sets a minimum spacing:

$$4\pi a_r < \ell_0$$

This condition can be realized for optical experiments, which often employ very small detectors. It is never encountered by microwave systems because the aperture of the receiver is always larger than the inner scale length.

**Large separations** The third region corresponds to inter-receiver separations that are greater than the outer scale length. One can use (5.1) in this range because the spatial covariance vanishes for large separations:

$$\kappa_0\rho > 1: \qquad \mathcal{D}_\varphi(\rho) = 2\langle\varphi^2\rangle$$

On substituting the expression for the phase variance (4.12) into this, we see that the result depends strongly on the outer scale length:

$$\kappa_0\rho > 1: \qquad \mathcal{D}_\varphi(\rho) = 0.073 R k^2 C_n^2 L_0^{\frac{5}{3}} \qquad (5.11)$$

One usually needs to know how rapidly the structure function approaches this asymptotic value which brings us to the second region.

**The similarity region** Microwave and optical experiments usually employ spacings that are intermediate between the inner scale and outer scale lengths. The 5/3 scaling law suggested in Figure 5.3 results if one makes two assumptions. The first is that the separation is *much greater* than the inner scale length so the product $\kappa_s\rho$ can be assumed to be infinite. The second is that the separation is *much less* than the outer scale length so that $\kappa_0\rho$ can be ignored. These assumptions lead to the following expression:

$$\text{Similarity:} \qquad \mathcal{D}_\varphi(\rho) = 2.606 R k^2 C_n^2 \rho^{\frac{5}{3}} \int_0^\infty \frac{dx}{x^{\frac{8}{3}}} \int_0^1 du \, [1 - J_0(xu)]$$

The integrals can be evaluated[1] and one can write the scaling law for spherical waves as follows [1]:

$$\text{Similarity:} \qquad \mathcal{D}_\varphi(\rho) = \rho^{\frac{5}{3}}(1.093 R k^2 C_n^2) \qquad (5.12)$$

This is often expressed as

$$\mathcal{D}_\varphi(\rho) = 2\left(\frac{\rho}{\rho_0}\right)^{\frac{5}{3}} \qquad (5.13)$$

where the *coherence radius* is defined by

$$\rho_0 = (0.546 R k^2 C_n^2)^{-\frac{3}{5}}. \qquad (5.14)$$

This parameter was introduced by Fried [3] and plays an important role in many descriptions of optical propagation through random media.[2] For

---

[1] Using a result from Appendix D,

$$\int_0^\infty \frac{dx}{x^{\frac{8}{3}}} \int_0^1 du \, [1 - J_0(xu)] = \int_0^1 du \, u^{\frac{5}{3}} \int_0^\infty \frac{dw}{w^{\frac{8}{3}}} [1 - J_0(w)] = 0.4194.$$

[2] Fried chose to write the structure function as

$$\mathcal{D}_\varphi(\rho) = 6.88\left(\frac{\rho}{r_0}\right)^{\frac{5}{3}}$$

and his coherence radius is related to our definition by $r_0 = 2.099\rho_0$.

instance, signals received by different points on a collecting mirror are not phase coherent if their separation is greater than $\rho_0$. That reality limits both the gain and the angular resolution of large telescopes. The coherence length is a few centimeters at optical wavelengths for horizontal paths. It is approximately 10 km at 10 GHz for typical path lengths.

**The exact solution** Many experiments are performed with inter-receiver spacings for which $\kappa_0\rho$ is not small. The previous analysis does not show how the 5/3 scaling law gives way to the saturation value. To explore this transition regime we return to the basic definition (5.9) and make the substitution $x = \rho\kappa_0 y$. The resulting upper limit $\kappa_s/\kappa_0$ is safely taken to infinity and the structure function can be expressed in terms of its asymptotic value:

$$\mathcal{D}_\varphi(\rho) = 2\langle\varphi^2\rangle\left(\frac{5}{3}\int_1^\infty \frac{dy}{y^{\frac{8}{3}}}\int_0^1 du\,[1 - J_0(yu\kappa_0\rho)]\right) \qquad (5.15)$$

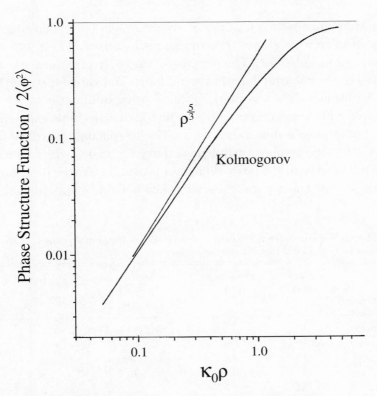

Figure 5.4: The phase structure function computed for a spherical wave in a random medium described by a Kolmogorov spectrum with an outer-scale cutoff. The 5/3 approximation is shown for comparison.

The quantity in large parentheses is plotted as a function of $\kappa_0\rho$ on logarithmic coordinates in Figure 5.4. The exact solution and the 5/3 approximation coincide for small values but they separate rapidly as $\kappa_0\rho$ increases. The slope of the exact solution is 5/3 only for $\kappa_0\rho < 0.01$. Above this threshold the slope falls steadily to zero as the curve approaches its asymptotic value.

Several conclusions emerge from this computation. The first point is that large eddies influence the wave structure function throughout the range of separations, a phenomenon first noted by Lutomirski and Yura [4]. The second point is that the 5/3 law for the phase structure function is *not synonymous* with the Kolmogorov spectrum; the two models are quite different over most of the range. The third conclusion is that the 5/3 scaling law is valid only for very small values of $\kappa_0\rho$ – a condition that is seldom met. One should expect departures from it in many situations and use computations based on (5.15) to interpret measurements.

### 5.1.1.3 Microwave Measurements

The phase difference between adjacent receivers has been measured over a wide range of microwave frequencies in various locations. These experiments are summarized in Table 5.1. The purpose of the early programs was to establish limits on accuracy for microwave guidance and range instrumentation systems. Additional data has been obtained using millimeter waves. These results now provide benchmarks against which structure-function models can be tested. Simultaneous data taken with different spacings is needed in order to do so because the level of turbulent activity $C_n^2$ varies widely during the day and from one day to the next. The data is usually plotted on logarithmic coordinates so that the separation scaling law is readily apparent.

Table 5.1: Microwave phase-difference measurements made on horizontal paths near the surface, where $N$ is the number of inter-receiver spacings employed

| Location | $R$ (km) | Frequency (GHz) | Inter-receiver Spacing (m) | $N$ | Ref. |
|---|---|---|---|---|---|
| Colorado Springs | 16 | 1.046 | 153, 433 and 586 | 3 | 5 |
| | 16 | 9.4 | 2–153 | 1 | 6 |
| Boulder | 15 | 9.4 | 380 | 1 | 7 |
| Maui | 25 | 9.4 | 1.25–790 | 8 | 8, 9 |
| Maui–Hawaii | 64 | 9.6 | 1.4–100 | 3 | 10, 11 |
| | 150 | 9.6 | 10 and 100 | 2 | 12 |
| Stanford | 28 | 35 | 3.4–24 | 8 | 13, 14 |
| Illinois | 1.4 | 173 | 1.43–10 | 6 | 15 |

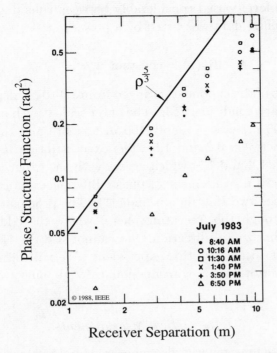

Figure 5.5: The phase structure function measured at 173 GHz by Hill, Bohlander, Clifford, McMillan, Priestley and Schoenfeld [15].

Three programs provided simultaneous phase-difference measurements. Confirmation of the predicted structure function came from a millimeter-wave experiment in Illinois. A reference signal at 173 GHz was transmitted over flat terrain on a 1.4-km path with an average height of 4 m [15]. The phase difference was measured for six inter-receiver spacings ranging from 1.43 to 10 m. Data for a single day is reproduced in Figure 5.5 and shows clearly that the structure function bends over as the separation increases. This behavior is consistent with the exact solution plotted in Figure 5.4 if the outer scale length was approximately 8 m. That value is twice the path height, but eddies near the surface have longer scale lengths in the horizontal propagation direction than they do in the vertical direction.

Additional confirmation comes from phase-difference data taken at 35 GHz on a 28-km path near Stanford [13][14]. Simultaneous measurements were made for eight inter-receiver separations ranging from 3.4 to 24 m. The slope on log–log coordinates was approximately constant for each run, although the level varied considerably between runs due to changes in $C_n^2$. The experimental data was fitted by a power law:

$$\mathcal{D}_\varphi(\rho) = \text{constant} \times \rho^\mu \qquad (5.16)$$

Measured values of the exponent $\mu$ ranged from 1.4 to 1.5 depending on the time of day and path conditions [13]. The outer scale length on this elevated path ($\mathsf{h} = 120$–$300$ m) was probably 100 m or more and the product $\kappa_0\rho$ therefore varied between 0.2 and 1.5. The exact solution indicates that $\mu$ should lie between 1.2 and 1.5, which agrees with the experimental results.

A third microwave measurement of phase difference was conducted on a 64-km path between two Hawaiian islands [11]. The propagation path traveled primarily through the free atmosphere, where the eddies are highly elongated in the horizontal direction. One cannot use the isotropic results developed thus far to analyze this experiment. We will return to it when we develop a description for transmission through anisotropic media in Section 5.1.3.

### 5.1.1.4  Optical Measurements

The phase structure function was also measured at optical wavelengths. The first experiment used a helium–neon laser with a Michelson interferometer on a 70-m path in Colorado [16]. The reference signal was a diverging wave and the phase structure function was measured simultaneously with four separa-

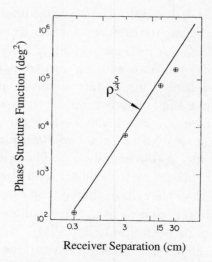

Figure 5.6: The phase structure function measured by Clifford, Bouricius, Ochs and Ackley using a helium–neon laser as the source, over a 70-m path [16].

tions between 3 mm and 30 cm. Temperature sensors gave estimates for $C_n^2$ and the structure function was divided by these values to obtain comparable scaling records. The outer scale length was estimated from temperature-sensor readings as 1–2 m, from which we infer that $0.01 < \kappa_0\rho < 2$. Averaged data from this experiment is reproduced in Figure 5.6 and the exact solution plotted in Figure 5.4 is a good match.

## 5.1.2 Plane Waves and Collimated Beams

We next examine the propagation of a plane wave near the surface. It is very difficult to generate a plane wave at microwave frequencies since beam spreading is a significant feature of short-range transmissions. On the other hand, one can make phase-difference measurements at optical wavelengths using large collimators. A collimated beam is indistinguishable from a plane wave if the receivers are placed sufficiently close to the line of sight. We will discuss a second technique for simulating a plane wave later on.

### 5.1.2.1 A Basic Description of the Phase Structure Function

The coordinates illustrated in Figure 5.7 define typical points along parallel ray paths. The scalar distance between typical points is written

$$D_{12} = |\mathbf{s}_1 - \mathbf{s}_2| = \sqrt{(x_1 - x_2)^2 + \rho^2} \tag{5.17}$$

and the basic phase-covariance expression (5.5) becomes

$$\langle \varphi_1 \varphi_2 \rangle = 4\pi k^2 \int_0^\infty d\kappa\, \kappa \Phi_n(\kappa) \int_0^R dx_1 \int_0^R dx_2\, \frac{\sin\left[\kappa\sqrt{(x_1 - x)^2 + \rho^2}\right]}{\sqrt{(x_1 - x_2)^2 + \rho^2}}.$$

Since the path integrations are symmetrical in $x_1$ and $x_2$ we can write

$$\mathcal{J} = 2\int_0^R dx_1 \int_0^{x_1} dx_2\, \frac{\sin\left[\kappa\sqrt{(x_1 - x_2)^2 + \rho^2}\right]}{\sqrt{(x_1 - x_2)^2 + \rho^2}}$$

Transforming to the difference coordinate $u = x_1 - x_2$ gives

$$\mathcal{J} = 2\int_0^R dx_1 \int_0^{x_1} du\, \frac{\sin\left(\kappa\sqrt{u^2 + \rho^2}\right)}{\sqrt{u^2 + \rho^2}}$$

$$= 2\int_0^R dx\,(R - x)\frac{\sin\left(\kappa\sqrt{x^2 + \rho^2}\right)}{\sqrt{x^2 + \rho^2}}$$

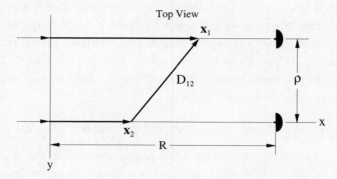

Figure 5.7: The geometry used to describe the phase correlation at adjacent receivers for a plane wave simulated by parallel rays.

or

$$\mathcal{J} = 2R \int_0^R dx\, \frac{\sin\left(\kappa\sqrt{x^2+\rho^2}\right)}{\sqrt{x^2+\rho^2}} + \frac{2}{\kappa}\left[\cos(\kappa\rho) - \cos\left(\kappa\sqrt{R^2+\rho^2}\right)\right]$$

The second term is proportional to the eddy size, which is much less than the path length for most terrestrial experiments. The integral term can be evaluated by letting

$$\rho\zeta = \sqrt{u^2+\rho^2}$$

so that

$$\int_0^R du\, \frac{\sin\left(\kappa\sqrt{u^2+\rho^2}\right)}{\sqrt{u^2+\rho^2}} = \int_1^{\zeta_{\max}} \frac{d\zeta\,\sin(\zeta\kappa\rho)}{\sqrt{\zeta^2-1}}$$

with

$$\zeta_{\max} = \sqrt{1 + \frac{R^2}{\rho^2}}.$$

The spacing is usually much less than the path length and one can take the upper limit to infinity. The resulting definite integral leads to the zeroth-order Bessel function

$$\mathcal{J} = \pi R J_0\left(\kappa\rho\right)$$

and the phase covariance for parallel rays becomes

$$\text{Plane:} \qquad \langle\varphi_1\varphi_2\rangle = 4\pi^2 R k^2 \int_0^\infty d\kappa\,\kappa\,\Phi_n(\kappa) J_0(\kappa\rho) \qquad (5.18)$$

The wavenumber weighting function $J_0(\kappa\rho)$ is small if the spacing is greater than the eddy size. This means that eddies smaller that the separation have little influence on the correlation, which confirms the qualitative argument developed from Figure 5.1.

The corresponding phase structure function follows directly from (5.18) and was first established by Tatarskii [17]:

$$\text{Plane:} \qquad \mathcal{D}_\varphi(\rho) = 8\pi^2 Rk^2 \int_0^\infty d\kappa \; \kappa \Phi_n(\kappa)[1 - J_0(\kappa\rho)] \qquad (5.19)$$

This expression describes horizontal propagation for irregularities that are isotropic and homogeneous. Notice that the structure functions for plane and spherical waves are related by a simple equation:

$$\mathcal{D}_{\text{sph}}(\rho) = \int_0^1 dw \; \mathcal{D}_{\text{pl}}(\rho w) \qquad (5.20)$$

For the same separation between receivers, the plane-wave structure function is larger than the spherical wave solution and behaves differently as it approaches the common asymptotic value.

### 5.1.2.2 Results for the Kolmogorov Spectrum

The inertial range should be influential because the weighting function in (5.19) vanishes for small wavenumbers. To evaluate the structure function we first use the Kolmogorov model, taking care to retain the inner- and outer-scale wavenumber limits:

$$\text{Plane:} \qquad \mathcal{D}_\varphi(\rho) = 2.606 Rk^2 C_n^2 \rho^{\frac{5}{3}} \int_{\kappa_0\rho}^{\kappa_s\rho} \frac{dx}{x^{\frac{8}{3}}}[1 - J_0(x)] \qquad (5.21)$$

The integral can be evaluated for the three regions identified in Figure 5.3 as follows [17][18]:

$$\kappa_s\rho < 1: \qquad \mathcal{D}_\varphi(\rho) = \rho^2(1.953 Rk^2 C_n^2 \kappa_s^{\frac{1}{3}})$$

$$\text{Similarity:} \qquad \mathcal{D}_\varphi(\rho) = \rho^{\frac{5}{3}}(2.914 Rk^2 C_n^2)$$

$$\kappa_0\rho > 1: \qquad \mathcal{D}_\varphi(\rho) = (0.073 Rk^2 C_n^2 L_0^{\frac{5}{3}}) \qquad (5.22)$$

It is important to note that the second result is valid only if $\kappa_0\rho$ is very small.

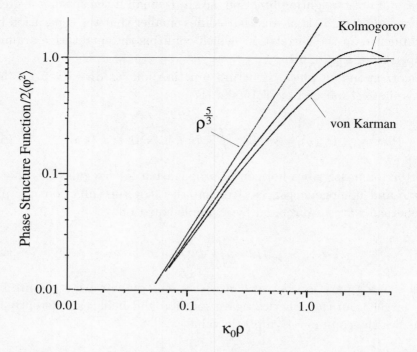

Figure 5.8: The phase structure function for a plane wave calculated using the von Karman model and the Kolmogorov spectrum with a cutoff at the outer-scale wavenumber. The 5/3 approximation is shown for reference.

One can establish a more accurate description by retaining the finite lower limit in (5.21). With the change of variable $x = \kappa_0 \rho y$ the structure function becomes

$$\text{Plane:} \qquad \mathcal{D}_\varphi(\rho) = 2\langle\varphi^2\rangle\left(\frac{5}{3}\int_1^\infty \frac{dy}{y^{\frac{8}{3}}}[1 - J_0(y\kappa_0\rho)]\right) \qquad (5.23)$$

The quantity in large parentheses is plotted in Figure 5.8. The slope is 5/3 for $\kappa_0\rho < 0.1$ but rapidly departs from that approximation. The curve approaches its asymptotic value only after *exceeding it*. This unrealistic behavior is a consequence of the sharp wavenumber cutoff assumed for the spectral model and provides further evidence of the profound influence that large eddies have on the phase structure function.

### 5.1.2.3 Results for the von Karman Spectrum

We want first of all to find out whether a smooth behavior for small wavenumbers gives a more realistic prediction for the structure function. Several models to describe the small-wavenumber region of the turbulence spectrum[3] have been proposed. Some are based on analysis of meteorological data but most are selected because they can be integrated to make analytical estimates. The von Karman model attempts to describe both the input and the inertial range. It assumes that the large eddies are isotropic but we will remove this restriction later on:

$$\mathcal{D}_{\varphi}(\rho) = 0.264\pi^2 R k^2 C_n^2 \int_0^{\infty} \frac{d\kappa\,\kappa}{\left(\kappa^2 + \kappa_0^2\right)^{\frac{11}{6}}} [1 - J_0(\kappa\rho)] \qquad (5.24)$$

The denominator is raised to exponential form with the Laplace-transform relation

$$\frac{1}{p^{1+\mu}} = \frac{1}{\Gamma(1+\mu)} \int_0^{\infty} d\zeta\,\zeta^{\mu} e^{-p\zeta}$$

so that

$$\mathcal{D}_{\varphi}(\rho) = \frac{0.264\pi^2}{\Gamma\left(\frac{11}{6}\right)} R k^2 C_n^2 \int_0^{\infty} d\zeta\,\zeta^{\frac{5}{6}} e^{-\zeta\kappa_0^2} \int_0^{\infty} d\kappa\,\kappa e^{-\zeta\kappa^2} [1 - J_0(\kappa\rho)]$$

$$= \frac{0.132\pi^2}{\Gamma\left(\frac{11}{6}\right)} R k^2 C_n^2 \int_0^{\infty} d\zeta\,\zeta^{-\frac{1}{6}} e^{-\zeta\kappa_0^2} \left[1 - \exp\left(-\frac{\rho^2}{4\zeta}\right)\right].$$

The first integral gives a gamma function and the second can be expressed as a MacDonald function:

$$\mathcal{D}_{\varphi}(\rho) = 1.563 R k^2 C_n^2 \kappa_0^{-\frac{5}{3}} \left(1 - \frac{2^{\frac{1}{6}}}{\Gamma\left(\frac{5}{6}\right)} (\rho\kappa_0)^{\frac{5}{6}} K_{\frac{5}{6}}(\rho\kappa_0)\right) \qquad (5.25)$$

The quantity in large parentheses is plotted in Figure 5.8 and approaches the asymptotic value smoothly. It bends over more rapidly than does the Kolmogorov model and departs from the 5/3 approximation even more dramatically. We will use this result to interpret experimental data because it acknowledges the existence of large eddies and because it avoids the asymptotic overshoot. Notice that it does not reduce to a quadratic form for very small separations because we have ignored the dissipation region of the

---

[3] Voitsekhovich [19] calculated the phase structure for three different models of the turbulence spectrum that have been proposed to describe the low-wavenumber region:

| | |
|---|---|
| von Karman: | $\Phi_n(\kappa) = 0.033 C_n^2 (\kappa^2 + \kappa_0^2)^{-\frac{11}{6}}$ |
| Greenwood–Tarazano: | $\Phi_n(\kappa) = 0.033 C_n^2 (\kappa^2 + \kappa_0\kappa)^{-\frac{11}{6}}$ |
| Exponential: | $\Phi_n(\kappa) = 0.033 C_n^2 \kappa^{-\frac{11}{3}} [1 - \exp(-\kappa^2/\kappa_0^2)]$ |

spectrum. The very-small-separation region is seldom measured and we need not carry out the difficult calculations required to describe it.[4]

### 5.1.2.4 *Optical Phase-difference Measurements*

Two techniques used to simulate an infinite plane wave at optical wavelengths are illustrated in Figure 5.9. The first method uses a telescope with a laser at its focus to generate a collimated beam. If the spacing between receivers is less than the diameter of the beam, the arriving signals are equivalent to a plane wave modulated by the random medium. The second approach uses two lasers to generate narrow beams aimed at the receivers. The arriving signals are equivalent to an infinite plane wave if the optical signals are phase locked [21]. In both approaches an optical interferometer measures the phase difference between the receivers.

Plane-wave expressions for the structure function were tested with a Mach–Zehnder interferometer in West Germany [22]. A Cassegrain telescope and krypton laser were used to generate a collimated beam with a diameter of 10 cm. The line of sight was 2.5 m above a level surface and data was taken for path lengths of 500, 1000, 1500 and 2000 m. The phase difference was measured for twelve separations between 0.5 and 60 mm. The outer scale length was close to 1 m, which suggests that $0.003 < \kappa_0\rho < 0.3$. The plots of structure-function data versus separation tended to agree with Figure 5.8. An interesting result of this experiment is that data taken 15 min apart followed curves corresponding to different values of the outer scale length. This suggests that conditions on the path changed significantly between runs. Rapid changes of large-eddy conditions are observed in measurements made with fine-wire thermometers near the surface [23]. One explanation is that an intermittent structure moved into the path.

A comprehensive series of optical phase-difference experiments was then conducted by Lukin and Pokasov in the USSR [21]. The small-separation region was explored with the two-laser method using photo receptors with 0.01-mm openings. A quadratic relation was confirmed for spacings less than 15 mm using both collimated and spherical reference signals on a 48-m path. Larger separations were studied using a variant of the two-laser method. Radiation from a single He–Ne laser traveled to two semitransparent beam splitters, which created parallel beams with a measured path difference. Phase-locked receivers measured the phase difference after recognizing the original path difference. These experiments were conducted on a 110-m path

---

[4] An early study [20] estimated the structure function for a von Karman spectrum modified by a Gaussian factor to characterize the dissipation region.

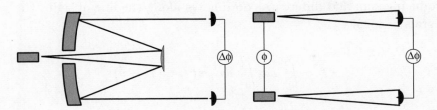

Figure 5.9: Two methods used to generate plane waves for measuring phase differences at optical wavelengths. On the left-hand side, a Cassegrain telescope collimates a laser signal and the receivers are connected as an interferometer. On the right-hand side, narrow beams from two phase-locked lasers intercept the coherent receivers to simulate an infinite plane wave.

and used separations between 5 and 150 cm. The 5/3 scaling law was found to describe the data *only* in the limited spacing range 20–40 cm. This is what one would expect from Figure 5.8, since the propagation was about 1 m above the terrain and the critical parameter fell in the range $0.3 < \kappa_0 \rho < 9$.

### 5.1.3 *The Effect of Atmospheric Anisotropy*

We have learned that the phase structure function is sensitive to large-scale irregularities. Refractometer and temperature measurements show that large eddies are not truly symmetrical – even near the surface. It is important to estimate the influence of this anisotropy on structure-function measurements. To do so we use the scaling technique devised by Zavorotny [24] to estimate the phase variance for anisotropic media. We first write the basic relationship (5.3) in rectangular coordinates and specialize it to describe horizontal propagation:

$$\mathcal{D}_\varphi(\boldsymbol{\rho}) = 2k^2 \int d^3\kappa \, \Phi_n(\boldsymbol{\kappa})[1 - \exp(\mathbf{i}\boldsymbol{\kappa} \cdot \boldsymbol{\rho})]$$
$$\times \int_0^R dx \int_0^R dx' \exp[i\kappa_x(x - x')] \tag{5.26}$$

We addressed the double path integration in Section 4.1.3 and made an effort to include the influence of the path length in the wavenumber integration through the weighting function (4.10). Our current interest is the *inter-receiver-spacing scaling law* and we can afford to be less precise. On introducing sum and difference coordinates along the line of sight, the path integrals become

$$\mathcal{J} = \int_0^R dx \int_0^R dx' \exp\left[i\kappa_x(x - x')\right]$$

$$= \int_0^R ds \int_{-s}^s du \, \exp(i\kappa_x u).$$

The exponential oscillates rapidly unless the product $\kappa_x u$ is small relative to unity. The distance along the path is typically quite large relative to the eddies that influence the phase difference. This means that one can send the limits to infinity and establish the following relationship:

$$\mathcal{J} = 2\pi R \delta(\kappa_x) \tag{5.27}$$

With this expression the structure function becomes

$$\mathcal{D}_\varphi(\boldsymbol{\rho}) = 4\pi R k^2 \int_{-\infty}^\infty d\kappa_x \, \delta(\kappa_x) \int_{-\infty}^\infty d\kappa_y$$

$$\int_{-\infty}^\infty d\kappa_z \, \Phi_n(\boldsymbol{\kappa})[1 - \exp(i\boldsymbol{\kappa} \cdot \boldsymbol{\rho})] \,. \tag{5.28}$$

We now use the stretched-coordinate description (2.87) for the spectrum of anisotropic media:

$$\mathcal{D}_\varphi(\boldsymbol{\rho}) = 4\pi R k^2 abc \int_{-\infty}^\infty d\kappa_y \int_{-\infty}^\infty d\kappa_z \, \Phi_n\left(\sqrt{b^2\kappa_y^2 + c^2\kappa_z^2}\right)$$

$$\times \left\{1 - \exp[i(\kappa_y\rho_y + \kappa_z\rho_z)]\right\}$$

The wavenumber coordinates are rescaled as follows:

$$\kappa_y = q_y/b \qquad \text{and} \qquad \kappa_z = q_z/c$$

which yields

$$\mathcal{D}_\varphi(\boldsymbol{\rho}) = 4\pi R k^2 a \int_{-\infty}^\infty dq_y \int_{-\infty}^\infty dq_z \, \Phi_n\left(\sqrt{q_y^2 + q_z^2}\right)$$

$$\times \left\{1 - \exp[i(q_y\rho_y/b + q_z\rho_z/c)]\right\} \,.$$

From this it is apparent that the structure function measured in the two directions normal to the line-of-sight path can be expressed in terms of the result for isotropic turbulence:

$$\text{Horizontal spacing:} \quad \mathcal{D}_\varphi^{\text{an}}(\rho_y) = a\mathcal{D}_\varphi^{\text{iso}}(\rho_y/b)$$

$$\text{Vertical spacing:} \quad \mathcal{D}_\varphi^{\text{an}}(\rho_z) = a\mathcal{D}_\varphi^{\text{iso}}(\rho_z/c) \qquad (5.29)$$

These results are independent of the spectral model used to describe the media, so long as the stretched-wavenumber-spectrum description (2.87) is valid. The same results were derived by Kon using a qualitative approach [25].

One can use these expressions to analyze the third microwave experiment identified earlier. A 9.6-GHz signal was transmitted over a 64-km path from Mount Haleakala (3040 m) on Maui to three receivers located at sea level on the island of Hawaii [11]. The phase differences were measured during April 1968 with horizontal separations of 1.4, 4.9 and 10.7 m. The experiment was repeated in November with spacings of 4.5, 10 and 100 m. If we approximate the transmitted signal by a plane wave, the relationship (5.29) for horizontal spacings suggests that one should replace $\rho_y$ by $\rho_y/a$. The guiding parameter $\kappa_0\rho$ for selecting the appropriate form of the phase structure function becomes

$$\kappa_0\rho \to \kappa_0\frac{\rho_y}{a} = 2\pi\frac{\rho_y}{aL_0}.$$

The average horizontal outer scale length on this elevated path was undoubtedly much larger than the inter-receiver spacings. One would expect the 5/3 scaling law to hold and this was observed experimentally.

It is interesting that the phase structure function has the same form both in isotropic and in anisotropic media. That observation suggests an experiment with which one can measure the degree of anisotropy. If receivers were deployed both in the vertical and in cross-path horizontal directions, one could measure the phase structure function in both directions simultaneously. By matching the two curves one can find separation pairs for which

$$\mathcal{D}_\varphi^{\text{an}}(\rho_y/b) = \mathcal{D}_\varphi^{\text{an}}(\rho_z/c)$$

and estimate the anisotropic ratio:

$$b/c = \rho_y/\rho_z$$

This would provide as many estimates of $b/c$ as there are separations between receivers. If the two curves cannot be made to match, one must question the ellipsoidal description of anisotropy.

### 5.1.4 Beam Waves on Horizontal Paths

The transmitted signal in optical experiments is often described accurately as a beam wave. In geometrical optics these signals are represented by straight rays going from various points on the transmitting aperture to a focal point. Typical points on these rays are identified by the rectangular coordinates in Figure 5.10:

$$\left(x_1, \frac{1}{2}\rho\frac{f_{\mathrm{L}} - x_1}{f_{\mathrm{L}} - R}\right) \qquad \text{and} \qquad \left(x_2, -\frac{1}{2}\rho\frac{f_{\mathrm{L}} - x_2}{f_{\mathrm{L}} - R}\right)$$

The phase structure function for isotropic irregularities depends on the scalar distance between these points:

$$D_{12}^2 = |s_1 - s_2|^2 = (x_1 - x_2)^2 + \rho^2\left(\frac{f_{\mathrm{L}} - \frac{1}{2}(x_1 + x_2)}{f_{\mathrm{L}} - R}\right)^2 \qquad (5.30)$$

On introducing sum and difference coordinates

$$u = x_1 - x_2 \qquad \text{and} \qquad v = \tfrac{1}{2}(x_1 + x_2)$$

the structure function becomes

$$\mathcal{D}_\varphi(\rho) = 8\pi k^2 \int_0^\infty d\kappa\, \kappa \Phi_n(\kappa) \int_0^R dv$$

$$\times \int_{-\infty}^\infty du \left(\frac{\sin(\kappa u)}{u} - \frac{\sin\left(\kappa\sqrt{u^2 + q^2}\right)}{\sqrt{u^2 + q^2}}\right)$$

where

$$q^2 = \rho^2\left(\frac{f_{\mathrm{L}} - v}{f_{\mathrm{L}} - R}\right)^2.$$

The difference integration has been extended to all positive and negative values because the integrand falls off rapidly with increasing $u$. This approximation is equivalent to the assumptions we have used before to simplify the structure functions for plane and spherical waves. The first integration on $u$ is elementary and the second is found in Appendix D:

$$\mathcal{D}_\varphi(\rho) = 8\pi^2 k^2 \int_0^\infty d\kappa\, \kappa \Phi_n(\kappa) \int_0^R dv \left[1 - J_0\left(\kappa\rho\frac{f_{\mathrm{L}} - v}{f_{\mathrm{L}} - R}\right)\right] \qquad (5.31)$$

The beam-wave phase-difference-weighting function reduces to the plane and spherical wave expressions for appropriate values of the focal length. A collimated beam corresponds to an infinite focal length and our expression

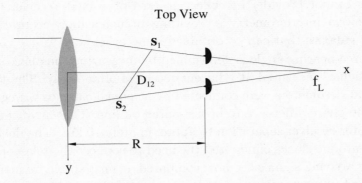

Figure 5.10: The coordinate system used to describe the phase correlation at adjacent receivers for a beam wave simulated by converging rays.

reduces to (5.19) for $f_{\mathrm{L}} = \infty$. If the focal length is taken to be zero, the rays go radially outward from the source and represent a spherical wave. Our prediction reduces to (5.8) in that situation.

The beam-wave result can be cast in a form that makes clear the roles played by eddies of different sizes. We let

$$w = \frac{f_{\mathrm{L}} - v}{f_{\mathrm{L}} - R}$$

to find

$$\mathcal{D}_\varphi(\rho) = 8\pi^2 k^2 (f_{\mathrm{L}} - R) \int_0^\infty d\kappa \, \kappa \Phi_n(\kappa) \left( \int_1^{(1 - R/f_{\mathrm{L}})^{-1}} dw \, [1 - J_0(\kappa \rho w)] \right). \tag{5.32}$$

This shows that only those eddies which are smaller than the spacing between receivers can have a significant influence on the phase structure function.

## 5.2 Microwave Interferometry

Optical astronomy has been augmented and enriched during the last fifty years by the development of microwave interferometry. Centimeter and millimeter radiation emitted by excited molecules in distant stars and gas clouds is measured by arrays of radio telescopes. When these signals are combined coherently the individual telescopes act like segments of a giant radio telescope *with respect to angular resolution*. Of course, a few small telescopes cannot provide the collecting area of a giant receiver. Nonetheless, microwave interferometry is providing an increasingly accurate portrait of distant galaxies that can be obtained in no other way.

The source-bearing angle is determined by measuring the phase difference between signals measured at adjacent receivers of an array. The telescopes of early interferometers were connected by a coaxial cable or waveguide that allowed the phase difference to be measured on-line. That arrangement limited the inter-receiver separation to approximately 10 km. The development of precise atomic clocks eliminated the need to connect the telescopes physically. The arriving signals are now compared against stable frequency references at each receiver. The signal phase is recorded digitally at each site and the records compared later at a common analysis center. Very large arrays (VLA) with baselines of up to 30 km have been used to make observations at frequencies ranging from 232 MHz to 100 GHz.

Using the same recording technique, it is possible to increase the inter-receiver separation and resolution dramatically in this way, making possible *very-long-baseline interferometry* (VLBI). Collaborations involving radio telescopes separated by hundreds or even thousands of kilometers have been established in the USA and in Europe. These combinations give angular resolutions of $10^{-3}$ arc seconds and provide unique insights into distant galactic structures. This approach has been extended by intercontinental cooperation to make the entire diameter of the earth available as a baseline. More recently, a satellite radio telescope operating at 1.7 and 5.0 GHz was placed in earth orbit by Japan. This satellite works in cooperation with a global network of ground-based radio telescopes and provides baselines as great as 30 000 km. In the future, combinations of satellite and ground stations will extend these baselines and steadily improve the accuracy with which distant galactic radio sources can be studied.

The angular resolution of these synthetic radio telescopes far surpasses that of optical telescopes. It is limited primarily by refractive-index fluctuations in the troposphere that distort the wavefront of an arriving plane wave.[5] Large horizontal eddies in the free atmosphere play a major role in creating this irregular phase front. Signals arriving at the individual telescopes have different random phase shifts impressed on them. The difference of these phase shifts produces angular errors when the signals of paired

---

[5] Below approximately 5000 MHz one must also consider the influence of plasma irregularities in the ionosphere and beyond. We ignore these complications here because the most precise observations of distant galaxies are being made at microwave and millimeter-wave frequencies well above the plasma threshold.

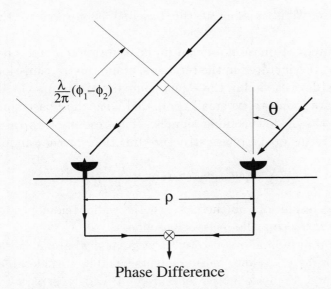

$$\frac{\lambda}{2\pi}(\phi_1 - \phi_2)$$

$\theta$

$\rho$

**Phase Difference**

Figure 5.11: A two-element interferometer indicating the extra path length traveled by one signal relative to the other. The signals were downconverted by mixing with a local oscillator and, in early versions, transmitted to a central point by a waveguide or coaxial cable. The two signals were then multiplied and the phase difference was extracted by passing the output signal through a phase detector.

telescopes are combined. The interferometric accuracy thus depends on the degree of phase coherence across the aperture.

One can describe this limitation by considering the simple two-element interferometer shown in Figure 5.11. If the source lies in the vertical plane defined by the two receivers, the angle $\theta$ is related to the total phase of the arriving signals by the *interferometer equation*:

$$\sin\theta = \frac{\lambda}{2\pi\rho}[\phi(R+\rho) - \phi(R)] \qquad (5.33)$$

The resolution of a microwave interferometer is equivalent to measurement errors in the bearing angle. It can be estimated by taking differentials of the preceding equation,

$$\delta\theta = \frac{\lambda}{2\pi\rho}\sec\theta\left[\delta\phi(R+\rho) - \delta\phi(R)\right] = \frac{\lambda}{2\pi\rho}\sec\theta\left[\varphi(R+\rho) - \varphi(R)\right]$$

and the angular variance becomes

$$\langle\delta\theta^2\rangle = \frac{1}{k^2\rho^2}\sec^2\theta\,\langle[\varphi(R+\rho) - \varphi(R)]^2\rangle. \qquad (5.34)$$

The factor $\sec^2\theta$ is missing if the baseline is normal to the plane of propagation.

Several sources of phase noise can limit an interferometer's performance. Phase errors are generated in the receivers, along the transmission lines or in the recording devices, and in the signal-comparison circuits. In almost every array these are found to be small compared with the phase shifts imposed along the paths of downcoming signals. The resolution is thus expressed primarily in terms of the phase structure function for the random medium:

$$\langle\delta\theta^2\rangle = \text{constant} \times \frac{\mathcal{D}_\varphi(\rho)}{k^2\rho^2} \tag{5.35}$$

The angular resolution of these synthetic radio telescopes is generally improved by increasing the receiver baseline.

Microwave interferometers also tell us a great deal about large structures in the troposphere. Their measurements exhibit considerable similarity between values of phase noise measured at widely separated receivers. The phase noise in the individual receivers is certainly uncorrelated and this similarity represents a *common response* to large-scale atmospheric irregularities. There is little doubt that such structures exist in the troposphere. Turbulent cloud patterns are observed in satellite photography and are routinely used by meteorologists for weather forecasting. Acoustic gravity waves also occur in the lower atmosphere. These formations have a broad range of horizontal dimensions. One can argue that they are part of a global weather pattern and should not be considered as components of the turbulence spectrum. On the other hand, they influence electromagnetic phase measurements directly. VLBI data therefore provides a unique insight into the world of large-scale tropospheric irregularities. Phase noise, which corrupts radio astronomical observations, is becoming the *signal* for geodesists and atmospheric scientists.

### 5.2.1 $\mathcal{D}_\varphi(\rho)$ *for Atmospheric Transmission*

To be more specific about these applications, we must develop an expression for the phase structure function that describes transmission through the entire atmosphere. Geometrical optics can be used to analyze astronomical observations made with microwave interferometers. Before we begin this process, it is helpful to identify the factors that are likely to be important. For this purpose we consider the overhead source illustrated in Figure 5.12. The separation $\rho$ of two rays reaching elements of an interferometer will certainly play a central role. We are primarily concerned here with establishing

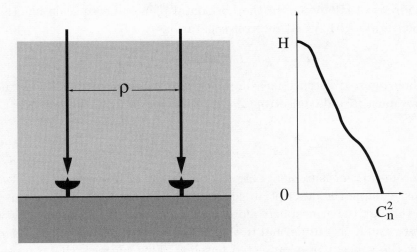

Figure 5.12: The use of a long-baseline microwave interferometer to measure the phase structure function for an overhead source. An idealized profile for the level of turbulent activity is shown on the right-hand side.

and confirming the scaling law which describes how the phase difference depends on this separation.

The next factor to consider is the limited extent of the random medium. Turbulent activity in the troposphere varies continuously along the down-coming rays, as suggested in Figure 5.12. This reminds us that there is a height $H$ above which turbulent activity can be ignored. The relationship between $\rho$ and $H$ is quite important for the interpretation of interferometric data.

A third factor is related to the way in which the data is processed. To reduce the errors in astronomical readings, phase-difference data must be averaged over a sufficient time to sample the full range of fluctuations. The sample length is usually expressed in terms of the *crossing time*:

$$T = \rho/v$$

This is the time required for eddies to travel from one receiver to another on prevailing winds. Sample lengths of an hour or more are therefore needed for large arrays. This influence will be analyzed after we have established the basic relationship between $\rho$ and $H$.

The propagation geometry of Figure 5.12 also reminds us that the down-coming rays pass through the entire atmosphere. The free atmosphere plays a strong role since it contains most of the ray path. Irregularities in this region are highly anisotropic and exhibit significant correlations over tens or hundreds of kilometers in the horizontal plane. These scale lengths are commensurate with VLBI receiver separations

$$\rho \approx L_0 \text{ (horizontal)}$$

and their phase-difference measurements are sensitive to them. This means that we must include anisotropy in our description from the outset.

### 5.2.1.1 $\mathcal{D}_\varphi(\rho)$ for Oblique Paths

Interferometric measurements are seldom made with overhead sources and we must enlarge the description suggested by Figure 5.12. The rays usually travel through the atmosphere along inclined paths. The separation between receivers usually is not normal to the direction of propagation and one must employ the concept of a projected baseline. This suggests that we use the propagation geometry defined in Figure 5.13 which neglects the curvature of the earth's surface. Plane waves provide a good description for signals arriving from distant sources. The general expression (5.3) for the phase correlation describes this situation if the upper limits for the path integrations are extended to infinity:

$$\langle \varphi_1 \varphi_2 \rangle = k^2 \int_0^\infty ds_1 \int_0^\infty ds_2 \int d^3\kappa \, \Phi_n\left(\boldsymbol{\kappa}, \frac{\mathbf{s}_1 + \mathbf{s}_2}{2}\right)$$
$$\times \exp[i\boldsymbol{\kappa} \cdot (\mathbf{s}_1 - \mathbf{s}_2 + \boldsymbol{\rho})] \tag{5.36}$$

We can usually separate the local properties of the random medium and the changing profile of turbulent activity as noted in Section 2.2.8. That separation allows us to write the phase structure functions as follows:

$$\mathcal{D}_\varphi(\rho) = 2k^2 \int_0^\infty ds_1 \int_0^\infty ds_2 \, C_n^2(r) \int d^3\kappa \, \Omega(\boldsymbol{\kappa})$$
$$\times \exp[i\boldsymbol{\kappa} \cdot (\mathbf{s}_1 - \mathbf{s}_2)] \, [1 - \exp(i\boldsymbol{\kappa} \cdot \boldsymbol{\rho})] \tag{5.37}$$

where

$$r = \tfrac{1}{2}(s_1 + s_2)$$

is the average slant range along the rays. The anisotropic nature of the medium is important and we use the stretched-wavenumber model (2.87) to describe the local turbulence spectrum.

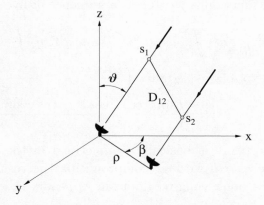

Figure 5.13: The geometry used to describe the phase structure function for a downcoming plane wave measured by an interferometer whose baseline makes an angle $\beta$ with the plane of propagation $(x, z)$.

The spatial vectors that occur in (5.37) must be expressed in the earth-based coordinates of Figure 5.13 which describe the anisotropic medium:

$$\mathbf{s}_1 = s_1(\mathbf{i}_x \sin \vartheta + \mathbf{i}_z \cos \vartheta)$$

$$\mathbf{s}_2 = s_2(\mathbf{i}_x \sin \vartheta + \mathbf{i}_z \cos \vartheta)$$

$$\boldsymbol{\rho} = \rho(\mathbf{i}_x \cos \beta + \mathbf{i}_y \sin \beta)$$

With these conventions the structure function is

$$\mathcal{D}_\varphi(\rho, \beta, \vartheta) = 2k^2 \int_0^\infty ds_1 \int_0^\infty ds_2 \, C_n^2(r) \int_{-\infty}^\infty d\kappa_x \int_{-\infty}^\infty d\kappa_y \int_{-\infty}^\infty d\kappa_z$$
$$\times abc\,\Omega\left(\sqrt{a^2 \kappa_x^2 + b^2 \kappa_y^2 + c^2 \kappa_z^2}\right)$$
$$\times \exp[i(s_1 - s_2)(\kappa_x \sin \vartheta + \kappa_z \cos \vartheta)]$$
$$\times \{1 - \exp[i\rho(\kappa_x \cos \beta + \kappa_y \sin \beta)]\}.$$

The turbulence spectrum becomes isotropic in the stretched coordinates

$$\kappa_x = \frac{1}{a} q \sin \psi \cos \omega \qquad \kappa_y = \frac{1}{b} q \cos \psi \qquad \kappa_z = \frac{1}{c} q \sin \psi \sin \omega$$

and the structure function is written as a sequence of five integrations:

$$\mathcal{D}_\varphi(\rho, \beta, \vartheta) = 2k^2 \int_0^\infty ds_1 \int_0^\infty ds_2\, C_n^2(r) \int_0^\infty dq\, q^2 \Omega(q) \int_0^\pi d\psi \sin\psi \int_0^{2\pi} d\omega$$

$$\times \exp\left[ iq(s_1 - s_2)\sin\psi \left( \frac{\sin\vartheta\cos\omega}{a} + \frac{\cos\vartheta\sin\omega}{c} \right) \right]$$

$$\times \left\{ 1 - \exp\left[ iq\rho\left( \frac{\sin\psi\cos\omega\cos\beta}{a} + \frac{\cos\psi\sin\beta}{b} \right) \right] \right\}$$

The lower atmosphere is basically symmetrical in the horizontal plane and one can take $a = b$ without compromising the result. The angular integrations can be done using results from Appendix D:

$$\mathcal{D}_\varphi(\rho, \beta, \vartheta) = 8\pi k^2 \int_0^\infty ds_1 \int_0^\infty ds_2\, C_n^2(r) \int_0^\infty dq\, q\Omega(q)$$

$$\times \left( \frac{\sin[q\mathsf{p}(s_1 - s_2)]}{\mathsf{p}(s_1 - s_2)} - \frac{\sin[qD_{12}]}{D_{12}} \right) \qquad (5.38)$$

where

$$D_{12} = \sqrt{ \mathsf{p}^2(s_1 - s_2)^2 + 2(s_1 - s_2)\rho\frac{\sin\vartheta\cos\beta}{a^2} + \frac{\rho^2}{a^2} }. \qquad (5.39)$$

The parameter $\mathsf{p}$ was introduced in (4.37) and describes inclined paths passing through anisotropic media. What remains is to select a spectrum model and complete the remaining integrations.

Let us pause to notice that the sequence of integrations followed here is quite different than the approach used to describe terrestrial links. In those applications we did the path integrations first, giving a delta function for the line-of-sight wavenumber component. That procedure is based on the assumption that the path length is large relative to the important eddy sizes. Microwave-interferometer measurements are sensitive to the horizontal outer scale length and to the atmospheric height. Because both are large, we are well advised to delay the path integrations until the last steps. Important physics would be lost if we followed the traditional strategy here.

When the characteristics of the random medium are known, these expressions allow one to estimate the phase structure function without approximation. That is usually a daunting task. One can also view (5.38) as an integral equation that connects phase-difference measurements to the profile of turbulent activity. Unfortunately it cannot be solved analytically. One must try to infer the profiles by introducing generic functions and identifying key parameters.

### 5.2.1.2 $\mathcal{D}_\varphi(\rho)$ Based on the Structure Function of $n$

There is an alternative way to describe the connection between the phase structure function and the random medium. Tropospheric experiments are often related to the refractive index structure function:

$$\mathcal{D}_n(\mathbf{R}, r) = \langle [\delta n(\mathbf{R} + \mathbf{r}) - \delta n(\mathbf{R})]^2 \rangle \tag{5.40}$$

When we use this approach, we gain additional insight into the relationship of interferometric measurements and atmospheric characteristics. The structure function is related to the spectrum of irregularities by the relationship established in Chapter 2 if the random medium is locally isotropic:

$$\mathcal{D}_n(R, r) = 8\pi \int_0^\infty d\kappa \, \kappa^2 \Phi_n(\kappa, r)\left(1 - \frac{\sin(\kappa R)}{\kappa R}\right)$$

$$= 8\pi \int_0^\infty d\kappa \, \kappa^2 C_n^2(r)\Omega(\kappa)\left(1 - \frac{\sin(\kappa R)}{\kappa R}\right)$$

If we associate the wavenumber $\kappa$ with the stretched coordinate $q$, we see that this combination of terms also occurs in (5.38). That observation allows us to construct an equivalent expression for the phase structure function of an astronomical signal:

$$\mathcal{D}_\varphi(\rho, \beta, \vartheta) = k^2 \int_0^\infty ds_1 \int_0^\infty ds_2 \, [\mathcal{D}_n(D_{12}, r) - \mathcal{D}_n(\mathsf{p}\,|s_1 - s_2|, r)] \tag{5.41}$$

The distance $D_{12}$ between points on the two ray paths is again given by (5.39). We can simplify this expression by introducing the sum and difference coordinates defined by (4.36):

$$\mathcal{D}_\varphi(\rho, \beta, \vartheta) = k^2 \int_0^\infty dr$$

$$\times \int_{-r}^r du \left[ \mathcal{D}_n\left(\sqrt{\mathsf{p}^2 u^2 + 2u\rho \frac{\sin\vartheta\cos\beta}{a^2} + \frac{\rho^2}{a^2}}, r\right) - \mathcal{D}_n(\mathsf{p}|u|, r) \right] \tag{5.42}$$

Several important features are retained in this description. The anisotropy of the atmosphere is included, provided that the ellipsoidal spectrum model (2.87) is a valid description. We have made no assumption about the relative size of the irregularities and the slant range. What remains is to evaluate (5.42) for particular models of $\mathcal{D}_n(R, r)$.

## 5.2.2  Separation Scaling for the 2/3 Law

In previous descriptions we have relied heavily on the Kolmogorov model
for the turbulence spectrum. The refractive-index structure function which
corresponds to this spectral model is described by the 2/3 power of the
separation distance:

$$\mathcal{D}_n(\mathbf{R}, r) = C_n^2(r)|\mathbf{R}|^{\frac{2}{3}} \tag{5.43}$$

The level of turbulent activity $C_n^2$ is allowed to change with position while the
second factor describes the local correlation of refractive irregularities. This
model is confirmed near the surface for separations less than the outer scale
length. In our examination of an interferometer's performance with (5.42)
we need a model that describes the free atmosphere and separations as great
as VLBI receiver baselines.

Early measurements of temperature fluctuations in the free atmosphere
were made with specially instrumented research aircraft flying level courses.
Those experiments showed that the structure function varies as the 2/3
power for separations as large as 10–100 km [26]. These observations were
extended by the Global Atmospheric Sampling Program which placed
velocity and temperature sensors on commercial aircraft. During countless
regular airline flights these sensors confirmed the 2/3 law for separations as
great as 300–400 km [27]. When we introduce this form into (5.42), the local
and inhomogeneous effects are separated:

$$\mathcal{D}_\varphi(\rho, \beta, \vartheta) = k^2 \int_0^\infty dr\, C_n^2(r)$$

$$\times \int_{-r}^r du \left[ \left( u^2\mathsf{p}^2 + 2u\rho\frac{\sin\vartheta\cos\beta}{a^2} + \frac{\rho^2}{a^2} \right)^{\frac{1}{3}} - \left( u^2\mathsf{p}^2 \right)^{\frac{1}{3}} \right]$$

$$\tag{5.44}$$

### 5.2.2.1  The Long-baseline Approximation

The separations between receivers for interferometric arrays are often much
greater than the effective height of the turbulent atmosphere. The argument
of the first term in (5.44) can then be approximated as

$$\sqrt{u^2\mathsf{p}^2 + 2u\rho\frac{\sin\vartheta\cos\beta}{a^2} + \frac{\rho^2}{a^2}} \simeq \frac{\rho}{a}$$

and the predicted scaling law for the baseline distance is simplified:

$$\rho \gg H: \qquad \mathcal{D}_\varphi(\rho, \beta, \vartheta) = \alpha \, \rho^{\frac{2}{3}} - \beta \tag{5.45}$$

The constants are independent of the orientation of the baseline:

$$\alpha = 2 \, k^2 \sec^2 \vartheta \int_0^\infty dz \, z C_n^2(z) a^{-\frac{2}{3}}$$

$$\beta = 2 \, k^2 \sec^{\frac{7}{3}} \vartheta \int_0^\infty dz \, z^{\frac{5}{3}} C_n^2(z) \mathsf{p}^{\frac{2}{3}}$$

This prediction is confirmed by interferometric phase measurements that we shall soon review. They provide strong support for using the 2/3 model to describe tropospheric irregularities.

### 5.2.2.2 The Baseline Normal to the Plane of Propagation

To develop our understanding further, it is useful to specialize the propagation geometry shown in Figure 5.13. The basic relationship (5.44) is simplified if the receivers are deployed perpendicular to the plane of the arriving signals, which is a preferred observational arrangement:

$$\mathcal{D}_\varphi\left(\rho, \frac{\pi}{2}, \vartheta\right) = 2k^2 \int_0^\infty dr \, C_n^2(r) \int_0^r du \left[\left(u^2\mathsf{p}^2 + \frac{\rho^2}{a^2}\right)^{\frac{1}{3}} - \left(\mathsf{p}^2 u^2\right)^{\frac{1}{3}}\right] \tag{5.46}$$

Notice that the same result describes an overhead source. When we change the variable of integration to $u = x\rho/(a\mathsf{p})$ and convert from slant range to altitude,

$$\mathcal{D}_\varphi\left(\rho, \frac{\pi}{2}, \vartheta\right) = 2k^2 \sec \vartheta \, \rho^{\frac{5}{3}} \int_0^\infty dz \, C_n^2(z) a^{-\frac{5}{3}} \mathsf{p}^{-1} K\left(\frac{za\mathsf{p} \sec \vartheta}{\rho}\right) \tag{5.47}$$

where the altitude weighting function

$$K(\eta) = \int_0^\eta d\zeta \left[(\zeta^2 + 1)^{\frac{1}{3}} - \zeta^{\frac{2}{3}}\right] \tag{5.48}$$

is plotted in Figure 5.14. In discussing satellite paths in the last chapter we noted that $\mathsf{p} \approx \cos\vartheta/c$ unless the source is very close to the horizon:

$$\vartheta < 85°: \qquad \mathcal{D}_\varphi\left(\rho, \frac{\pi}{2}, \vartheta\right) \simeq 2k^2 \sec^2 \vartheta \, \rho^{\frac{5}{3}} \int_0^\infty dz \, C_n^2(z) ca^{-\frac{5}{3}} K\left(\frac{za}{\rho c}\right) \tag{5.49}$$

The function $K(\eta)$ weights the contributions of different levels in the lower atmosphere. It depends on the inter-receiver separation $\rho$ and the ratio $a/c$.

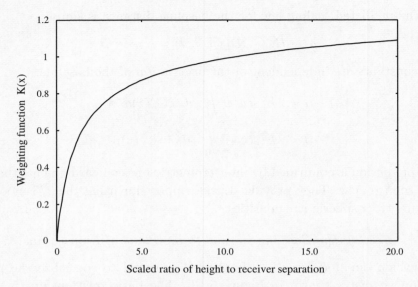

Figure 5.14: The function which weights the contributions of anisotropic turbulent irregularities for interferometric measurements using signals that have passed through the troposphere.

The values of $z$ are bounded by the effective height $H$ of the atmosphere. Using small- and large-argument expressions,[6] one finds different scaling laws for short and long baselines relative to the height of the troposphere:

$$\rho \ll \frac{a}{c} H: \qquad \mathcal{D}_\varphi(\rho, \vartheta) = \rho^{\frac{5}{3}} \left( 2.914 k^2 \sec^2 \vartheta \int_0^\infty dz\, C_n^2(z) c a^{-\frac{5}{3}} \right)$$
(5.50)

$$\rho \gg \frac{a}{c} H: \qquad \mathcal{D}_\varphi(\rho, \vartheta) = \rho^{\frac{2}{3}} \left( 2 k^2 \sec^2 \vartheta \int_0^\infty dz\, z C_n^2(z) a^{-\frac{2}{3}} \right) \qquad (5.51)$$

These scaling laws were first established by Stotskii [28] assuming an isotropic atmosphere. The second result agrees with the long-baseline approximation (5.45), which is valid for any orientation. For even larger separations, $\mathcal{D}_\varphi(\rho, \vartheta)$ approaches a constant value equal to twice the single-path phase variance. The separation at which this saturation occurs is determined by the horizontal outer scale length and by atmospheric curvature.

The transition region between the two scaling laws is of great interest for the design of instruments. We note first that the transition spacing is related to the *effective height* of the troposphere $aH/c$. Since the elongation ratio is large in the free atmosphere, the height limit on turbulent conditions $H$ need not be great to have a significant influence on the performance of microwave arrays.

[6] From (5.47) we have $\lim_{\eta \to 0} [K(\eta)] = \eta$ and from Appendix B $\lim_{\eta \to \infty} [K(\eta)] = 1.4572$.

The change in scaling laws can best be understood on physical grounds. When the two ray paths are separated by a distance that is considerably smaller than the effective height, the signals encounter many independent eddies. A three-dimensional description is appropriate in this case and this leads to the 5/3 law. The irregularities are essentially two-dimensional when they are sampled by signals whose separation is large relative to the depth of the troposphere. That condition leads naturally to the 2/3 scaling law.

To establish predictions that can be tested against observations, we need to consider this transition more carefully. The plot of $K(\eta)$ suggests that the scaling law changes from 5/3 to 2/3 when the primary contribution to the integration of $C_n^2$ comes from an altitude below

$$z \approx 5 \frac{c}{a} \rho.$$

The elongation ratio $a/c$ is about 100 in the free atmosphere. This means that a change from one scaling law to the other should occur at about $\rho = 10$ km if the main contributions to the $C_n^2$ integration occur below 500 m. A breakpoint at this spacing is observed experimentally. A numerical simulation used a *slab model* of the troposphere[7] and assumed that its irregularities are isotropic [29]. Those results showed in detail how one scaling law gives way to another for several values of the zenith angle and baseline orientation.

If the underlying assumptions are correct, there should be a marked change in the scaling law at a separation that is related to the effective height of the troposphere and the elongation of its irregularities. To test that prediction against experimental data, we must include a second effect that has an important influence on the phase structure function.

### 5.2.3 The Influence of the Sample Length

In examining the factors which influence single-path phase measurement, we found that phase trends and the length of the data sample both play important roles. We noted that phase trends usually disappear when one measures phase differences between nearby paths. That is usually the experimental situation for terrestrial links. It is also a good assumption for astronomical paths if the interferometer receivers are less than a kilometer apart. This characterizes optical interferometers and some millimeter-wave arrays. For large separations, trends in the refractive index influence the paths differently

---

[7] This model assumes that the level of turbulent activity is uniform up to a fixed height $H$ and zero above that level.

and one must treat them as single-path phase measurements. In this case one must attempt to detrend the data or use the Allan variance which is introduced in Chapter 6.

Phase-difference measurements are quite sensitive to the sample length even when the phase trends cancel out. Astronomical data is often averaged over large time spans to reduce observational errors. That integration is limited by the sample length. At the beginning of this section, we noted that the sample length must be at least as great as the time required for eddies to travel from one receiver to another.

To interpret astronomical observations, we need to describe the combined effect of the inter-receiver separation and the sample length on the phase difference measured between adjacent receivers. Two conventions are commonly used to represent the influence of a finite sample length and they tend to explore different aspects of the effect. The first is the ensemble average of the squared phase difference [30]:

$$M^2 = \left\langle \left( \frac{1}{T} \int_0^T dt \, \Delta\varphi(t) \right)^2 \right\rangle \tag{5.52}$$

This should approach zero as the sample length goes to infinity if the process is stationary. Its departure from zero is thus an indication of the influence of the sample length.

The second approach examines fluctuations about the sample mean. This measure is described by the variance of the phase difference averaged over the sample length:

$$\overline{\langle \Delta\varphi^2 \rangle} = \frac{1}{T} \int_0^T dt \left\langle \left( \Delta\varphi(t) - \frac{1}{T} \int_0^T dt' \, \Delta\varphi(t') \right)^2 \right\rangle \tag{5.53}$$

One can write this out as follows:

$$\overline{\langle \Delta\varphi^2 \rangle} = \frac{1}{T} \int_0^T dt \, \langle \Delta\varphi^2(t) \rangle - \frac{1}{T^2} \int_0^T dt \int_0^T dt' \, \langle \Delta\varphi(t) \, \Delta\varphi(t') \rangle$$

The first term is independent of the sample length and is identical to the single-path phase structure function. This means that the two measures are simply connected and one can calculate one from the other:

$$\overline{\langle \Delta\varphi^2 \rangle} = \mathcal{D}_\varphi(\boldsymbol{\rho}) - M^2 \tag{5.54}$$

We will develop the subject here using the phase-difference-variance convention employed previously in Section 4.1.7. On substituting

$$\Delta\varphi(t) = \varphi(\mathbf{r}, t) - \varphi(\mathbf{r} + \boldsymbol{\rho}, t)$$

into the basic definition (5.53), we find that

$$\overline{\langle\Delta\varphi^2\rangle} = \mathcal{D}_\varphi(\boldsymbol{\rho}) - \frac{1}{T^2}\int_0^T dt \int_0^T dt' \, \mathcal{Q}(t, t', \boldsymbol{\rho}) \tag{5.55}$$

where

$$\mathcal{Q}(t, t', \boldsymbol{\rho}) = \langle[\varphi(\mathbf{r}, t) - \varphi(\mathbf{r} + \boldsymbol{\rho}, t)][\varphi(\mathbf{r}, t') - \varphi(\mathbf{r} + \boldsymbol{\rho}, t')]\rangle. \tag{5.56}$$

This can be related to the phase structure function using the algebraic identity

$$2(a - b)(c - d) = (a - d)^2 + (b - c)^2 - (a - c)^2 - (b - d)^2 \tag{5.57}$$

and letting $\tau = t - t'$:

$$\begin{aligned}
2\mathcal{Q}(\tau, \boldsymbol{\rho}) = & \langle[\varphi(\mathbf{r}, t) - \varphi(\mathbf{r} + \boldsymbol{\rho}, t + \tau)]^2\rangle \\
& + \langle[\varphi(\mathbf{r} + \boldsymbol{\rho}, t) - \varphi(\mathbf{r}, t + \tau)]^2\rangle \\
& - \langle[\varphi(\mathbf{r}, t) - \varphi(\mathbf{r}, t + \tau)]^2\rangle \\
& - \langle[\varphi(\mathbf{r} + \boldsymbol{\rho}, t) - \varphi(\mathbf{r} + \boldsymbol{\rho}, t + \tau)]^2\rangle
\end{aligned}$$

Taylor's hypothesis relates separations in time and space by assuming that the random medium is frozen and blown past the receivers on a steady wind so that $\rho = v\tau$. This is a reasonable assumption for small baselines but must be abandoned for large arrays. When the wind blows parallel to the receiver baseline the expression above simplifies to

$$2\mathcal{Q}(\tau, \rho) = \mathcal{D}_\varphi(|\rho - v\tau|) + \mathcal{D}_\varphi(|\rho + v\tau|) - 2\mathcal{D}_\varphi(v\tau). \tag{5.58}$$

With this result the variance of the phase difference averaged over a finite sample can be written as a time integral of phase structure functions:

$$\begin{aligned}
\overline{\langle\Delta\varphi^2\rangle}\frac{1}{T^2}\int_0^T d\tau \, (T - \tau)[2\mathcal{D}_\varphi(\rho) + 2\mathcal{D}_\varphi(v\tau) \\
- \mathcal{D}_\varphi(|\rho - v\tau|) - \mathcal{D}_\varphi(|\rho + v\tau|)] \tag{5.59}
\end{aligned}$$

The next step is to introduce the expression for $\mathcal{D}_\varphi(R)$ developed for transmission through the atmosphere. If we assume that the wind speed does not vary over the altitude region of significant phase disturbance, we can use (5.46) to describe the temporal structure function. The resulting calculation involves two integrations on height and one on time:

$$\overline{\langle\Delta\varphi^2\rangle} = 2k^2 \sec^2\vartheta \int_0^\infty dz\, C_n^2(z) \int_0^z du\, \frac{1}{T^2} \int_0^T d\tau\, (T-\tau)\Lambda(u,t,\rho) \quad (5.60)$$

where

$$\Lambda(u,t,\rho) = 2\left(u^2\mathsf{p}^2 + \frac{\rho^2}{a^2}\right)^{\frac{1}{3}} + 2\left(u^2\mathsf{p}^2 + \frac{(vt)^2}{a^2}\right)^{\frac{1}{3}} - 2(\mathsf{p}u)^{\frac{2}{3}}$$

$$- \left(u^2\mathsf{p}^2 + \frac{(\rho+vt)^2}{a^2}\right)^{\frac{1}{3}} - \left(u^2\mathsf{p}^2 + \frac{(\rho-vt)^2}{a^2}\right)^{\frac{1}{3}}. \quad (5.61)$$

These integrals are difficult to complete even when the $C_n^2$ profile is represented by simple models. For a slab model of the atmosphere the integrations can be expressed in terms of the hypergeometric functions described in Appendix H but are more easily evaluated numerically.

One can approximate the sample-length effect by disconnecting the time and altitude integrations. To do so, we represent the structure functions in (5.59) by their asymptotic expansions developed in (5.50) and (5.51). This *ad hoc* approach is a substantial simplification but allows the calculation to be completed in a way that preserves the essential physics [30].

The first case to analyze is that for inter-receiver spacings considerably less than the effective height of the turbulent atmosphere. We abbreviate the phase structure function (5.50) as follows:

$$\rho \ll \frac{a}{c}H: \qquad \mathcal{D}_\varphi(\tau) = \mathsf{Q}\rho^{\frac{5}{3}} \quad (5.62)$$

Here $\mathsf{Q}$ depends on the wavelength, the zenith angle, the profile of $C_n^2$ and the anisotropy scaling parameters $a$ and $c$. The joint dependence on the inter-receiver spacing and sample length emerges when one substitutes these expressions into (5.59):

$$\rho \ll \frac{a}{c}H: \qquad \overline{\langle\Delta\varphi^2(\rho,T)\rangle} = \mathsf{Q}\rho^{\frac{5}{3}}\mathsf{q}\!\left(\frac{\rho}{vT}\right) \quad (5.63)$$

The function

$$\mathsf{q}(x) = 1 + \frac{9}{44}x^{-2}\left(1 + x^{\frac{11}{3}} - \frac{1}{2}|x+1|^{\frac{11}{3}} + \frac{1}{2}|x-1|^{\frac{11}{3}}\right) \quad (5.64)$$

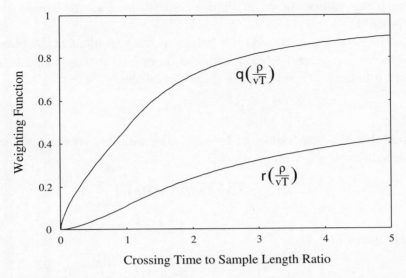

Figure 5.15: Plots of the functions $q(x)$ and $r(x)$ which describe the joint dependence of the phase structure function on the inter-receiver separation and sample length for interferometric measurements using signals that have passed through the troposphere.

is plotted in Figure 5.15. Expanding this expression allows one to describe very large and very small sample lengths relative to the crossing time:

$$\rho \ll \frac{a}{c}H: \qquad vT > \rho: \qquad \overline{\langle \Delta \varphi^2(\rho, T) \rangle} = Q\rho^{\frac{5}{3}}\left[1 - \left(\frac{\rho}{vT}\right)^{\frac{1}{3}}\right]$$

$$vT < \rho: \qquad \overline{\langle \Delta \varphi^2(\rho, T) \rangle} = \frac{9}{44}Q(vT)^{\frac{5}{3}} \qquad (5.65)$$

It is significant that the first asymptotic form is approached very slowly. When the sample length exceeds the crossing time by a large margin, one recovers the assumed model, since

$$\overline{\langle \Delta \varphi^2(\rho, \infty) \rangle} = \mathcal{D}_\varphi(\rho).$$

By contrast, the short-sample result is independent of the separation. This suggests that the variation of $\overline{\langle \Delta \varphi^2 \rangle}$ with $\rho$ is similar to the generic structure function presented in Figure 5.3. In this analogy, $vT$ plays the role of an *effective* outer scale length above which the variance of the phase difference saturates. That makes physical sense because a finite sample length excludes eddies larger than $vT$ from the measurement. In doing so, it plays the role of $L_0$ in cutting off small-wavenumber contributions to the structure function. One can trace the evolution of the distance dependence from $\rho^{\frac{5}{3}}$

to $\rho^0$ with the values shown in Figure 5.15, so long as the approximation (5.62) is valid.

When the separation exceeds the height of the turbulent atmosphere, we use the second asymptotic expansion to describe the transmission phase structure function. We summarize (5.51) as follows:

$$\rho \gg \frac{a}{c}H: \qquad \mathcal{D}_\varphi(\rho) = \mathsf{R}\rho^{\frac{2}{3}} \tag{5.66}$$

On combining this expression with (5.59), the sample-averaged variance of the phase difference becomes

$$\rho \gg \frac{a}{c}H: \qquad \overline{\langle \Delta\varphi^2(\rho, T)\rangle} = \mathsf{R}\rho^{\frac{2}{3}}\mathsf{r}\left(\frac{\rho}{vT}\right) \tag{5.67}$$

where the function

$$\mathsf{r}(x) = 1 + \frac{9}{20}x^{-2}\left(1 + x^{\frac{8}{3}} - \frac{1}{2}|x+1|^{\frac{8}{3}} - \frac{1}{2}|x-1|^{\frac{8}{3}}\right) \tag{5.68}$$

is also plotted in Figure 5.15. Expanding this function gives complementary expressions for very large and very small sample lengths relative to the crossing time:

$$\rho \gg \frac{a}{c}H: \qquad vT > \rho: \qquad \overline{\langle \Delta\varphi^2(\rho, T)\rangle} = \mathsf{R}\rho^{\frac{2}{3}}\left[1 - \left(\frac{\rho}{vT}\right)^{\frac{4}{3}}\right]$$

$$vT < \rho: \qquad \overline{\langle \Delta\varphi^2(\rho, T)\rangle} = \frac{9}{20}\mathsf{R}(vT)^{\frac{2}{3}} \tag{5.69}$$

Intermediate cases can be constructed with the values given in Figure 5.15. The concept of an effective outer scale length $vT$ is again appropriate. Several experiments have used sample lengths large enough for $vT$ to exceed the horizontal outer scale length in the free atmosphere. This means that the sample length provides the practical transition to saturation for phase-difference measurements.

Let us pause to reflect on these analytical results and ask how they are likely to be reflected in experimental data. If the interferometer baselines are small relative to the thickness of the atmosphere, one could measure a scaling law with an exponent anywhere between 0 and 5/3 depending on the sample length. This variability is compounded by the transition from 5/3 to 2/3 which reflects the relationship between the spacing and the effective height of the random medium. Our approximate description does not capture this transition and deals only with the end-point behaviors of $\mathcal{D}_\varphi(\rho)$. We expect that the 2/3 scaling for large spacings will become flat when the sample length exceeds the crossing time by a comfortable margin. Evidently one must be quite careful in analyzing such measurements and in drawing conclusions from them about the random medium.

### 5.2.4 Microwave-interferometer Measurements

Considerable effort has been devoted to measuring phase structure functions for atmospheric transmission. The initial motivation was to understand the inherent limits of existing microwave interferometers. A second reason was to support the design of new arrays to exploit millimeter-wave sources. This data is now being applied to establish error limits for differential GPS position measurements used in surveying and geodesy.

It is significant that the troposphere is nondispersive in the range of interest. This means that the refractive index does not depend on the carrier frequency[8] and one can combine phase differences measured at different wavelengths with a simple frequency scaling. Stotskii attempted to do so using data taken at different times at the NRAO, Cambridge and the TAO in the USSR [32]. The measurements covered a wavelength range from 4 to 11 cm and inter-receiver spacings from 100 m to 400 km. Using this combined data set, he suggested that a transition from 5/3 to 2/3 occurs for a separation of 5.6 km. One must be cautious in accepting this estimate, since we believe that large structures in the troposphere change with location, season and time of day.

One would prefer to analyze simultaneous data taken at the same site with a wide range of inter-receiver separations. In doing so, it is desirable to use a sample length that exceeds the crossing time. Table 5.2 provides a summary of recent phase-difference experiments. In combination with the predictions developed above, these measurements now provide a reasonably complete description for the phase structure function of microwave signals that pass through the atmosphere.

#### 5.2.4.1 VLA Structure-function Measurements

A pioneering experiment by Armstrong and Sramek used the Very Large Array at the National Radio Astronomy Observatory in New Mexico to measure the phase structure function [34]. This facility has twenty-seven receiving dishes with adjustable spacings. Signals at 4.885 MHz were monitored on near-vertical paths over a period spanning two and a half years.

---

[8] By contrast, the imaginary part of the tropospheric dielectric constant which controls attenuation of the signal changes dramatically with frequency over this range [31].

Table 5.2: Microwave-interferometer measurements of atmospheric phase structure function ($N$ is the number of receivers used and $D$ is their diameter)

| Location and altitude | Frequency (GHz) | $N$ | $D$ (m) | Inter-receiver separations | Sample length | Ref. |
|---|---|---|---|---|---|---|
| Cambridge, 17 m | 5 | 2 | 13 | 36 m to 4.6 km | 12 h | 33 |
| NRAO, 2124 m | 5 | 27 | 25 | 1–35 km | 5 min | 34 |
| | 5 & 15 | 27 | 25 | 50 m to 35 km | 2 h | 35 |
| | 22 | 27 | 25 | 100–3500 m | 16 min | 36 |
| | 22 | 27 | 25 | 200 m to 20 km | 90 min | 37 |
| Nobeyama, 1350 m | 22 | 5 | 10 | 27–540 m | 1–20 min | 38 |
| | 19.5 | 2 | 1.8 | 35 m | 5 min | 39 |
| Hat Creek, 1024 m | 86 | 3 | 6 | 6–100 m | 15 min | 40 |
| | 86 | 3 | 6 | 6–300 m | 100 s | 40 |
| | 86 | 6 | 6 | 6–846 m | 15 min | 41 |
| IRAM, 2550 m | 86 | 3 | 15 | 24–288 m | 15 min | 42 |
| VLBI | 100 | 3 | | 400, 800 and 1200 km | 100 s | 40 |

The phase difference was measured with spacings ranging from 1 to 10 km and the data fitted to a simple scaling law:

$$\mathcal{D}_\varphi(\rho) = A\rho^\mu + B \tag{5.70}$$

The thirteen measured exponents were spread uniformly across the range $0.84 < \mu < 1.95$. It is curious that a transition to a slope of $2/3$ was not observed. On the other hand, the sample length of 5 min is much less than the crossing time and this limitation would have eliminated the contributions of large energy-bearing eddies which are important to the scaling law for the phase difference.

This problem was corrected by Sramek in a second experiment in which he used sample lengths between 2 and 4 h [35]. Signals at 4.9 and 15 GHz were measured during a two-year campaign at the NRAO for sources above $60°$. The range of inter-receiver separations was enlarged to include spacings as small as 50 m and as large as 35 km. The scaling law exponents were distributed over a wide range, $0 < \mu < 1.8$. The specific slope depended on the combination of inter-receiver spacings employed, as the data reproduced in Figure 5.16 makes clear. The first panel shows phase-difference measurements for separations less than 1 km and gives $\mu = 1.44$, in fair agreement with the prediction $5/3$ for a long data sample. The second panel shows data for spacings larger than 1 km and gives $\mu = 0.68$. This is remarkably close to the value of $2/3$ predicted for long data samples *and* inter-receiver

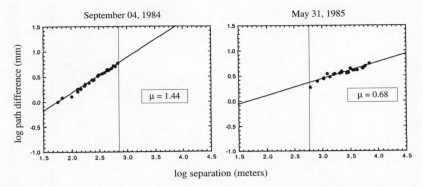

Figure 5.16: The phase difference as a function of the inter-receiver separation measured by Sramek at 4.885 and 15 GHz with the NRAO array on different days [35]. The sources were at high elevations and the sample length was 2 h.

separations that exceed the effective height of the atmosphere. Such data suggests a transition spacing of approximately 1 km. An important reservation is that the data was not all taken on the same day because many of the antennas had to be moved to establish the second set of spacings.

Sramek later used the VLA at the NRAO to obtain nineteen *simultaneous* measurements at 22 GHz by restricting the inter-receiver separations to less than 3500 m [36]. The structure function was inferred from measurements of the Allan variance using expressions that will be derived in the next chapter. This conversion is possible because the use of 16-min data samples ensured that $vT > \rho$. The distribution of scaling-law exponents was bimodal. Three measurements clustered around $\mu = 0.6$ and the remaining values fell in the range $0.9 < \mu < 1.84$. The smaller value agrees with the prediction 2/3 and suggests that the effective height of the troposphere was small when these measurements were performed. The larger values suggest a transition to 5/3 and a relatively high atmosphere. Had one simply averaged the results, one would have found $\mu = 1.2 \pm 0.2$ and thereby missed important physics.

The question of scaling for $\mathcal{D}_\varphi(\rho)$ was finally laid to rest by simultaneous measurements of phase difference taken at 22 GHz with the NRAO Very Large Array [37]. A wide range of antenna spacings was possible at this facility, namely 200 m $< \rho <$ 20 km. The basic data was averaged for 90 min to generate rms estimates of the phase difference for each separation and these results are reproduced as open circles in Figure 5.17. A constant electronic phase noise error was noted in the equipment. This error was subtracted in quadrature from the raw data and the corrected points are

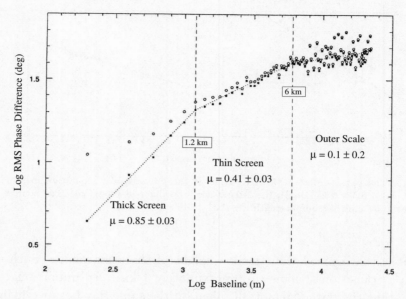

Figure 5.17: The phase difference measured at 22 GHz by Carilli and Holdaway using the Very Large Array at the NRAO [37]. The open circles are measured points averaged for 90 min. The solid circles are the same data corrected by subtracting in quadrature a constant *10°* phase error that was detected in the measuring equipment.

plotted in Figure 5.17 as solid circles. This second data set should represent pure propagation phase noise. Numerous measurements of the rms phase difference were thus obtained in three important regions: (1) a thick screen, (2) a thin screen and (3) the saturation region where outer-scale effects dominate. The results of this comparison are presented in Table 5.3. The agreement is remarkably good and leaves little doubt that the foregoing predictions provide a valid description of microwave-interferometric observations.

Phase-difference measurements were also made at the Nobeyama Radio Observatory in Japan using 22-GHz signals [38]. Five antennas in this array provided ten spacings between 27 and 540 m. The sample length was varied between 1 and 20 min. The Allan variance was measured as a function of baseline and converted into the phase structure function. Fitted values of the scaling-law index for various sample lengths are reproduced in Figure 5.18. The inferred index increases gradually with the sample length, as one would expect from Figure 5.15. The average values agree with the small-spacing prediction 5/3. Two smaller antennas at this facility with a spacing of 35 m monitored the much stronger 19.5-GHz signal from a geostationary satellite.

Table 5.3: Rms-phase-difference scaling-law exponents predicted by turbulence theory compared with values measured at 22 GHz by Carilli and Holdaway using the VLA [37]

| Region | Spacing Range (km) | Prediction | Measurement |
|---|---|---|---|
| Thick screen | $0 < \rho < 1.2$ | $\mu = 0.83$ | $\mu = 0.85 \pm 0.03$ |
| Thin screen | $1.2 < \rho < 6$ | $\mu = 0.33$ | $\mu = 0.41 \pm 0.03$ |
| Saturation | $6 < \rho < 20$ | $\mu = 0$ | $\mu = 0.1 \pm 0.2$ |

The measured index coincided with the curve established previously using stellar sources [39].

Several experiments have measured the phase structure function at 86 GHz using the Berkeley–Illinois–Maryland Association Millimeter Array at Hat Creek, California. The array has progressively been expanded from three to ten receivers. The first series of measurements used baselines of less than 100 m with an averaging time of 15 min [40]. The scaling-law exponents fitted to this data fell in the range $\mu = 1.4 \pm 0.4$, which agrees with the C-band data in Figure 5.16. This result is reasonably close to the prediction 5/3 for these small spacings and relatively large sample lengths.

The phase structure function was also measured at 86 GHz in France using the IRAM interferometer on the Plateau de Bure [42]. The separations of its three receivers were varied between 24 and 288 m during a 49-day observing period. Distributions of the scaling-law exponent were measured for different times of day and sample lengths. Typical values in the morning were $\mu = 1.85 \pm 0.04$, in agreement with the 5/3 model; at night they were $\mu = 1.11 \pm 0.02$. In both experiments, the limited number of receivers meant that $\mathcal{D}_\varphi(\rho)$ could be measured *simultaneously* for only a few baselines.

A comprehensive experiment was later conducted at Hat Creek by Wright using 86-GHz signals [41]. Six receivers provided spacings between 6 and 846 m. The phase structure function was measured as a function of the inter-receiver spacing, sample length (averaging interval) and elevation angle. This data provides the most complete picture we have for separations less than 1 km. The measurements were fitted to the following general expression:

$$\mathcal{D}_\varphi(\rho, T, \vartheta) = \text{constant} \times \rho^{\mu(T)} (\sin E)^{-\gamma} + \text{noise} \qquad (5.71)$$

The reported data was processed with four sample lengths: $T = 5, 15, 45$ and 135 min. Measured values of $\mu$ increased gradually with sample length as one would expect from Figure 5.18. Results from the best data set are reproduced in Table 5.4. This summary shows that $\mu$ and $\gamma$ assume different values for spacings less than and greater than 300 m. Values of $\mu$ for larger

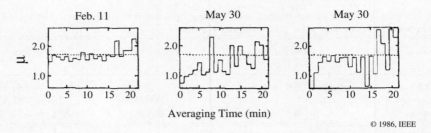

Figure 5.18: The separation scaling-law index for the phase difference as a function of sample length measured at 22 GHz by Kasuga, Ishiguro and Kawabe [38].

separations agree with the 2/3 prediction and suggest that the transition spacing at this location is close to 300 m – although no clear break was observed in the data. This implies that phase modulation is imposed by irregularities quite close to the surface. The measured exponents below this threshold are large but did not reach the 5/3 limit. There is thus a gap between these observations and our theory.

Increases in wind speed from 2 to 6 m s$^{-1}$ drove the exponent to 5/3 for baselines less than 40 m. A seasonal variation in $\mu$ was observed and probably reflects changes in the amount of turbulent activity.[9] The elevation-angle scaling indicated by Table 5.4 should be compared with the sec$^2 \vartheta$ prediction both of (5.50) and of (5.51) which was driven by the anisotropic nature of the troposphere. The measured values for $\gamma$ are consistent with that prediction but may also tell a more complicated story when they are compared with a fuller description of the propagation.

### 5.2.4.2  VLBI Structure-function Measurements

VLBI phase difference was measured at 100 GHz for very large separations in a collaborative effort involving the groups at Hat Creek, Owens Valley and Kitt Peak [40]. Distances between these facilities gave spacings of 400, 800 and 1200 km. The Allan variance was measured as a function of separation for 100-s samples and then converted into the structure function. Within experimental error, the measured phase structure function was the same for all three spacings. This is what one would predict for separations that are very much larger than the travel distance associated with the short averaging time. A companion measurement at Hat Creek used separations

---

[9]  This experiment found the following mean values:

$$\text{Winter:} \quad \mu = 0.88 \pm 0.22$$
$$\text{Summer:} \quad \mu = 1.34 \pm 0.38$$

Table 5.4: Scaling-law exponents for structure-function measurements made at 86 GHz by Wright using the BIMA millimeter array [41]

| Inter-receiver spacings (m) | Distance scaling | | Elevation scaling |
|---|---|---|---|
| | $T = 15$ min | $T = 45$ min | |
| $6 < \rho < 300$ | $\mu = 1.16 \pm 0.06$ | $\mu = 1.18 \pm 0.21$ | $\gamma = 1.34 \pm 0.40$ |
| $300 < \rho < 846$ | $\mu = 0.48 \pm 0.10$ | $\mu = 0.58 \pm 0.10$ | $\gamma = 1.60 \pm 0.44$ |

less than 300 m and the same sample length of 100 s. The results were later scaled to 100 GHz and compared with the VLBI data to suggest a baseline exponent $\mu = 0.3 \pm 0.1$. One must regard this result with some care because it is likely that the physics for signals received 300 m apart is different than that for those at 1000 km. The horizontal outer scale length might have been exceeded by these large separations. The spherical nature of the atmosphere imposes a further restriction on signals passing through it, as noted in Chapter 4.

### 5.2.4.3 Ionospheric Influences

Phase differences have been measured with large arrays in the Netherlands and in China using signals in the range 232–608.5 MHz [43]. This data is more difficult to interpret because the ionosphere is involved in an important way. The propagation plane of the downcoming waves, the receiver baseline and the orientation of the magnetic field each present important vector directions that must be recognized in the analysis of such experiments.

### 5.2.4.4 A Perspective

The measurements described above provide the best description currently available for the phase structure function induced by atmospheric transmission. The predictions that had previously been developed agree with this data in several respects but fail to do so in others. The 2/3 spacing law predicted by the Kolmogorov model (5.43) is observed for separations larger than a threshold that lies between 300 and 1000 m. This suggests that phase errors are imposed close to the surface and is consistent with the rapid decline of $C_n^2$ with height that is measured. The structure function should be independent of spacing for much larger separations and this too is observed.

The prediction 5/3 for small baselines is confirmed in some measurements but not in others. This mismatch between theory and experiment is due to the crude approximations that were made in developing the scaling laws. Even the numerical simulation of separation scaling laws [29] replaced the $C_n^2$ profile by a simple slab, which is a major simplification. The important influence of sample-length averaging was introduced in an *ad hoc* manner, using the asymptotic expressions (5.50) and (5.51) as a starting point. These, in turn, were replaced by asymptotic expansions. The theoretical description needs to be refined before one can expect it to match the measured data with precision. Even with a complete analysis in hand, one must remember that phase measurements are largely controlled by meteorology, which is extraordinarily difficult to describe and predict.

## 5.3  Optical Interferometry

The idea of using interference phenomenon to make precise astronomical observations originated with Fizeau in 1868. His concept was to place a mask with two holes at the entrance plane of a telescope and exploit the interference fringes generated by waves passing through the two openings. This idea was first implemented by Michelson using the Lick 12-inch reflector. This technique allowed him to measure the size of Jupiter's four Galilean satellites for the first time. The same approach was then used with the 100-inch Mount Wilson reflector to resolve Capella as a binary star and to measure the angular width of Betelgeuse – the first star size other than that of our own sun to be so defined. The technique then went into hibernation for several decades.

After the Second World War, microwave interferometers became extraordinarily successful at resolving point sources to reveal the structures of distant galaxies. Their success stimulated the development of comparable optical interferometers, which are fundamentally more capable than Fizeau's original concept of segmenting the aperture of a single telescope. The revised plan was to connect separated telescopes by optical beams or fiber-optic cables [44]. This capability has evolved rapidly in the last decade as electro-optical technology has improved and financial support for a wide range of scientific and military applications has become available. A large number of prototype or test-bed interferometers has been built in Europe, Australia and the USA, operating both in the visible and at infrared wavelengths, as noted in Table 5.5. These instruments have also made significant astronomical observations.

Table 5.5: Optical interferometers currently being used and developed for astronomical observations in the visible and infrared bands (the number of telescopes employed is $N$ and $D$ is their diameter; the inter-receiver spacings are measured between the centers of the individual receivers)

| Instrument and location | $N$ | Spacings (m) | $D$ (cm) | $\lambda$ ($\mu$m) | First light | Ref. |
|---|---|---|---|---|---|---|
| SOIRDETE, Plateau de Calern, France | 2 | 15 | 100 | 11 | 1978 | 45 |
| GI2T, Plateau de Calern, France | 2 | 13–65 | 152 | 1.25–2.2 | 1985 | 46 |
| Mark III, Mount Wilson | 5 | 3–31.5 | 2.5 | 0.4–0.9 | 1986 | 47 |
| IRMA, Irma Flatts, Wyoming, USA | 2 | 2.5–19.5 | 20 | 2.2 | 1986 | 48 |
| ISI, Mount Wilson | 2 | 4–34 | 165 | 11 | 1988 | 49 |
| COAST, Cambridge, UK | 5 | 4–100 | 40 | 0.65–0.95 and 1.3–2.2 | 1991 | 50 |
| IOTA, Mount Hopkins, Arizona, USA | 3 | 5–38 | 45 | 0.45–0.75 and 1.25–2.0 | 1993 | 51 |
| I2T+ASSI, Plateau de Calern, France | 2 | 12 | 26 | 0.78–1.05 | 1994 | 52 |
| NPOI, Flagstaff, Arizona, USA | 10 | 2–437 | 50 | 0.45–0.85 | 1994 | 53 |
| PTI, Palomar, California, USA | 3 | 86 and 110 | 40 | 1.5–2.4 | 1995 | 54 |
| SUSI, New South Wales, Australia | 11 | 5–640 | 20 | 0.4–0.9 | 1991 | 55 |
| CHARA, Mount Wilson, California, USA | 6 | 34–331 | 100 | 1.6 and 2.2 | 2000 | 56 |
| Keck, Mauna Kea, Hawai, USA | 2 | 85 | 1000 | 1.5–5 and 10 | 2001 | 57 |
| LBT, Mount Graham, Arizona, USA | 2 | 14.4 | 840 | 0.3–2.5 | 2002 | 58 |
| VLT, Cerro Paranal, Chile | 4 | 57–129 | 820 | 0.4–25 | 2006 | 59 |

### 5.3.1 *Optical Interferometer Installations*

The purpose of the early interferometers was to provide the foundation for a new generation of high-resolution interferometric observatories, which are now being constructed and are identified in Table 5.5. The individual telescopes in these new arrays have large primaries so they can capture faint signals from distant galaxies and gas clouds. The telescopes are optically connected with baselines ranging from 14 to 120 m, which translates into an angular resolution of $10^{-3}$ arc second. That compares with an accuracy of 0.5 arc second for excellent ground-based seeing and a resolution of $50 \times 10^{-3}$ arc second for the Hubble space telescope. It is hoped that large interferometers will eventually be able to capture images of planets that are circling nearby stars.

The generic design for an interferometer operating at visible wavelengths is suggested in Figure 5.19. Two or more telescopes are used to measure the phase difference of signals collected by the individual receivers. The signals are routed by mirrors inside a vacuum enclosure to a common detector, where they are multiplied. To generate optically stable fringes, the individual optical paths must be adjusted by moving the pairs of reflecting mirrors which act as variable delay lines. Laser beams provide the metrology required to measure the path difference with great precision. The relative delay of the two optical signals then gives the bearing angle through the interferometer equation (5.33).

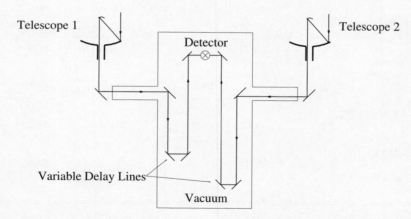

Figure 5.19: The typical arrangement of an optical interferometer designed for astronomical observations. Light from each telescope is routed to a common point by a series of mirrors within a vacuum enclosure. The signal path lengths are varied so that fringes can be counted by the detector.

The design and operation of interferometers are a good deal simpler at long infrared wavelengths. The telescope signals can often be passed through tunnels that need not be evacuated, because these frequencies are less sensitive to propagation conditions along the combining routes. A few installations have exploited heterodyne techniques using $CO_2$ lasers as local oscillators to convert the infrared signals into microwave frequencies, whereupon they can be transmitted and combined with traditional radio methods.

## 5.3.2 Limitations on the Accuracy of Optical Interferometers

The fundamental limit on the accuracy of these instruments is imposed by turbulent irregularities in the troposphere. The phenomenon is similar to those we examined in connection with microwave interferometry. The structure function is the natural measure of this limitation and provides an estimate of the angular error through (5.35). One can approach this problem theoretically or with actual experimental data. In view of the complexity of the lower atmosphere, we begin with the measurements.

### 5.3.2.1 Optical Measurements of $\mathcal{D}_\varphi(\rho)$

A considerable number of time-structure-function measurements has been made with single telescopes. In those experiments Taylor's hypothesis is invoked to convert time delays into equivalent spatial separations. This approximation and its implications for $\mathcal{D}_\varphi(\rho)$ are considered in Chapter 6. We are concerned here only with phase-difference measurements made by a number of receivers *simultaneously*. Most interferometers have only two receivers. They provide only one point on the structure-function curve for each instant and the available data is therefore quite limited.

The first optical structure-function measurements were made using a mask in front of a single telescope. Breckinridge used the 1.5-m solar telescope at Kitt Peak to measure the fringe amplitude, which he related to the phase difference [60]. The points agreed with the 5/3 inertial scaling law for spacings 20 cm $< \rho <$ 150 cm but some softening of the fit was found for the largest separation. The same basic technique was used to measure the phase structure function with the 4.2-m William Herschel Telescope in the Azores soon after its commissioning. This early experiment explored the range 15 cm $< \rho <$ 220 cm but these measurements were not made simultaneously. It found some evidence of saturation toward the largest spacing and an outer scale of 2 m was suggested [61]. That experiment was repeated using simultaneous data taken with six baselines in the range 55 cm $< \rho <$ 350 cm. In this second experiment the 5/3 scaling law was a good fit to the data over the entire range and there was no evidence that the outer scale length exerted any influence [62]. While the results are interesting and supportive of the simplest approximation to $\mathcal{D}_\varphi(\rho)$, these single-aperture experiments do not address the performance of interferometric telescopes separated by tens or hundreds of meters.

The first structure-function measurements for longer baselines were made with the I2T Interferometer at CERGA using 2.2-μm radiation [63]. Two telescopes provided phase-difference measurements for separations of 8, 12.5 and 16 m. The data could not be taken simultaneously because one telescope had to be moved to change the baseline. Nonetheless, the experimenters reported difficulty in fitting their data to the presumed 5/3 scaling law. In an apparent elaboration using data attributed to the CERGA experiment, Coulman and Vernin revealed clear evidence of a breakdown of the Kolmogorov scaling law for spacings greater than 10 m [64].

A decisive experiment was conducted with the eleven-element Sydney University Stellar Interferometer [65]. The individual receivers operate at visible wavelengths and are operated in a manner comparable to the technique explained in Figure 5.19. Nearly simultaneous measurements of the phase-difference variance were obtained on a single night, which was judged to be typical. Three baselines of 5, 20 and 80 m were employed and the experimental results are reproduced in Figure 5.20. The structure function begins to approach saturation at about $\rho = 5$ m. The 5/3 approximation is plotted for reference and there is little doubt that one is seeing the influence of a finite outer scale length. This result is consistent with temporal structure-function measurements made on Mount Wilson with the Mark III interferometer [47].

### 5.3.2.2 A Theoretical Description

The description of microwave interferometry presented in Section 5.2 provides the framework for analyzing optical measurements. We judge that it is important to include anisotropy in this analysis and we use (5.42) as our point of departure. We do so with the understanding that the turbulence parameters that occur in the refractive-index structure function are those appropriate for optical propagation. We make one concession to analytical simplicity and assume that the source direction is perpendicular to the receiver baseline. That is valid for sources near the zenith and this geometry

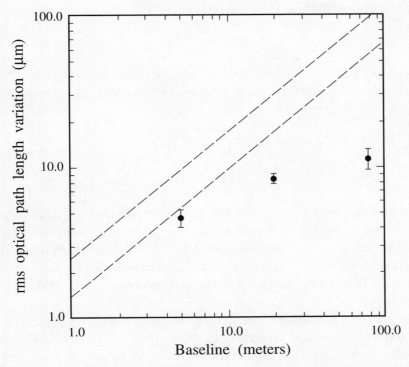

Figure 5.20: The rms path difference measured for three baselines at visible wavelengths by Davis, Lawson, Booth, Tango and Thorvaldson [65]. This data was taken on 27 April 1994 under typical nighttime conditions with the Sydney University Stellar Interferometer. The straight lines are the 5/3 scaling for two different values of the Fried coherence radius.

is often used for other elevations because it provides the longest effective baseline:

$$\mathcal{D}_\varphi\left(\rho, \frac{\pi}{2}, \vartheta\right) = 2k^2 \int_0^\infty dr \int_0^r du \left[ \mathcal{D}_n\left( \sqrt{\mathsf{p}^2 u^2 + \frac{\rho^2}{a^2}}, r \right) - \mathcal{D}_n(\mathsf{p}u, r) \right]$$
(5.72)

We convert from slant range to altitude using $r = z \sec \vartheta$ and make the substitution $u = v\rho/(a\mathsf{p})$ to find that

$$\mathcal{D}_\varphi\left(\rho, \frac{\pi}{2}, \vartheta\right) = 2k^2 \sec \vartheta \int_0^\infty dz \, \frac{\rho}{a\mathsf{p}}$$

$$\times \int_0^M dv \left[ \mathcal{D}_n\left( \frac{\rho}{a} \sqrt{v^2 + 1}, z \right) - \mathcal{D}_n\left( \frac{\rho}{a} v, z \right) \right]$$
(5.73)

where the upper limit on the $v$ integration is

$$M = z\frac{a\mathsf{p}}{\rho}\sec\vartheta.$$

In evaluating (5.47) we noted that $\mathsf{p} \approx \cos\vartheta/c$ unless the source is very close to the horizon, but that situation is invariably avoided. This approximation gives the following expression:

$$\mathcal{D}_\varphi\left(\rho, \frac{\pi}{2}, \vartheta\right) = 2k^2\sec^2\vartheta\int_0^\infty dz\left(\frac{c}{a}\rho\right)$$

$$\times \int_0^{\frac{az}{c\rho}} dv\left[\mathcal{D}_n\left(\frac{\rho}{a}\sqrt{v^2+1},\, z\right) - \mathcal{D}_n\left(\frac{\rho}{a}v,\, z\right)\right]$$

Notice that we have not limited the height integration by introducing an effective height for the troposphere, as we did in discussing long-baseline microwave arrays. The spacings employed in optical interferometers are quite modest relative to the height of the atmosphere and we doubt that a transition from three- to two-dimensional behavior is important for these instruments.

What remains is to select a model for the refractive-index structure functions and complete the integrations. The data in Figure 5.20 suggests that outer-scale effects are important for telescope separations greater than 5 m. This compels us to choose a model for the refractive-index structure functions that includes a finite $L_0$. We select the von Karman model as a reasonable description of the small-wavenumber region of the turbulence spectrum where energy is fed into the cascade. We can use the expression (2.79) for $\mathcal{D}_n(R, z)$ if we regard $C_n^2$ and $\kappa_0$ as functions of altitude:

$$\mathcal{D}_n(R,z) = 1.0468\,C_n^2(z)[\kappa_0(z)]^{-\frac{2}{3}}\left(1 - \frac{2^{\frac{2}{3}}}{\Gamma\left(\frac{1}{3}\right)}(R\kappa_0(z))^{\frac{1}{3}}K_{\frac{1}{3}}(R\kappa_0(z))\right)$$

$$(5.74)$$

If we suppress the height dependence of $\kappa_0$ for notational simplicity, we find

$$\mathcal{D}_\varphi\left(\rho, \frac{\pi}{2}, \vartheta\right) = 2.0972k^2\sec^2\vartheta\int_0^\infty dz\,C_n^2(z)\left(\frac{c}{a}\rho\right)\mathcal{J}\left(\frac{az}{c\rho}, \frac{\kappa_0\rho}{a}\right)$$

$$(5.75)$$

where

$$\mathcal{J}\left(\frac{az}{c\rho}, \eta\right) = \int_0^{\frac{az}{c\rho}} dv\left[(v\eta)^{\frac{1}{3}}K_{\frac{1}{3}}(v\eta) - \left(\eta\sqrt{v^2+1}\right)^{\frac{1}{3}}K_{\frac{1}{3}}\left(\eta\sqrt{v^2+1}\right)\right]. \quad (5.76)$$

The parameter $\eta$ is expressed in terms of the spacing and the horizontal outer scale length:

$$\eta = \frac{\kappa_0 \rho}{a} = 2\pi \frac{\rho}{L_0} \qquad (5.77)$$

One could evaluate the integral (5.76) numerically for various values of $\eta$, just as we did in treating the integral in (5.48) which corresponds to an infinite outer scale length. One need not go to this trouble since the upper limit is quite large because the ratio $a/c$ is large and increases as one ascends. Moreover, the spacing $\rho$ is small compared with the height of maximum turbulent activity. This means that one can take

$$\frac{az}{c\rho} \rightarrow \infty$$

and evaluate the weighting function analytically with two results that can be found at the end of Appendix D:

$$\mathcal{D}_\varphi\left(\rho, \frac{\pi}{2}, \vartheta\right) = 1.563 k^2 \sec^2 \vartheta \, c \int_0^\infty dz \, \frac{C_n^2(z)}{(\kappa_0)^{\frac{5}{3}}} \left[1 - \frac{2^{\frac{1}{6}}}{\Gamma\left(\frac{5}{6}\right)} \left(\frac{\kappa_0 \rho}{a}\right)^{\frac{5}{6}} K_{\frac{5}{6}}\left(\frac{\kappa_0 \rho}{a}\right)\right]$$

$$(5.78)$$

Notice that this expression reduces to the plane-wave result (5.25) if $C_n^2$ and $\kappa_0$ are constant along the path and the anisotropy parameters $a$ and $c$ are both unity. One can introduce experimental profiles for $C_n^2(z)$ and the outer-scale wavenumber to complete the remaining height integration numerically.

A reasonable estimate is established by ignoring the variation with altitude of the outer scale length:

$$\mathcal{D}_\varphi\left(\rho, \frac{\pi}{2}, \vartheta\right) = \frac{1.563 k^2 \sec^2 \vartheta \, c}{(\kappa_0)^{\frac{5}{3}}} \left[1 - \frac{2^{\frac{1}{6}}}{\Gamma\left(\frac{5}{6}\right)} \left(\frac{\kappa_0 \rho}{a}\right)^{\frac{5}{6}} K_{\frac{5}{6}}\left(\frac{\kappa_0 \rho}{a}\right)\right] \int_0^\infty dz \, C_n^2(z)$$

$$(5.79)$$

This result is almost the same as the terrestrial expression in (5.25) with the single replacement

$$R C_n^2 \rightarrow \int_0^\infty dz \, C_n^2(z).$$

One can therefore use the numerical values for the quantity in square brackets in (5.79) which are plotted in Figure 5.8 for the von Karman model. The graceful bend over to a saturation value portrayed there agrees with the data reproduced in Figure 5.20. We judge that the approximation (5.79) provides a satisfactory description for optical interferometric measurements of the phase noise. One can extrapolate phase-difference measurements made at one spacing to another with a reasonable estimate for the horizontal outer scale length. The analysis can also be expanded to include the influence of the finite length of data samples [66].

## 5.4 Problems

### Problem 1

Regard the phase-correlation expression (5.7) as an integral equation connecting the turbulence spectrum to phase data measured with a variety of receiver baselines. Show that it can be inverted as follows [67]:

$$\Phi_n(\kappa) = \frac{1}{4\pi^2 R k^2} \int_0^\infty d\rho \, \rho [\kappa\rho J_1(\kappa\rho) - J_0(\kappa\rho)] \langle \varphi(r)\varphi(r+\rho) \rangle$$

In principle, reconstruction of the spectrum is limited only by the number of inter-receiver spacings employed and the precision of the phase measurements. Unfortunately this is a serious problem. The kernel which weights the data increases with $\kappa\rho$ and emphasizes phase measurements taken with large separations. There is a practical limit to the number of receivers that can be deployed in a given experiment. This implies a maximum separation and one cannot explore the spectral influence of eddies larger than this distance.

Bias errors in the phase detectors cause a second problem. These can produce artificial correlations, even though the actual phase of the signal is uncorrelated between the receivers. Such problems mean that it is unlikely that one can exploit this analytical solution to reconstruct $\Phi_n(\kappa)$ from spaced-receiver experiments.

### Problem 2

In the real atmosphere an exponential model describes the decrease of $C_n^2$ with altitude better than does the slab model. In this case, show that

$$\mathcal{D}_\varphi(\rho) = 2k^2 H \int_0^\infty dz \, e^{-z/H} \left[ \mathcal{D}_n\left( \sqrt{\rho^2 + z^2} \right) - \mathcal{D}_n(z) \right].$$

Can you find an analytical solution for this integral equation that expresses $\mathcal{D}_n(R)$ in terms of the phase structure function?

# References

[1] D. L. Fried, "Limiting Resolution Looking Down Through the Atmosphere," *Journal of the Optical Society of America*, **56**, No. 10, 1380–1384 (October 1966).

[2] D. L. Fried, "Propagation of a Spherical Wave in a Turbulent Medium," *Journal of the Optical Society of America*, **57**, No. 2, 175–180 (February 1967).

[3] D. L. Fried, "Optical Resolution Through a Randomly Inhomogeneous Medium for Very Long and Very Short Exposures," *Journal of the Optical Society of America*, **56**, No. 10, 1372–1379 (October 1966).

[4] R. F. Lutomirski and H. T. Yura, "Wave Structure Function and Mutual Coherence Function of an Optical Wave in a Turbulent Atmosphere," *Journal of the Optical Society of America*, **61**, No. 4, 482–487 (April 1971).

[5] J. W. Herbstreit and M. C. Thompson, "Measurements of the Phase of Radio Waves Received over Transmission Paths with Electrical Lengths Varying as a Result of Atmospheric Turbulence," *Proceedings of the IRE*, **43**, No. 10, 1391–1401 (October 1955).

[6] A. P. Deam and B. M. Fannin, "Phase-Difference Variations in 9,350 Megacycle Radio Signals Arriving at Spaced Antennas," *Proceedings of the IRE*, **43**, No. 10, 1402–1404 (October 1955).

[7] H. B. Janes and M. C. Thompson, "Errors Induced by the Atmosphere in Microwave Range Measurements," *Radio Science, Journal of Research NBS/USNC-URSI*, **68D**, No. 11, 1229–1235 (November 1964).

[8] K. A. Norton, J. W. Herbstreit, H. B. Janes, K. O. Hornberg, C. F. Peterson, A. F. Barghausen, W. E. Johnson, P. I. Wells, M. C. Thompson, M. J. Vetter and A. W. Kirkpatrick, *An Experimental Study of Phase Variations in Line-of-Sight Microwave Transmissions* (National Bureau of Standards Monograph 33, U.S. Government Printing Office, Washington, 1 November 1961).

[9] M. C. Thompson and H. B. Janes, "Measurements of Phase-Front Distortion on an Elevated Line-of-Sight Path," *IEEE Transactions on Aerospace and Electronic Systems*, **AES-6**, No. 5, 645–656 (September 1970).

[10] H. B. Janes, M. C. Thompson and D. Smith, "Tropospheric Noise in Microwave Range-Difference Measurements," *IEEE Transactions on Antennas and Propagation*, **AP-21**, No. 2, 256–260 (March 1973).

[11] H. B. Janes and M. C. Thompson, "Comparison of Observed and Predicted Phase-Front Distortion in Line-of-Sight Microwave Signals," *IEEE Transactions on Antennas and Propagation*, **AP-21**, No. 2, 263–266 (March 1973).

[12] M. C. Thompson, H. B. Janes, L. E. Wood and D. Smith, "Phase and Amplitude Scintillations at 9.6 GHz on an Elevated Path," *IEEE Transactions on Antennas and Propagation*, **AP-23**, No. 6, 850–854 (November 1975).

[13] R. R. Rogers, R. W. Lee and A. T. Waterman, "Measurements of the Phase Structure Function at 35 GHz on a 28-km Path," USNC/URSI Spring Meeting in Seattle, Washington: Session F-1 (18–22 June 1979). These measurements are also described briefly in the Stanford Electronics Laboratory report by R. W. Lee and A. T. Waterman, "Wave-Front Observations at 35 GHz on a Line-of-Sight, Radio-Wave Propagation Path", AFCRL-68-0096 and SU-SEL-68-015 (December 1967).

[14] R. W. Lee and J. C. Harp, "Weak Scattering in Random Media, with Applications to Remote Probing," *Proceedings of the IEEE*, **57**, No. 4, 375–406 (April 1969).

[15] R. J. Hill, R. A. Bohlander, S. F. Clifford, R. W. McMillan, J. T. Priestley and W. P. Schoenfeld, "Turbulence-Induced Millimeter-Wave Scintillation Compared with Micrometeorological Measurements," *IEEE Transactions on Geoscience and Remote Sensing*, **26**, No. 3, 330–342 (May 1988).

[16] S. F. Clifford, G. M. B. Bouricius, G. R. Ochs and M. H. Ackley, "Phase Variations in Atmospheric Optical Propagation," *Journal of the Optical Society of America*, **61**, No. 10, 1279–1284 (October 1971).

[17] V. I. Tatarskii, *Wave Propagation in a Turbulent Medium*, translated by R. A. Silverman (Dover, New York, 1967), 111.

[18] V. A. Krasil'nikov, "On Fluctuations of the Angle-of-Arrival in the Phenomenon of Twinkling of Stars," *Doklady Akademii Nauk SSSR, Seriya Geofizicheskaya*, **65**, No. 3, 291–294 (1949) and "On Phase Fluctuations of Ultrasonic Waves Propagating in the Layer of the Atmosphere Near the Earth," *Doklady Akademii Nauk SSSR, Seriya Geofizicheskaya*, **88**, No. 4, 657–660 (1953). (These references are in Russian and no translations are currently available.)

[19] V. V. Voitsekhovich, "Outer Scale of Turbulence: Comparison of Different Models," *Journal of the Optical Society of America A*, **12**, No. 6, 1346–1353 (June 1995).

[20] Y. V. Vaysleyb, Y. R. Milyutin and B. Y. Frezinskiy, "Asymptotic Behavior of the Structure Functions of Fluctuations of the Amplitude and Phase of a Plane Optical Wave in a Turbulent Atmosphere with a Modified Karman Spectrum," *Radiotekhnika i Elektronika (Radio Engineering and Electronic Physics)*, **30**, No. 10, 116–123 (October 1985).

[21] V. P. Lukin and V. V. Pokasov, "Optical Wave Phase Fluctuations," *Applied Optics*, **20**, No. 1, 121–135 (1 January 1981).

[22] G. K. Born, R. Bogenberger, K. D. Erben, F. Frank, F. Mohr and G. Sepp, "Phase-Front Distortion of Laser Radiation in a Turbulent Atmosphere," *Applied Optics*, **14**, No. 12, 2857–2863 (December 1975).

[23] G. R. Ochs, private communication on 9 June 1997.

[24] V. U. Zavorotny, private communication on 30 July 1998.

[25] A. I. Kon, "Qualitative Theory of Amplitude and Phase Fluctuations in a Medium with Anisotropic Turbulent Irregularities," *Waves in Random Media*, **4**, No. 3, 297–306 (July 1994). (Notice that there is an error in equation (32) of this paper.)

[26] L. R. Tsvang, "Measurements of the Spectrum of Temperature Fluctuations in the Free Atmosphere," *Izvestiya Akademii Nauk SSSR, Seriya Geofizicheskaya (Bulletin of the Academy of Sciences of the USSR, Geophysical Series)*, No. 1, 1117–1120 (January 1960).

[27] G. D. Nastrom and K. S. Gage, "A Climatology of Atmospheric Wavenumber Spectra of Wind and Temperature Observed by Commercial Aircraft," *Journal of the Atmospheric Sciences*, **42**, No. 9, 950–960 (1 May 1985).

[28] A. A. Stotskii, "Concerning the Fluctuation Characteristics of the Earth's Troposphere," *Izvestiya Vysshikh Uchebnykh Zavedenii, Radiofizika (Soviet Radiophysics)*, **16**, No. 5, 620–622 (May 1973).

[29] R. N. Treuhaft and G. E. Lanyi, "The Effect of the Dynamic Wet Troposphere on Radio Interferometric Measurements," *Radio Science*, **22**, No. 2, 251–265 (March–April 1987).

[30] A. F. Dravskikh and A. M. Finkelstein, "Tropospheric Limitations in Phase and Frequency Coordinate Measurements in Astronomy," *Astrophysics and Space Science*, **60**, No. 2, 251–265 (February, 1979).

[31] B. R. Bean and E. J. Dutton, *Radio Meteorology* (National Bureau of Standards Monograph 92, U.S. Government Printing Office, Washington, 1 March 1966), 271–280.

[32] A. A. Stotskii, "Fluctuation Characteristics of the Electric Component of the Troposphere," *Radiotekhnika i Electronika (Radio Engineering and Electronic Physics)*, **17**, No. 11, 1827–1833 (November 1972).

[33] G. Pooley, "Atmospheric Fluctuations Measured with a Short-Baseline Interferometer," in *Radioastronomical Seeing*, edited by J. E. Baldwin and W. Shouguan (Pergamon Press, London, 1990), 64–65.

[34] J. W. Armstrong and R. A. Sramek, "Observations of Tropospheric Phase Scintillations at 5 GHz on Vertical Paths," *Radio Science*, **17**, No. 6, 1579–1586 (November–December 1982).

[35] R. A. Sramek, "Atmospheric Phase Stability at the VLA," in *Radioastronomical Seeing*, edited by J. E. Baldwin and W. Shouguan (Pergamon Press, London, 1990), 21–30.

[36] R. A. Sramek, "VLA Phase Stability at 22 GHz on Baselines of 100 m to 3 km," VLA Test Memo 143, NRAO (20 October 1983). See especially Table II on page 11.

[37] C. L. Carilli and M. A. Holdaway, "Tropospheric Phase Calibration in Millimeter Interferometry," *Radio Science*, **34**, No. 4, 817–840 (July–August 1999).

[38] T. Kasuga, M. Ishiguro and R. Kawabe, "Interferometric Measurement of Tropospheric Phase Fluctuations at 22 GHz on Antenna Spacings of 27 to 540 m," *IEEE Transactions on Antennas and Propagation*, **AP-34**, No. 6, 797–803 (June 1986).

[39] M. Ishiguro, T. Kanzawa and T. Kasuga, "Monitoring of Atmospheric Phase Fluctuations Using Geostationary Satellite Signals," in *Radioastronomical Seeing*, edited by J. E. Baldwin and W. Shouguan (Pergamon Press, London, 1990), 60–63.

[40] M. C. H. Wright and W. J. Welch, "Interferometer Measurements of Atmospheric Phase Noise at 3 mm," in *Radioastronomical Seeing*, edited by J. E. Baldwin and W. Shouguan (Pergamon Press, London, 1990), 71–74.

[41] M. C. H. Wright, "Atmospheric Phase Noise and Aperture-Synthesis Imaging at Millimeter Wavelengths," *Publications of the Astronomical Society of the Pacific*, **108**, No. 724, 520–534 (June 1996).

[42] L. Olmi and D. Downes, "Interferometric Measurement of Tropospheric Phase Fluctuations at 86 GHz on Antenna Spacings of 24 m to 288 m," *Astronomy and Astrophysics*, **262**, No. 2, 634–643 (September 1992).

[43] T. A. Th. Spoelstra and Y. Yi-Pei, "Ionospheric Scintillation Observations with Radio Interferometry," *Journal of Atmospheric and Terrestrial Physics*, **57**, No. 1, 85–97 (January 1995).

[44] P. R. Lawson, Editor, *Selected Papers on Long Baseline Stellar Interferometry* (SPIE Optical Engineering Press, Bellingham, WA 98227-0010, 1997).

[45] P. Assus, H. Choplin, J. P. Corteggiani, E. Cuot, J. Gay, A. Journet,
     G. Merlin and Y. Rabbia, "L'interféromètre infrarouge du C.E.R.G.A.,"
     *Journal d'Optique*, **10**, No. 6, 345–350 (1979).

[46] D. Mourard, I. Tallon-Bosc, A. Blazit, D. Bonneau, G. Merlin,
     F. Morand, F. Vakili and A. Labeyrie, "The G12T Interferometer on Plateau
     de Calern," *Astronomy and Astrophysics*, **23,** No. 2, 705–713 (1994).

[47] D. F. Buscher, J. T. Armstrong, C. A. Hummel, A. Quirrenbach,
     D. Mozurkewich, K. J. Johnston, C. S. Denison, M. M. Colavita and M.
     Shao, "Interferometric Seeing Measurements on Mt. Wilson: Power Spectra
     and Outer Scales," *Applied Optics*, **34**, No. 6, 1081–1096 (February 1995).

[48] H. M. Dyck, J. A. Benson and S. T. Ridgway, "IRMA: A Prototype Infrared
     Michelson Stellar Interferometer," *Publications of the Astronomical Society of
     the Pacific*, **105**, 610–615 (June 1993).

[49] M. Bester, W. C. Danchi and C. H. Townes, "Long Baseline Interferometer for
     the Mid-infrared," *Proceedings of the SPIE*, **1237**, 40–48 (1990).

[50] J. E. Baldwin, R. C. Boysen, G. C. Cox, C. A. Haniff, J. Rogers, P. J. Warner,
     D. M. A. Wilson and C. D. Mackay, "Design and Performance of COAST,"
     *Proceedings of the SPIE*, **2200**, 118–128 (June 1994).

[51] N. P. Carleton, W. A. Traub, M. G. Lacasse, P. Nisenson, M. R. Pearlman,
     R. D. Reasenberg, Xingi Xu, C. Coldwell, A. Panasyuk, J. A. Benson,
     C. Papaliolios, R. Predmore, F. P. Schloerb, H. M. Dyck and D. Gibson,
     "Current Status of the IOTA Interferometer," *Proceedings of the SPIE*, **2200**,
     152–165 (June 1994).

[52] S. Robbe, B. Sorrente, F. Cassaing, Y. Rabbia and B. Rousset, "Performance
     of the Angle of Arrival Correction System of the I2T + ASSI Stellar
     Interferometer," *Astronomy and Astrophysics*, **125**, 367–380 (1997).

[53] J. T. Armstrong, D. Mozurkewich, L. J. Rickard, D. J. Hutter, J. A. Benson,
     P. F. Bowers, N. M. Elias II, C. A. Hummel, K. J. Johnston,D. F. Buscher,
     J. H. Clark III, L. Ha, L. C. Ling, N. M. White and  R. S. Simon, "The Navy
     Prototype Optical Interferometer," *The Astrophysical Journal*, **496**, 550–571
     (March 1998).

[54] M. M. Colavita, J. K. Wallace, B. E. Hines, Y. Gursel, F. Malbet,
     D. L. Palmer, X. P. Pan, M. Shao, J. W. Yu, A. F. Boden, P. J. Dumont,
     J. Gubler, C. D. Koresko, S. R. Kulkarni, B. F. Lane, D. W. Mobley and
     G. T. van Belle, "The Palomar Testbed Inter- ferometer", *The Astrophysical
     Journal*, **510**, 505–521 (January 1999).

[55] M. M. Colavita, A. F. Boden, S. L. Crawford, A. B. Meinel, M. Shao,
     P. N. Swanson, G. T. van Belle, G. Vasisht, J. M. Walker, J. K. Wallace and
     P. L. Wizinowich, "The Keck Interferometer," *Proceedings of the SPIE*,
     **3350**, 776–784 (March 1998).

[56] J. Davis, W. J. Tango, A. J. Booth, T. A. ten Brummelaar, R. A. Minard and
     S. M. Owens, "The Sydney University Stellar Interferometer – I. The
     Instrument," *Monthly Notices of the Royal Astronomical Society*, **303**,
     773–782 (1999).

[57] H. A. McAlister, W. G. Bagnuolo, T. ten Brummelaar, W. I. Hartkopf,
     M. A. Shure, L. Sturmann and N. H. Turner, "Progress on the CHARA
     Array," *Proceedings of the SPIE*, **3350**, 947–950 (March 1998).

[58] J. M. Hill and P. Salinari, "The Large Binocular Telescope Project,"
     *Proceedings of the SPIE*, **3352**, 23–33 (1998).

[59] D. Enard, "E.S.O. VLT Project I: A Status Report," *Proceedings of the SPIE*, **1236**, 63–70 (1990).

[60] J. B. Breckinridge, "Measurement of the Amplitude of Phase Excursions in the Earth's Atmosphere," *Journal of the Optical Society of America*, **66**, No. 2, 143–144 (February 1976).

[61] N. S. Nightingale and D. F. Buscher, "Interferometric Seeing Measurements at the La Palma Observatory," *Monthly Notices of the Royal Astronomical Society*, **251**, 155–166 (1991).

[62] C. A. Haniff, J. E. Baldwin and P. J. Warner, "Atmospheric Phase Fluctuation Measurement: Interferometric Results from the WHT and COAST Telescopes," *Proceedings of the SPIE*, **2200**, 407–417 (June 1994).

[63] J. M. Mariotti and G. P. Di Benedetto, "Pathlength Stability of Synthetic Aperture Telescopes: The Case of the 25 cm CERGA Interferometer," in *Proceedings of the International Astronomical Union Colloquium No. 79: Very Large Telescopes, the Instrumentation and Programs* (Garching, 9–12 April 1984), 257–265.

[64] C. E. Coulman and J. Vernin, "Significance of Anisotropy and the Outer Scale of Turbulence for Optical and Radio Seeing," *Applied Optics*, **30**, No. 1, 118–126 (January 1991). (See especially Figure 2 of this paper.)

[65] J. Davis, P. R. Lawson, A. J. Booth, W. J. Tango and E. D. Thorvaldson, "Atmospheric Path Variations for Baselines up to 80 m Measured with the Sydney University Stellar Interferometer," *Monthly Notices of the Royal Astronomical Society*, **273**, L53–L58 (1995).

[66] M. M. Colavita, M. Shao and D. H. Staelin, "Atmospheric Phase Measurements with the Mark III Stellar Interferometer," *Applied Optics*, **26**, No. 19, 4106–4112 (October 1987).

[67] A. D. Wheelon, "Relation of Radio Measurements to the Spectrum of Tropospheric Dielectric Fluctuations," *Journal of Applied Physics*, **28**, No. 6, 684–693 (June 1957).

# 6

# The Temporal Variation of Phase

The microwave measurement of the single-path phase reproduced in Figure 4.1 shows that there are significant variations with time. Similar variability is observed in the phase difference measured between adjacent receivers. The phase of an electromagnetic signal varies with time because the dielectric constant is changing along the path. These phase variations influence the performance and design of astronomical instruments. They suggest appropriate exposure times for telescope measurements. They often indicate suitable data-processing approaches for GPS distance and interferometric angular measurements. The spectrum of phase variations is needed to design wave-front correctors for large telescopes and fringe trackers for interferometers. Temporally variable phase fluctuations define the accuracy limit for angular rates that can be extracted from microwave tracking systems. For the scientist, they reveal the movements of the random medium through which a signal has passed.

## 6.1 Atmospheric Variability

Phase variability is related to changing dielectric fluctuations along the path through the basic relation (4.1). In this chapter we shall consider only propagation in the lower atmosphere. This shifts our attention to the refractive index and we set $\Delta\epsilon = 2\,\delta n$ in the basic relation. The time-shifted phase covariance is the basic building block for describing temporal variability:

$$\langle \varphi(t)\varphi(t+\tau) \rangle = k^2 \int_0^R dx \int_0^R dx' \, \langle \delta n(x,t)\, \delta n(x',t+\tau) \rangle \qquad (6.1)$$

The ensemble-averaged product of two refractive-index irregularities measured at different times and places is evidently the key to describing phase

240

variability. It is determined primarily by winds in the troposphere and we pause to discuss that meteorological matter briefly.

### 6.1.1 Atmospheric Wind Fields

Temperature and water vapor are the primary determinants of the refractive index. Both are passive scalar quantities, which means that $\delta n$ is determined by the diffusion equation

$$\frac{\partial}{\partial t}\delta n + \mathbf{v}(\mathbf{r},t) \cdot \boldsymbol{\nabla}\ \delta n = D\,\nabla^2 \delta n \qquad (6.2)$$

The central role played by the wind vector $\mathbf{v}(\mathbf{r},t)$ is made plain in this relation. It provides the turbulent motion which creates the irregularities and the mechanisms by which they are moved and rearranged.

Meteorological measurements show that the wind vector is almost always horizontal. It can be represented by a constant wind-speed vector and a variable component as indicated in Figure 6.1.

$$\mathbf{v}(\mathbf{r},t) = \mathbf{v}_0(\mathbf{r}) + \delta\mathbf{v}(\mathbf{r},t) \qquad (6.3)$$

The steady component varies with location and height in the troposphere. During the time periods used for most propagation experiments it can be considered constant. Surface wind values are often substituted for it in making estimates but the height profile of wind vectors is important when

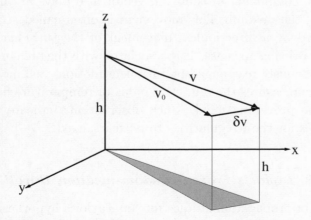

Figure 6.1: The horizontal wind vector at height h which is decomposed into a steady component and a random component.

we consider signals from distant sources that transit the entire atmosphere. Wind profiles are routinely measured by instrumented balloon flights up to altitudes as great as 30 km. Over a thousand such flights are made each day at various points around the world as part of an international meteorological cooperation among ninety-two nations [1]. These profiles can now be recorded more frequently with electromagnetic sounding techniques using radar, lidar and GPS signals.

The variable wind component $\delta\mathbf{v}(\mathbf{r}, t)$ is usually associated with turbulent conditions. It is small relative to the average wind speed because only a small fraction of the latter's energy is converted into turbulence. It is usually about 10% of the steady wind speed and its primary effect is to generate random changes in the total wind speed. It can change the wind direction when it is comparable to the steady component.

The differential equation which describes the generation and evolution of refractive-index irregularities emerges from the combination of (6.2) and (6.3):

$$\frac{\partial}{\partial t}\delta n + [\mathbf{v}_0(\mathbf{r}) + \delta\mathbf{v}(\mathbf{r}, t)] \cdot \boldsymbol{\nabla}\,\delta n = D\,\nabla^2\delta n \qquad (6.4)$$

This equation cannot be solved generally since the three components of $\delta\mathbf{v}$ are stochastic functions of position and time. On the other hand, it allows one to identify the primary roles played by the two wind components. The average wind vector is primarily responsible for moving the irregularities. This is called *advection*. The variable component is the only term with random properties. It is responsible for creating the irregularities and moving them in combination with the mean wind.

The variable component of the wind vector also plays an important role in rearranging the medium. This movement is often called *self motion* and is characterized by a directionless movement of the turbulent atmosphere relative to the drifting air mass. It is associated with the breakup of eddies as they are continuously rearranged by turbulent motion. Self motion becomes apparent if one measures the refractive index or temperature from a balloon drifting on the prevailing wind. Such observations indicate that it takes several minutes for the irregularities to be rearranged.

### 6.1.2 Taylor's Frozen-random-medium Hypothesis

The majority of propagation studies rely on Taylor's hypothesis to describe the temporal changes in a random medium [2]. This approximation makes two important assumptions. It postulates that the entirety of the turbulent medium is frozen during the measurement interval. It also assumes that one can ignore the variable component of velocity $\delta \mathbf{v}(\mathbf{r}, t)$ so the wind velocity is constant at a given location. In combination, these two assumptions imply that the entire air mass is transported horizontally at constant speed without being deformed. When this model is applied to electromagnetic propagation Figure 6.2 shows that this bulk motion of the random medium is equivalent to moving the ray path parallel to itself through a stationary atmosphere. The phase covariance measured at $t$ and $t + \tau$ should be identical to the spatial correlation between parallel rays separated by a vector

$$\boldsymbol{\rho} = \mathbf{v}\tau. \tag{6.5}$$

This intuitive equivalence can be demonstrated analytically by making the coordinate translation $\mathbf{r} = \mathbf{r}' - \mathbf{v}t$ in the diffusion equation (6.4). One can write the temporal and spatial covariance of refractive-index variations in terms of the wavenumber spectrum by adding this additional displacement to (4.2) and neglecting the variation of the spectrum with height:

$$\langle \delta n(\mathbf{r}, t) \, \delta n(\mathbf{r} + \mathbf{R}, t + \tau) \rangle = \int d^3 \kappa \, \Phi_n(\boldsymbol{\kappa}) \exp[\, i\boldsymbol{\kappa} \cdot (\mathbf{R} + \tau\mathbf{v})] \tag{6.6}$$

This expression is consistently used in propagation studies and we will also rely heavily on it in our developments.

It is important to note that this approach is only an approximation to what happens in the real atmosphere. It clearly breaks down for large time delays because the random medium is gradually rearranged by the turbulent velocity field. Taylor suggested this simple model, but it was left to Tatarskii to explore the conditions under which it is a valid description [3]. Its accuracy depends on the type of measurement that is being made.

Taylor's hypothesis provides an accurate description for temperature and refractivity measurements made with airborne sensors flown on straight-line flight paths [4][5]. In these situations the sensor passes through the medium at the aircraft's speed, which is usually more than ten times the wind speed. A reasonable volume of the atmosphere is sampled before it can be rearranged by turbulent motion, which allows these experiments to be described by (6.6) over large distance scales.

Ground-based measurements require more detailed consideration since atmospheric winds now transport the frozen irregularities. This situation can be visualized using Figure 6.2 if one imagines that a temperature sensor or refractometer is co-located with the receiver. For Taylor's hypothesis to

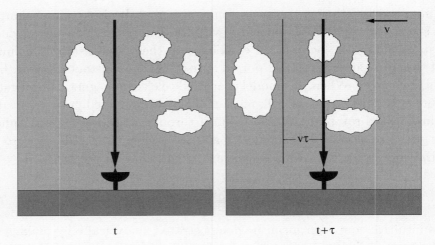

t                                    t+τ

Figure 6.2: A frozen random medium carried past the line of sight on the prevailing wind, illustrating Taylor's hypothesis.

be valid, the volume of air being sampled must pass over the sensor before it can evolve significantly. Tatarskii examined this question using dimensional analysis [3]. For eddies that are smaller than the outer scale length, he noted that the characteristic eddy speed is related to its size by

$$v(\ell) = (\mathcal{E}\ell)^{\frac{1}{3}}$$

where $\mathcal{E}$ is the dissipation rate for turbulent energy in the cascade process. This suggests that the intrinsic lifetime or *evolution time* of an eddy is represented by the following ratio:

$$T_{\text{ev}} = \frac{\ell}{v(\ell)} = \frac{\ell^{\frac{2}{3}}}{\mathcal{E}^{\frac{1}{3}}} \tag{6.7}$$

The random medium can be considered frozen only if the evolution time is much greater than the crossing time. The latter is set by the average wind speed:

$$\text{Single point}: \quad T_{\text{cr}} = \ell/v_0$$

The time comparison can be expressed in terms of the velocity components:

$$\text{Taylor's Hypothesis}: \quad v_0 \gg (\mathcal{E}\ell)^{\frac{1}{3}} = v(\ell) \tag{6.8}$$

This is satisfied for eddies smaller than the outer scale length and the condition can also be expressed in terms of scale lengths:

$$\text{Taylor's Hypothesis:} \quad \ell < L_0 \tag{6.9}$$

Taylor's model can therefore be used both in the inertial and in the dissipation ranges of the turbulence spectrum.

That prediction was confirmed in experiments performed by Tsvang [6]. He measured the power spectrum of temperature variations 70 m above the ground using one sensor mounted on a tower. An identical sensor was carried by an aircraft that flew nearby at the same height. The power spectra of these two data sets were identical for scale sizes as large as 1000 m, thereby confirming Taylor's hypothesis for an even greater range of eddy sizes than that suggested above. Gossard made a similar comparison between temperature sensors and microwave refractometers carried on a tethered balloon and identical instruments carried on a passing aircraft at altitudes up to 1000 m [5]. These results give considerable confidence that expression (6.6) can be used to describe the spatial and temporal covariance for single-point measurements.

The same approximation has also been used consistently to describe microwave interferometers. However, in these applications one must ask whether the random medium will *stay frozen* as it travels from one receiver to another. This is a difficult test because the crossing time is now determined by the separation between receivers.

$$\text{Interferometer:} \quad T_{\text{cr}} = \rho/v_0$$

For the Very Large Array at the NRAO the largest spacing is 35 km and the crossing time can be as long as an hour. This gives the irregularities ample time to evolve as they make the transit. On the other hand, we learned in Chapter 5 that the angular resolution of these large arrays is determined primarily by eddies whose horizontal size is comparable to the inter-receiver separation. Only those large structures can influence both signals simultaneously and their evolution time provides the condition

$$\frac{\rho}{v(\rho)} = \frac{\rho^{\frac{2}{3}}}{\mathcal{E}^{\frac{1}{3}}} \gg \frac{\rho}{v_0}.$$

This leads again to (6.8) with $v(\ell)$ replaced by $v(\rho)$. Taylor's model ought to describe such data so long as the mean wind speed is a good deal greater than the turbulent velocity of the very large horizontal eddies. That requirement focuses our attention on the input range of the turbulent velocity spectrum, for which we have no reliable physical theory. The speed of these very large eddies is not well defined and we cannot be sure that

$$v_0 \gg v(\rho) \approx v(L_0) \qquad (6.10)$$

A second problem is that the mean wind speed itself can be different at the individual receivers of a large array and can change during the long sample lengths used to make astronomical observations. For these reasons, one should use the locally frozen approximation presented below combined with actual wind data to interpret such experiments.

By contrast, optical interferometers employ baselines less than 100 m because it is difficult to transmit and compare coherent optical signals over large distances. The crossing time for these short baselines is usually less than the evolution time for the relevant eddies and Taylor's hypothesis can be used to analyze such data. The equivalence of spatial and temporal phase correlations has an important consequence for observations with optical interferometers. One can duplicate the spatial correlation that would be measured with a large array of receivers by using temporally displaced data taken on a single path. The economy of this approach indicates why temporal measurements are often preferred.

It is important to note that the accuracy of Taylor's model improves as the mean wind speed increases, since the blobs are transported more rapidly over the receivers in an interferometer and therefore have less time to be rearranged. It evidently falls apart when the mean wind drops to zero. This happens on still days and one cannot then neglect the variable wind component.

### 6.1.3 The Locally Frozen Random-medium Approximation

Tatarskii introduced the concept of a *locally frozen medium* to provide a more accurate description of atmospheric variability [7]. This approach abandons both assumptions that were used in framing Taylor's hypothesis and recognizes the variable wind component in (6.3) – which Taylor ignored. It assumes only that a small volume element remains frozen in time, whereas the entire random medium was fixed in the simpler model. Typical trajectories for these frozen eddies are illustrated in Figure 6.3.

One must first reconcile the variable wind component with the assumption that the local refractive-index structure remains fixed. To do this we

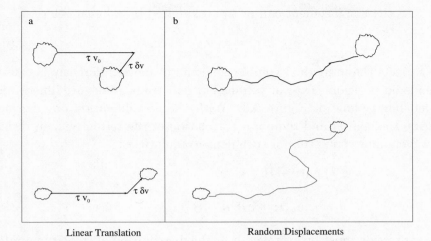

a

b

Linear Translation

Random Displacements

Figure 6.3: The locally frozen random-medium approximation illustrated for (a) the linear translation expected for small time delays, and (b) the more complicated trajectories that occur for larger time delays.

assume that the random velocity component and refractive-index fluctuations are statistically independent and uncorrelated.[1] In that case the turbulent velocity can move the eddies by advection but need not change their configuration. The frozen eddies follow complicated curvilinear paths as suggested in the right-hand panel of Figure 6.3. These trajectories are determined by the velocity at each point along the path. The assumption that the eddy remains frozen as it moves along this trajectory means that the values of $\delta n$ at successive times are the same:

$$\delta n(\mathbf{r}, t) = \delta n[\mathbf{r} + \boldsymbol{\varrho}(\mathbf{r}, t, \tau), t + \tau] \qquad (6.11)$$

The random vector $\boldsymbol{\varrho}(\mathbf{r}, t, \tau)$ describes the displacement of a volume element in the time interval $\tau$. Our task is to relate this displacement to the velocity components.

The first applications of this method assumed that the delay time was small relative to the time during which the velocity changes appreciably. In

---

[1] The lack of correlation can be proven if the fluid is incompressible (i.e. $\boldsymbol{\nabla} \cdot \delta \mathbf{v} = 0$) and both $\delta n$ and $\delta \mathbf{v}$ are isotropic [8]. Notice that independence of these two quantities assures their decorrelation since the mean values of both variables vanish. We accept their independence for now as an assumption. It would be useful to confirm this with simultaneous measurements of $\delta n$ and $\delta \mathbf{v}$ in the real atmosphere, especially in view of their relationship, which is evident in the diffusion equation (6.4).

that case the displacement can be approximated by a linear relation:

$$\boldsymbol{\varrho}(\mathbf{r}, t, \tau) = \tau[\mathbf{v}_0 + \delta\mathbf{v}(\mathbf{r}, t)] \tag{6.12}$$

One expects the individual volume elements to move differently because $\delta\mathbf{v}$ depends on position. Typical volume-element translations are illustrated in the left-hand panel of Figure 6.3. With this simplification one can again describe the spatial and temporal correlation of the refractivity in terms of the wavenumber spectrum of turbulent irregularities:

$$\langle \delta n(\mathbf{r}, t)\, \delta n(\mathbf{r} + \mathbf{R}, t + \tau) \rangle$$

$$= \int d^3\kappa\, \Phi_n(\boldsymbol{\kappa}) \exp\{i\boldsymbol{\kappa} \cdot [\mathbf{R} + \tau\mathbf{v}_0 + \tau\, \delta\mathbf{v}(\mathbf{r}, t)]\} \tag{6.13}$$

This expression must be averaged over the distribution of velocity variations before it can be used to describe refractivity measurements. This averaging involves only the exponential term:

$$\langle \delta n(\mathbf{r}, t)\, \delta n(\mathbf{r} + \mathbf{R}, t + \tau) \rangle = \int d^3\kappa\, \Phi_n(\boldsymbol{\kappa}) \exp\{i\boldsymbol{\kappa} \cdot [\mathbf{R} + \tau\mathbf{v}_0]\}$$

$$\times\, \langle \exp[i\tau\boldsymbol{\kappa} \cdot \delta\mathbf{v}(\mathbf{r}, t)] \rangle_{\delta v} \tag{6.14}$$

The last term is the *characteristic function* for the random velocity components. We have no universal theory that can define this function and must resort to models to represent it.

A three-dimensional Gaussian distribution with spherical symmetry has been assumed for the probability density function of the variable velocity components in several studies [7][9]:

$$\mathsf{P}(\delta\mathbf{v}) = \frac{1}{(2\pi\sigma_v^2)^{\frac{3}{2}}} \exp\left(-\frac{1}{2\sigma_v^2}\left(\delta v_x^2 + \delta v_y^2 + \delta v_z^2\right)\right) \tag{6.15}$$

This representation is reasonable so long as the average wind speed is much larger than the random component. That condition is usually satisfied in the lower atmosphere since the energy transferred to turbulent motion is but a fraction of that transferred by the average wind field. It is a simple matter to calculate the characteristic function with (6.15) using integrals that can be found in Appendix C:

$$\langle \exp[i\tau\boldsymbol{\kappa} \cdot \delta\mathbf{v}(\mathbf{r}, t)] \rangle_{\delta v} = \exp\left(-\frac{1}{2}|\boldsymbol{\kappa}|^2\tau^2\sigma_v^2\right) \tag{6.16}$$

On combining this with (6.14) we find that the temporally and spatially shifted covariance is given by the following expression:

$$\langle \delta n(\mathbf{r}, t)\, \delta n(\mathbf{r} + \mathbf{R}, t + \tau) \rangle = \int d^3\kappa\, \Phi_n(\boldsymbol{\kappa}) \exp[\, i\boldsymbol{\kappa} \cdot (\mathbf{R} + \tau \mathbf{v}_0)]$$

$$\times \exp\left(-\frac{1}{2}|\boldsymbol{\kappa}|^2 \tau^2 \sigma_v^2\right) \tag{6.17}$$

The Gaussian distribution of turbulent velocity components was validated by measurements of temperature made in the USSR [10]. The spectrum and co-spectrum were calculated from (6.17) and compared with the same quantities measured for various separations of sensors. The agreement was generally good with only a small indication of eddy evolution. We will use this model when we need to address problems that are sensitive to velocity fluctuations.

### 6.1.4 More General Descriptions

The linear approximation for the volume displacement (6.12) breaks down as the time delay increases. The random velocity component carries the volume element first this way and then that, in which case movements of the frozen cells are better described by the curvilinear trajectories illustrated in the second panel of Figure 6.3. The random displacement vector is now expressed as an integral of the wind velocity evaluated at all prior positions along the trajectory:

$$\varrho(\mathbf{r}, t, \tau) = \int_t^{t+\tau} dt'\, \{\mathbf{v}_0 + \delta \mathbf{v}[\mathbf{r} + \varrho(\mathbf{r}, t, t'), t + t']\} \tag{6.18}$$

In writing (6.12) we neglected changes in the velocity fluctuations that occur during the movement of the frozen eddies. We need not have done so. If we ignore changes in the frozen volume's displacement $\varrho(\mathbf{r}, t, t')$ inside the time integral of (6.18), we can write

$$\varrho(\mathbf{r}, \tau) = \tau \mathbf{v}_0 + \int_0^\tau dt\, \delta \mathbf{v}(\mathbf{r}, t). \tag{6.19}$$

The experimental support for (6.15) indicates that the components of $\delta\mathbf{v}$ are distributed as Gaussian random variables. The integration of these components over time is represented by linear combinations of these values and the components of the second term should also be Gaussian random variables.

With this observation, we can take the ensemble average of the velocity fluctuations to write

$$\left\langle \exp\left( i\boldsymbol{\kappa} \cdot \int_0^\tau dt'\, \delta\mathbf{v}(\mathbf{r}, t') \right) \right\rangle_{\delta v}$$

$$= \exp\left( -\frac{1}{2} |\boldsymbol{\kappa}|^2 \sigma^2 \int_0^\tau dt' \int_0^\tau dt'\, C(|t - t'|) \right) \qquad (6.20)$$

where $C(t)$ is the temporal correlation of the $\delta v_i$ and $\sigma^2$ is their common variance. Although there is no reliable physical model for the temporal correlation, we have a general impression of its behavior. We believe that it is close to unity for temporal separations less than a *decorrelation time* T that describes the lack of similarity between values of the turbulent velocity measured at the same point at different times. The velocity correlation should fall rapidly to zero for temporal separations greater than the decorrelation time. This is enough information to suggest the general behavior of the characteristic function for time delays that are considerably smaller and larger than T:

$$\left\langle \exp\left( i\boldsymbol{\kappa} \cdot \int_0^\tau dt'\, \delta\mathbf{v}(\mathbf{r}, t') \right) \right\rangle_{\delta v} = \begin{cases} \tau \ll \mathrm{T}: & \exp(-\frac{1}{2} |\boldsymbol{\kappa}|^2 \sigma^2 \tau^2) \\[2mm] \tau \gg \mathrm{T}: & \exp(- |\boldsymbol{\kappa}|^2 \sigma^2 \tau\, \mathrm{T}) \end{cases} \qquad (6.21)$$

Both forms have the same wavenumber dependence but their variations with the delay time are quite different. This change should be reflected in temperature and refractivity measurements through (6.14).

Nalbandyan carried this analysis to the next level by retaining $\varrho(\mathbf{r}, t, t')$ in the temporal integration of (6.18). He used functional derivatives to express the temporal and spatial covariance of the refractivity in terms of the turbulent velocity field's characteristics [11]. This more refined description is quite new and is just beginning to be applied to studies of electromagnetic propagation.

## 6.2  The Single-path Variability of Phase

We consider first an electromagnetic signal that travels along a single ray path connecting the transmitter and the receiver. Its phase varies in response to the changing combination of irregularities along the path. There are several ways to describe time-varying phase fluctuations. Different approaches are chosen to deal with the variety of problems encountered in making these difficult measurements. We address the techniques separately.

There is a considerable body of single-path phase measurements in which temporal variability has been investigated. In the vast majority of these experiments terrestrial paths with the signal traveling near the surface were used. Most of this data comes from the microwave experiments identified in Table 4.1. Important optical measurements using laser signals over relatively short paths have also been made. One can generally ignore variations of the spectrum along the path in these situations.

A growing number of experiments are using GPS satellite signals to measure phase variations induced by the troposphere. These signals travel through the entire atmosphere, which requires recognition of the variability of $C_n^2$ and $\kappa_0$ with altitude in order to interpret them.

### 6.2.1 Autocorrelation of Phase

The *autocorrelation* function is one way to describe the variability of phase fluctuations. It is simply the phase covariance normalized by the mean square value:

$$\mu(\tau) = \frac{\langle \varphi(t+\tau)\varphi(t)\rangle}{\langle \varphi^2(t)\rangle} \tag{6.22}$$

The phase covariance is calculated by combining the basic expressions (6.1) and (6.6). As a first approximation we assume that the wind vector and turbulence spectrum do not change with time and position:

$$\langle \varphi(t)\varphi(t+\tau)\rangle = k^2 \int d^3\kappa \, \Phi_n(\boldsymbol{\kappa}) e^{i\tau\boldsymbol{\kappa}\cdot\mathbf{v}} \int_0^R dx \int_0^R dx' \, \exp\left[i\kappa_x(x-x')\right]$$

The double path integration was evaluated in (5.27) to give a delta function that drives the line-of-sight wavenumber component to zero:

$$\langle \varphi(t)\varphi(t+\tau)\rangle = 2\pi k^2 R \int d^3\kappa \, \Phi_n(\boldsymbol{\kappa})\delta(\kappa_x) \exp(i\tau\boldsymbol{\kappa}\cdot\mathbf{v}) \tag{6.23}$$

The wind vector is naturally expressed in the rectangular coordinates of Figure 4.2:

$$\boldsymbol{\kappa}\cdot\mathbf{v} = \kappa_x v_x + \kappa_y v_y + \kappa_z v_z$$

For links near the surface we can assume that the irregularities are isotropic. In this case, it is convenient to use spherical wavenumber coordinates centered on the line-of-sight axis to express the wavenumber integration:

$$\langle\varphi(t)\varphi(t+\tau)\rangle = 2\pi Rk^2 \int_0^\infty d\kappa\; \kappa^2 \Phi_n(\kappa) \int_0^\pi d\psi \sin\psi\, \delta(\kappa\cos\psi)$$
$$\times \int_0^{2\pi} d\omega \exp[\,i\tau\kappa\sin\psi\,(v_y\sin\omega + v_z\cos\omega)]$$

The $\psi$ integration collapses and the $\omega$ integral gives a Bessel function:

$$\langle\varphi(t)\varphi(t+\tau)\rangle = 4\pi^2 Rk^2 \int_0^\infty d\kappa\; \kappa\Phi_n(\kappa) J_0\!\left(\kappa\tau\sqrt{v_y^2 + v_z^2}\right) \qquad (6.24)$$

This result depends on the velocity components that are normal to the line of sight. The wind component along the path has no influence. The wind blows horizontally in most situations of interest:

$$\langle\varphi(t)\varphi(t+\tau)\rangle = 4\pi^2 Rk^2 \int_0^\infty d\kappa\; \kappa\Phi_n(\kappa) J_0(\kappa\tau v_y) \qquad (6.25)$$

For notational simplicity, we often drop the subscript on the horizontal speed and replace $v_y$ by $v$. Notice that we could have established the same result by substituting (6.5) into the spatial correlation for parallel rays given by (5.18).

To proceed further one must know more about the random medium. The small-wavenumber portion of the turbulence spectrum is important for temporal correlations – as it was for the phase variance. In previous chapters we have used the von Karman model to provide a soft transition to the large-eddy portion of the spectrum and we will do so again here. The temporal covariance for this model can be inferred from (5.25) by setting $\rho = v\tau$:

$$\langle\varphi(t)\varphi(t+\tau)\rangle = 0.782Rk^2 C_n^2 \kappa_0^{-\frac{5}{3}} \left(\frac{2^{\frac{1}{6}}}{\Gamma\!\left(\frac{5}{6}\right)}(\kappa_0 v\tau)^{\frac{5}{6}} K_{\frac{5}{6}}(\kappa_0 v\tau)\right) \qquad (6.26)$$

The function in large parentheses defines the autocorrelation $\mu(\tau)$ and is the solid curve plotted in Figure 6.4.

### 6.2.1.1 Optical Measurements of Autocorrelation

The autocorrelation of optical phase fluctuations was measured in the USSR [12][13][14] using the experimental arrangement shown in Figure 4.3. Phase

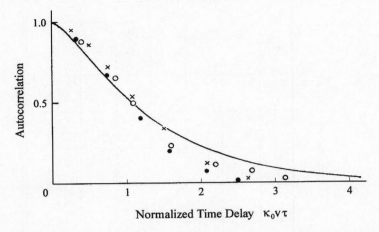

Figure 6.4: Predicted and measured values of the autocorrelation for single-path fluctuations experienced on a terrestrial path. The solid curve corresponds to the von Karman model. The experimental data was taken with a helium–neon laser over 95-m (crosses) and 200-m (circles) paths by Lukin, Mironov, Pokasov and Khmelevtsov [12].

histories were recorded for path lengths of 95 and 200 m. Time delays of up to 0.5 s were used to create the autocorrelation from these histories. Time was converted to the dimensionless variable $\kappa_0 v \tau$ by (*a*) using the measured wind speed $v = 1.2$ m s$^{-1}$ and (*b*) estimating that $\kappa_0 = 5$ m$^{-1}$ from the path height. The resulting data is plotted in Figure 6.4 and agrees with the prediction in the range $\kappa_0 v \tau < 1$. For larger values the measured values fall faster than does the prediction and suggest that the data could become negative.[2] Negative values would contradict the prediction (6.26), which is everywhere positive.

### *6.2.1.2 Microwave Measurements of Autocorrelation*

Careful measurements of $\mu(\tau)$ were also made using 9.4-GHz signals [15][16] over the 25-km elevated path shown in Figure 4.9. The phase autocorrelation was measured at four different times within a 24-h period and the data is reproduced in Figure 6.5. A significant diurnal variability, which corresponds to changing turbulent conditions and wind vectors along the path, is evident. These measurements give *both* positive and negative values, which contradicts the prediction of (6.26).

---

[2] The first data published from this series of experiments [13] showed the autocorrelation going negative at about $\kappa_0 v \tau = 3$. Later publications omitted negative values and gave only the data reproduced in Figure 6.4.

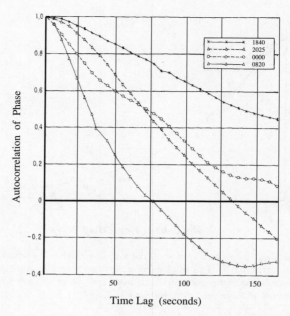

Figure 6.5: Autocorrelations of single-path phase fluctuations measured at 9.4 GHz on Maui by Norton, Herbstreit, Janes, Hornberg, Peterson, Barghausen, Johnson, Wells, Thompson, Vetter and Kirkpatrick [15].

This mismatch between prediction and measured data is caused by our neglect of the important influence of the finite sample length which was used in processing the data. Good agreement between theory and experiment is achieved when the sample length or integration interval is properly included. To explain this effect, we recall how the time histories of phase were converted into autocorrelation in this experiment. Data samples between 15 and 22 min long were first selected from the data stream. These samples were detrended by subtracting apparent phase trends as explained in Section 6.2.5. An average value was computed for each sample using

$$\overline{\varphi} = \frac{1}{T} \int_0^T dt \, \varphi(t).$$

The integrations were approximated by averaging phase values read every 0.2 s. The time-shifted autocorrelation was estimated from

$$\mu(\tau, T) = \frac{1}{\mathcal{N}(T - \tau)} \int_0^{T-\tau} dt \, \langle [\varphi(t + \tau) - \overline{\varphi}] \cdot [\varphi(t) - \overline{\varphi}] \rangle \qquad (6.27)$$

where the normalization is just the phase variance:

$$\mathcal{N} = \frac{1}{T} \int_0^T dt \, \langle [\varphi(t) - \overline{\varphi}]^2 \rangle \qquad (6.28)$$

The foreshortened integration interval $T - \tau$ was used to ensure that the original and delayed phase histories contained the same points. By combining these expressions we can write

$$\mu(\tau, T)\mathcal{N} = \langle \varphi(t+\tau)\varphi(t) \rangle + \frac{1}{T^2} \int_0^T dt \int_0^T dt' \, \langle \varphi(t) \, \varphi(t') \rangle$$

$$- \frac{1}{T(T-\tau)} \int_0^T dt \int_0^{T-\tau} dt' \, [\langle \varphi(t)\varphi(t'+\tau) \rangle + \langle \varphi(t)\varphi(t') \rangle].$$

The ensemble averages that occur in this expression correspond to an infinitely long sample length and we can replace them by their wavenumber representation (6.25). The following expression emerges after reversing wavenumber and time integrations:

$$\mu(\tau, T) = \frac{4\pi^2 R k^2}{\mathcal{N}} \int_0^\infty d\kappa \, \kappa \Phi_n(\kappa)\Lambda(\kappa, \tau, T) \qquad (6.29)$$

where the weighting function is defined by

$$\Lambda(\kappa, \tau, T) = J_0(\kappa v \tau) - \frac{2\tau}{T^2(T-\tau)} \int_0^T d\zeta \, (T - \zeta)J_0(\kappa v \zeta)$$

$$+ \frac{2}{T(T-\tau)} \int_0^\tau d\zeta \, (\tau - \zeta)J_0(\kappa v \zeta)$$

$$- \frac{2}{T(T-\tau)} \int_0^{T-\tau} d\zeta \, (T - \tau - \zeta)J_0(\kappa v \zeta). \qquad (6.30)$$

The sample-length-modified autocorrelation was computed from these expressions using the von Karman spectral model. The results are presented in Figure 6.6, in which the autocorrelation is plotted as a function of the time shift $\tau$ divided by the sample length $T$. Curves for several values of the dimensionless parameter $\kappa_0 v T$ are provided. It is clear that a finite sample length drives the autocorrelation to negative values in each case.

To relate these results to the microwave data reproduced in Figure 6.5 we note that the maximum time shift was 160 s, so our attention is focused on the region $\tau < 0.2\,T$. The curve corresponding to $\kappa_0 v T = 10$ goes negative

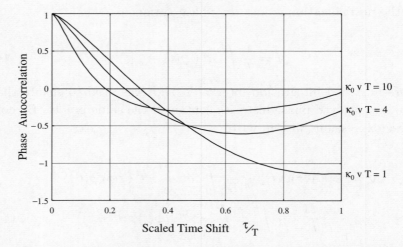

Figure 6.6: The effect of a finite sample length on the autocorrelation of phase fluctuations expressed as a function of the time delay divided by the sample length. These curves were computed with the von Karman model for several values of the dimensionless parameter $\kappa_0 v T$.

in this range whereas the others do not. Comparing this with the data taken at 0820 and 2025 suggests that

$$10 < \kappa_0 v T < 15.$$

The sample length was approximately 20 min and the cross-path wind speed was about 10 m s$^{-1}$, providing the following estimate for the horizontal outer scale length:

$$L_0 \approx 5-8 \text{km}$$

This is consistent with the values measured by airborne refractometers that were discussed in Chapter 2.

### 6.2.2 The Temporal Structure Function

The spatial structure function was introduced in Chapter 5 to describe the phase difference measured between spaced receivers. This convention removes the influence of large eddies and phase trends, which are often common to adjacent paths. It is tempting to use the same approach to describe time-displaced phase measurements:

$$\mathcal{D}_\varphi(\tau) = \langle [\, \varphi(t+\tau) - \varphi(t)]^2 \rangle \tag{6.31}$$

An explicit expression emerges when we substitute the time-shifted covariance (6.26) into this definition:

$$\mathcal{D}_\varphi(\tau) = 1.564 R k^2 C_n^2 \kappa_0^{-\frac{5}{3}} \left( 1 - \frac{2^{\frac{1}{6}}}{\Gamma\left(\frac{5}{6}\right)} (\kappa_0 v \tau)^{\frac{5}{6}} K_{\frac{5}{6}}(\kappa_0 v \tau) \right) \tag{6.32}$$

When $\kappa_0 v \tau$ is considerably less than unity the small-argument expansion for the MacDonald function found in Appendix D shows that the structure function is independent of the outer scale length:

$$\kappa_0 v \tau \ll 1: \qquad \mathcal{D}_\varphi(\tau) = \left( 2.914 R k^2 C_n^2 v^{\frac{5}{3}} \right) \tau^{\frac{5}{3}} \tag{6.33}$$

This prediction was confirmed in a pioneering experiment performed by Bouricius and Clifford [17]. They used a helium–neon laser with a Michelson interferometer. The laser signal was divided into two components by a beam splitter. One component acted as a reference, while the other traveled 25 m through the atmosphere and was reflected back to the transmitter. The reference signal was retarded and then compared with the returning signal to provide interference fringes from which the phase difference was extracted. The time difference was varied by changing the retardation of the reference signal. The 5/3 scaling law was confirmed for time delays of less than 1 s. The structure function started to bend over for larger delays, as one would expect from Figure 5.3. Similar experiments were performed in the USSR using the apparatus illustrated in Figure 4.3. That data confirmed the 5/3 law for delays less than 0.2 s [13][14]. The difference in bend-over times was probably due to different wind speeds and outer scale lengths being encountered at the widely separated sites and times.

### 6.2.3 The Phase Power Spectrum

The power spectrum is the most common description of phase fluctuations. It emerges naturally when the stream of phase data is passed through a spectrum analyzer. The power spectrum is intimately connected to the temporal covariance by the Wiener–Khinchine theorem:[3]

$$W_\varphi(\omega) = \int_{-\infty}^{\infty} d\tau \, e^{i\omega\tau} \langle \varphi(t)\varphi(t+\tau) \rangle \tag{6.34}$$

---

[3] Some authors [18] used the convention

$$W_\varphi(\omega) = \int_0^\infty d\tau \, \cos(\omega\tau) \, \langle \varphi(t)\varphi(t+\tau) \rangle$$

because early experiments often analyzed data in this format. The development of fast Fourier transforms means that (6.34) is a natural convention for modern data reduction. It is also more consistent with symmetrical descriptions of Fourier integrals.

The covariance and power spectrum provide equivalent descriptions of the phase fluctuations. One can calculate the spectrum for all frequencies if one knows the correlation for all temporal separations. Conversely, one can calculate the covariance as the inverse Fourier integral if one knows the phase spectrum:

$$\langle \varphi(t)\varphi(t+\tau) \rangle = \frac{1}{2\pi} \int_{-\infty}^{\infty} d\omega \, e^{-i\omega\tau} W_\varphi(\omega) \qquad (6.35)$$

The power spectrum has an important advantage over the covariance and the temporal structure function. The following observation makes this point clear: "*Phase noise in a power spectrum measurement is uncorrelated between different frequency points, whereas the noise on the measured [temporal] structure function is highly correlated between adjacent time lags*" [19].

### 6.2.3.1 A Basic Description of the Power Spectrum

In reducing experimental data, the phase spectrum is often estimated by taking the Fourier transform of the covariance expression. For analytical purposes, it is simpler to substitute the wavenumber representation of the covariance (6.25) and reverse the order of integration. By doing so one can relate the phase spectrum directly to $\Phi_n(\kappa)$ and avoid a complicated integral:

$$W_\varphi(\omega) = \int_{-\infty}^{\infty} d\tau \, e^{i\omega\tau} \left( 4\pi^2 R k^2 \int_0^\infty d\kappa \, \kappa \Phi_n(\kappa) J_0(\kappa v \tau) \right)$$

$$= 8\pi^2 R k^2 \int_0^\infty d\kappa \, \kappa \Phi_n(\kappa) \int_0^\infty d\tau \cos(\omega\tau) \, J_0(\kappa v \tau) \qquad (6.36)$$

The cosine transform of the Bessel function is discontinuous,

$$\int_0^\infty d\tau \cos(\omega\tau) \, J_0(\kappa v \tau) = \begin{cases} \left( \kappa^2 v^2 - \omega^2 \right)^{-\frac{1}{2}} & \kappa v > \omega \\ 0 & \kappa v < \omega \end{cases}$$

and a general expression for the *propagation power spectrum* emerges in this way:

$$W_\varphi(\omega) = 8\pi^2 R k^2 \int_{\omega/v}^\infty \frac{d\kappa \, \kappa \Phi_n(\kappa)}{\left( \kappa^2 v^2 - \omega^2 \right)^{\frac{1}{2}}} \qquad (6.37)$$

The restricted wavenumber integration has a simple physical interpretation. An eddy of size $\ell$ is carried completely past the line of sight in a time $\ell/v$. It can influence frequencies that are smaller than the reciprocal transit time:

$$\omega < v/\ell$$

This means that the spectrum at $\omega$ is determined by wavenumbers greater than $\omega/v$. That is what our result requires.

We know that single-path phase fluctuations depend primarily on large refractive-index irregularities. This is the region for which we have no universal physical theory. To complete the calculation, however, we must make an assumption about the spectrum for small wavenumbers. In doing so we hope to validate or reject the assumed model by comparing its consequences with experimental data. The von Karman model gives

$$W_\varphi(\omega) = 2\,\pi^2 \times 0.132 R k^2 C_n^2 \int_{\omega/v}^{\infty} \frac{d\kappa\,\kappa}{\left(\kappa^2 + \kappa_0^2\right)^{\frac{11}{6}}\left(\kappa^2 v^2 - \omega^2\right)^{\frac{1}{2}}}$$

which can be evaluated by setting

$$\kappa = \frac{\omega}{v}\sqrt{\zeta^2 + 1}$$

to yield the following expression:

$$W_\varphi(\omega) = 2.606 R k^2 C_n^2 \frac{v^{\frac{5}{3}}}{\omega^{\frac{8}{3}}} \int_0^{\infty} \frac{d\zeta}{\left(\zeta^2 + 1 + \kappa_0^2 v^2/\omega^2\right)^{\frac{11}{6}}}$$

The remaining integral can be found in Appendix B:

$$W_\varphi(\omega) = \frac{2.192 R k^2 C_n^2 v^{\frac{5}{3}}}{\left(\omega^2 + \kappa_0^2 v^2\right)^{\frac{4}{3}}} \tag{6.38}$$

Most of the fluctuation energy is contained in the low-frequency portion of the spectrum, corresponding to the passage of large eddies through the line of sight. These take longer to cross the path than do small eddies. The low-frequency portion of the power spectrum is determined primarily by large eddies. The low-frequency measurements of the phase spectrum should therefore provide unique insight into the small-wavenumber portion of the turbulence spectrum [20]. In this range the power spectrum approaches a constant value:

$$\omega < \omega_0: \qquad W_\varphi(\omega) = 2.192 R k^2 C_n^2 v^{-1} \kappa_0^{-\frac{8}{3}} \tag{6.39}$$

One should thus be able to measure the combination $C_n^2 \kappa_0^{-\frac{8}{3}}$ from the level of the low-frequency spectrum given the local wind speed.

The predicted spectrum (6.38) changes its behavior markedly above the *threshold frequency*:

$$\omega_0 = \kappa_0 v \ \text{rad s}^{-1} \qquad \text{or} \qquad f_0 = \frac{v}{L_0} \ \text{Hz}$$

For typical wind speeds, this frequency should be a few hertz for optical links close to the surface, where most experiments are performed. By contrast, the threshold frequency is less than $10^{-2}$ Hz for paths in the free atmosphere, which were used in many microwave experiments. The spectrum changes to an inverse power law above this threshold:

$$\omega > \omega_0: \qquad W_\varphi(\omega) = \omega^{-\frac{8}{3}}\left(2.192Rk^2C_n^2v^{\frac{5}{3}}\right) \qquad (6.40)$$

It is this prediction that is usually compared with experimental results.

The distribution of energy in the phase spectrum (6.38) changes as the wind speed varies. Spectral energy shifts from low to high frequencies when the wind picks up. That agrees with our physical intuition. It is also evident in the predictions (6.39) and (6.40). This readjustment takes place in a way that keeps the area under the power spectrum constant:

$$\frac{1}{2\pi}\int_{-\infty}^{\infty} d\omega\, W_\varphi(\omega) = \frac{1}{\pi}\int_0^{\infty} d\omega\, \frac{2.192Rk^2C_n^2v^{\frac{5}{3}}}{(\omega^2 + \kappa_0^2v^2)^{\frac{4}{3}}} = 0.782Rk^2C_n^2 \qquad (6.41)$$

The outcome is the phase variance as we should expect. This conservation holds true even though the low-frequency components of $W_\varphi(\omega)$ increase without bound as the wind speed drops. That suggests an important issue to which we shall return later.

### 6.2.3.2 The Influence of the Sample Length
### on the Power Spectrum

We found that the sample length exerts a strong influence on autocorrelation measurements. One should also ask how it affects the power spectrum of phase fluctuations. We could take the Fourier transform of (6.29), but we would miss an important insight in doing so. The sample-length effect is expressed in a very natural way in the frequency domain. To illustrate this we represent the phase history as a Fourier transform:

$$\varphi(t) = \frac{1}{2\pi}\int_{-\infty}^{\infty} d\omega\, F(\omega)\exp(-i\omega t) \qquad (6.42)$$

The complex weighting functions $F(\omega)$ are stochastic functions of frequency since $\varphi(t)$ is a random function of time. With this expression we can represent the phase variance in the following form:

$$\langle\varphi^2\rangle = \frac{1}{4\pi^2}\int_{-\infty}^{\infty} d\omega \int_{-\infty}^{\infty} d\omega'\, \langle F(\omega)F(\omega')\rangle \exp[-it(\omega + \omega')]$$

Implicit in our discussions has been the assumption that the phase fluctuations represent a *stationary random process*. When this is so, Tatarskii has shown [21] that the ensemble average of two weighting functions vanishes if the frequencies are different and becomes infinite if they are the same:

$$\langle F(\omega)F(\omega')\rangle = 2\pi\delta(\omega + \omega')W_\varphi(\omega) \tag{6.43}$$

Notice that this relation leads to the previous expression for the phase variance in terms of the power spectrum on setting $\tau = 0$ in (6.35).

Suppose that an average phase value is estimated using a data sample of length $T$. If that average is subtracted from the data stream, the following expression describes the phase variance that would be measured from the residual points:

$$\langle |\varphi(t) - \overline{\varphi}|^2\rangle = \frac{1}{T}\int_0^T dt \left\langle \varphi^2(t)\right\rangle - \frac{1}{T^2}\int_0^T dt \int_0^T dt' \left\langle \varphi(t)\varphi(t')\right\rangle$$

Introducing the Fourier integral expression (6.42) for the phase history,

$$\langle |\varphi(t) - \overline{\varphi}|^2\rangle = \frac{1}{4\pi^2}\int_{-\infty}^{\infty} d\omega \int_{-\infty}^{\infty} d\omega' \left\langle F(\omega)F(\omega')\right\rangle \mathcal{K}(\omega, \omega')$$

where

$$\mathcal{K}(\omega, \omega') = \frac{1}{T}\int_0^T dt \, \exp[-i(\omega + \omega')t]$$
$$- \frac{1}{T^2}\int_0^T dt \int_0^T dt' \, \exp[-i(\omega t + \omega' t')]. \tag{6.44}$$

One of the frequency integrations collapses when the ensemble average is replaced by (6.43) and the time-like integrals in $\mathcal{K}(\omega, \omega')$ are easily performed:

$$\langle |\varphi(t) - \overline{\varphi}|^2\rangle = \frac{1}{2\pi}\int_{-\infty}^{\infty} d\omega \, W_\varphi(\omega)\left(1 - \frac{\sin^2(\omega T/2)}{(\omega T/2)^2}\right) \tag{6.45}$$

The factors which influence $\langle \varphi^2\rangle$ appear as a product of terms in the frequency domain. The first term represents the influence of the turbulent atmosphere and is simply the propagation power spectrum described by (6.38). The second represents the result of processing the data with a finite sample length and eliminates frequencies smaller than $1/T$. That seems reasonable when we note that phase changes associated with lower frequencies cannot be detected in a time $T$. The sample size therefore sets a lower limitation on the spectrum of phase fluctuations which can be measured. As a practical matter, this means that the zero-frequency limit (6.39) is seldom reached.

Other influences appear as multiplicative terms in the frequency domain. Similar factors describe the *sampling interval* of the data and the *frequency response* of the phase-measuring equipment. They each multiply the propagation spectrum in representing phase-variance measurements.[4] This *stacking feature* makes the power spectrum a favorite approach.

### 6.2.3.3 Microwave Measurements of the Power Spectrum

The spectrum of single-path phase fluctuations has been measured many times at microwave frequencies using the technique of downconverting and retransmission described in Section 4.1.1. Running records of phase are usually passed through a spectrum analyzer to give the power spectrum. A summary of these experiments is presented in Table 6.1. Most of this data was taken above the threshold frequency and therefore provides tests for the high-frequency prediction (6.40).

Table 6.1: Microwave measurements of single-path phase and phase-difference spectra made on terrestrial paths

| Location | $R$ (km) | Frequency (GHz) | $W_\varphi$ | $W_{\Delta\varphi}$ | Ref. |
|---|---|---|---|---|---|
| Colorado Springs | 16 | 1.046 | x | x | 22 |
| | 16 | 9.4 | | x | 23 |
| Boulder | 15 | 9.4 | x | x | 24, 25 |
| Maui | 25 | 9.4 | x | x | 15, 26, 27 |
| Maui–Hawaii | 64 | 9.6 and 34.52 | x | x | 28 |
| | 64 | 9.6 | x | x | 29 |
| | 64 | 9.6, 19.1, 22.2, 25.4 and 33.3 | x | | 30 |
| | 97 | 9.6 | x | | 31 |
| | 150 | 9.6 | x | x | 32 |
| Stanford | 28 | 35 | | x | 33, 34 |

---

[4] Notice that the sample-length factor depends on the type of measurement being made. The frequency factor which represents the sample-length effect for the *temporal covariance* of phase is different than that given in (6.45) and is indicated in Problem 3.

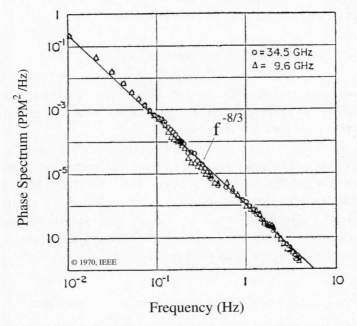

Figure 6.7: Power spectra for single-path phase fluctuations measured in Hawaii at 9.6 and 34.52 GHz on a 64-km path by Janes, Thompson, Smith and Kirkpatrick [28]. The straight line is the high-frequency prediction (6.40).

An early microwave program measured the phase spectrum for two radio frequencies simultaneously [28]. Signals at 9.6 and 34.5 GHz were transmitted over a 64-km path between Maui and Hawaii. The observed spectra reproduced in Figure 6.7 indicate a slope of $-8/3$ in striking agreement with the prediction. As Tatarskii observed [20] *"This result hardly could have been predicted beforehand."* A second experiment on the same path measured phase spectra at 9.55 and 22.2 GHz and the data fitted the following scaling law [30]:

$$W_\varphi(\omega) = \text{constant} \times \omega^{-j} \qquad (6.46)$$

The measured values of $j$ ranged from 2.37 to 2.91 but their midpoint value was 2.67, which is exactly the prediction. This result has consistently been verified by other campaigns. An experiment to determine whether aperture averaging influences the phase spectrum was done but no measurable effect was found [35].

It is significant that the high-frequency prediction (6.40) is generated by any model of turbulence that has the following form:

$$\Phi_n(\kappa) = \text{constant} \times \kappa^{-\frac{11}{3}}$$

The Kolmogorov and von Karman models share this characteristic in the inertial range. The *low-frequency* region of the phase power spectra may reflect a difference between them.

Several microwave experiments probed the low-frequency range of the phase spectrum. The 9.4-GHz experiments conducted on Maui indicated that there is a gradual saturation below $10^{-2}$ Hz [15]. A similar experiment in Colorado confirmed the $-8/3$ relationship down to $10^{-4}$ Hz and sometimes $10^{-5}$ Hz before saturating [26]. These experiments provide clear evidence that very large refractive-index structures are present in the free atmosphere and influence such transmissions.

### 6.2.3.4 Optical Measurements of the Power Spectrum

The situation is different for optical links close to the surface, where the outer scale length is a few meters. The threshold frequency is a few hertz in this case and one should be able to explore the low-frequency region with collimated laser signals. This was done in the USSR [13][14] using the experimental approach shown in Figure 4.3. Single-path phase records were converted to power spectra using the relationship (6.34). The prediction (6.38) was expressed in terms of $2\pi f/(\kappa_0 v)$, multiplied by $f$ and then divided by $\langle \varphi^2 \rangle$ to provide a dimensionless function that could be compared with the data:

$$\frac{fW_\varphi(f)}{\langle \varphi^2 \rangle} = 0.466 \left( \frac{2\pi f}{\kappa_0 v} \right) \left[ 1 + \left( \frac{2\pi f}{\kappa_0 v} \right)^2 \right]^{-\frac{4}{3}} \tag{6.47}$$

This combination is plotted in Figure 6.8 together with measurements for two path lengths approximately 1.5 m above the surface. The agreement between theory and experiment is remarkably good.

The linear rise for small frequencies evident in Figure 6.8 suggests that the power spectrum itself is flat in this range. That conclusion is consistent with the prediction (6.38) based on the von Karman model. This gives one some confidence in the ability of that model to describe the large-eddy region of the turbulence spectrum. If further observations verify that the

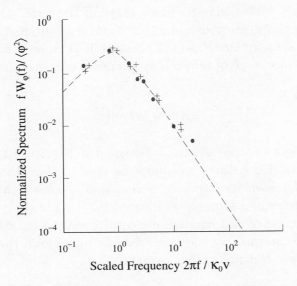

Figure 6.8: The power spectrum of single-path phase fluctuations measured with a collimated laser signal near the surface for distances of 95 m (crosses) and 200 m (circles) by Lukin, Mironov, Pokasov and Khmelevtsov [12]. The dotted line is the von Karman prediction (6.47).

power spectrum is flat for small frequencies, one can make a more general statement about the model of turbulence. The power-spectrum expression (6.37) can be evaluated in a manner that is independent of the model selected to represent $\Phi_n(\kappa)$:

$$\lim_{\omega \to 0}[W_\varphi(\omega)] = 8\pi^2 R k^2 \lim_{\omega \to 0}\left(\int_{\omega/v}^{\infty} \frac{d\kappa\ \kappa \Phi_n(\kappa)}{(\kappa^2 v^2 - \omega^2)^{\frac{1}{2}}}\right)$$

$$= \frac{8\pi^2 R k^2}{v}\int_0^{\infty} d\kappa\ \Phi_n(\kappa)$$

One can then assert that the integrated turbulence spectrum must be constant:

$$\int_0^{\infty} d\kappa\ \Phi_n(\kappa) = \text{constant} \tag{6.48}$$

and this places an important limit on the admissible models of turbulence. We need to consider other explanations before embracing this conclusion.

The observed flatness could also be due to a stand-off between the rising sample-length factor (6.45) and a falling power spectrum. Phase trends that have not carefully been removed also influence the spectrum in this spectral region. More fundamentally, notice that low frequencies correspond to very large irregularities that may fill much or all of the path. One cannot then simplify the path integrations using (5.27) and the convenient delta function in (6.23) is absent. In that case the phase shift is more nearly described by the instantaneous value of refractivity averaged over the path,

$$\varphi(t) = kR\,\overline{\delta n(t)}$$

which is analogous to the situation discussed in Section 4.1.6. This means that the Central Limit theorem cannot be used to show that phase fluctuations obey Gaussian statistics. It is clear that the low-frequency region of the phase spectrum is strongly influenced by the boundary between well-developed turbulence and nominal meteorological conditions. One must be cautious when interpreting such experiments but rejoice that they provide some insight into this mysterious region.

### 6.2.4  Wind-speed Variations

The discussion so far has assumed that a constant wind carries eddies across the path. We need to test that assumption against meteorological data. A sequence of surface wind vectors measured every 30 s is reproduced in Figure 6.9. It is clear that the direction and magnitude of the wind are changing with time. Although there are many occasions when the wind is steady, we judge that there is usually a variable component that is quite significant.

These measurements compel us to consider the variability of the cross-path wind. This changing situation is naturally described by a probability density function for the normal wind speeds, which we denote by $\mathsf{P}(v)$. If the phase spectrum is measured by taking samples of data over a long period,

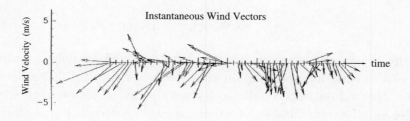

Figure 6.9: Successive surface wind vectors averaged over 30 s that were measured in connection with a millimeter-wave experiment by Otto, Hill, Sarma, Wilson, Andreas, Gosz and Moore [36].

the result becomes an average of (6.38) taken over the distribution of wind values:

$$\langle W_\varphi(\omega) \rangle = 2.192 k^2 C_n^2 \int_{-\infty}^{\infty} dv \; \frac{\mathsf{P}(v)|v|^{\frac{5}{3}}}{(\omega^2 + \kappa_0^2 v^2)^{\frac{4}{3}}} \qquad (6.49)$$

In writing this we have replaced $v$ by $|v|$ since the result is indifferent to which way the wind blows across the path. One should replace $\mathsf{P}(v)$ by the measured wind distribution when it is available. Only by doing so can one obtain accurate estimates for $C_n^2$ from low-frequency data.

### 6.2.4.1 The Gaussian Distribution of Wind Speeds

The distribution of wind vectors and their path profile are seldom available for analyzing propagation experiments. This is always the case when one tries to predict the performance of a system because one cannot forecast the weather with confidence. It is therefore helpful to have a model for cross-path wind distributions. This situation can be described by a constant cross-path wind plus a variable component:

$$v = v_0 + \delta v \qquad (6.50)$$

The Gaussian distribution for velocity fluctuations (6.15) suggests that

$$P(v) = \frac{1}{\sigma_v \sqrt{2\pi}} \exp\!\left( -\frac{1}{2\sigma_v^2} (v - v_0)^2 \right) \qquad \text{with} \quad \sigma_v^2 = \langle \delta v^2 \rangle \qquad (6.51)$$

should provide a reasonable description for wind values. When this model is combined with (6.49) one finds that the speed integration must be evaluated numerically. By setting $v = v_0 u$ we find that

$$\langle W_\varphi(\omega) \rangle = \frac{2.192 k^2 C_n^2}{v_0 (\kappa_0)^{\frac{8}{3}}} \mathcal{X}\!\left( \frac{\omega}{\kappa_0 v_0}, \frac{v_0}{\sigma_v} \right) \qquad (6.52)$$

where

$$\mathcal{X}(\xi, p) = \frac{\eta}{\sqrt{2\pi}} \int_{-\infty}^{\infty} du \; |u|^{\frac{5}{3}} \frac{\exp[-\frac{1}{2}\xi^2 (u-1)^2]}{(u^2 + p^2)^{\frac{4}{3}}} \qquad (6.53)$$

The normalized spectrum defined by $\mathcal{X}(\xi, p)$ is plotted logarithmically in Figure 6.10 for two values of $v_0/\sigma_v$, which identifies the variability of the wind speed relative to its average value. The curves share a $-8/3$ high-frequency behavior. Their low-frequency behavior is different and becomes quite large for $\omega = 0$. This occurs because the Gaussian distribution includes the possibility that the combined wind speed *vanishes*. The chance that the

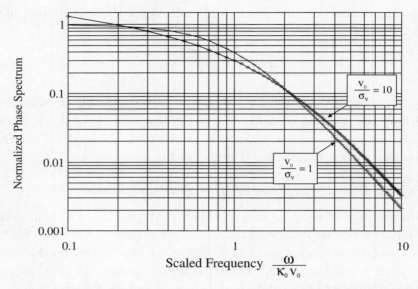

Figure 6.10: The predicted phase power spectrum for a Gaussian distribution of wind speeds for two values of the average wind speed divided by the rms value of the random component.

variable component is equal and opposite to the average speed is small but finite. It depends on the ratio $v_0/\sigma$ and the low-frequency behavior of the two curves reflects this sensitivity.

### 6.2.4.2 The Zero-wind Problem

This discussion reminds us that we should consider the case when the wind stops blowing entirely. There are certainly calm days when the wind speed goes to zero. *What happens to the phase spectrum in that case?* We might have asked the same question in connection with the constant-wind solution (6.38). We noted before that the spectral energy is shifted to lower frequencies as the horizontal wind speed decreases. Expression (6.39) suggests that the low-frequency spectrum grows without bound as the wind speed goes to zero. On the other hand, we know from field measurements that $C_n^2$ increases as the wind picks up and decreases as it dies down. This suggests that the product $C_n^2 v$ could be finite or even zero. That idea is comforting but it masks a deeper question. *Are time-varying phase fluctuations created if no wind blows across the path?*

The average velocity field normally plays two roles in establishing the phase spectrum. In the first role it provides the horizontal transportation needed to move irregularities across the path, which is expressed by Taylor's hypothesis. When this familiar vehicle is absent, self motion becomes the dominant agent for rearranging refractive irregularities.

The second role is more profound. It involves the mechanism by which kinetic energy from the steady wind field is converted into turbulent motion. This question lies at the heart of turbulence theory and is an important unsolved problem in physics. We have no universal theory that can describe the conversion of steady flow to turbulent conditions in the atmosphere.

We believe that turbulent velocity fields are solutions of the Navier–Stokes equations of hydrodynamics but we cannot solve these equations by analytical techniques. On the other hand, numerical simulations of the solutions to these equations exhibit many of the properties which are measured in wind tunnels. Dimensional analysis provides another type of solution. In this approach, a fraction of the energy in the ambient wind field is converted to turbulent energy at large scale lengths – by mechanisms that are not specified. That input energy is redistributed in a cascade process to smaller and smaller scales until it is eventually dissipated as heat. This raises a basic question: *are turbulent velocity fluctuations created if there is no steady wind from which to draw energy?*

Different components of the turbulent velocity field mix passive scalars like temperature and humidity. In doing so they generate eddies of refractive index. This secondary process is also described by dimensional analysis and gives the wavenumber spectrum which we have often used:

$$\Phi_n(\kappa) = 0.033 C_n^2 \kappa^{-\frac{11}{3}}$$

This raises a third question: *are refractive-index irregularities created in the absence of turbulent velocity components?* The chain of reasoning outlined above suggests a simple answer to our two questions. *No wind, no turbulence, no refractive-index variations and hence $C_n^2 = 0$.* Beyond this conjecture, we must confess that we do not know how $C_n^2$ depends on the wind speed – especially for calm days.

Fortunately we can approximate the effect of small random velocities on the phase spectrum without knowing the details of how they are generated. To do so we separate the wind vector into a constant horizontal speed and a variable component, as we did in (6.50). Small random velocity components persist even when the average speed goes to zero. These vectors are associated with self motion or vertical convection driven by temperature gradients. It is significant that they can be oriented in any direction. In calm conditions we replace the cross-path velocity components in the covariance expression (6.24) by these surviving random components:

$$\sqrt{v_y^2 + v_z^2} \rightarrow \sqrt{\delta v_y^2 + \delta v_z^2}$$

To take advantage of its random orientation, the random velocity vector is described in terms of spherical coordinates referenced to the line of sight:

$$\delta \mathbf{v} = \delta v \left( \mathbf{i}_x \cos\theta + \mathbf{i}_y \sin\theta \cos\phi + \mathbf{i}_z \sin\theta \sin\phi \right)$$

We insert the cross-path velocity components into (6.24) and average over $\delta v$, $\theta$ and $\phi$. In doing so we follow the description (6.51) and assume that each component is distributed as a Gaussian random variable:

$$\overline{\langle \varphi(t+\tau)\varphi(t)\rangle}\Big|_{\theta,\phi,\delta v} = \frac{\sqrt{2}}{\sigma_v^3 \sqrt{\pi}} \int_{-\infty}^{\infty} d(\delta v)(\delta v)^2 \exp\left(-\frac{(\delta v)^2}{2\sigma_v^2}\right) \frac{1}{4\pi} \int_0^{2\pi} d\phi$$

$$\times \int_0^{\pi} d\theta \sin\theta \left( 4\pi^2 R k^2 \int_0^{\infty} d\kappa \, \kappa \Phi_n(\kappa) J_0(\kappa\tau \, \delta v \sin\theta) \right)$$

The integration on random speeds $\delta v$ is to be found in Appendix C and the polar integration is a special case of a result to be found in Appendix D:

$$\overline{\langle \varphi(t+\tau)\varphi(t)\rangle}\Big|_{\theta,\phi,\delta v} = 4\pi^2 R k^2 \int_0^{\infty} d\kappa \, \kappa \Phi_n(\kappa) \exp\left(-\frac{\tau^2 \kappa^2 \sigma_v^2}{2}\right) \qquad (6.54)$$

The Bessel function which appeared in the constant-wind expression (6.25) has been replaced here by a Gaussian term that depends on the dispersion $\sigma_v$ of the random velocity components. This change has a striking effect on the behavior for large time delays.

To complete the calculation we use the von Karman model as we did in treating the constant-wind case. The result can be expressed in terms of the *second* type of Kummer's function described in Appendix G:

$$\overline{\langle \varphi(t+\tau)\varphi(t)\rangle}\Big|_{\theta,\phi,\delta v} = 0.6514 R k^2 C_n^2 \kappa_0^{-\frac{5}{3}} U(1, \tfrac{1}{6}, \tfrac{1}{2}\tau^2 \kappa_0^2 \sigma_v^2) \qquad (6.55)$$

This reduces to the variance (4.12) for small time delays since $U(1, \tfrac{1}{6}, 0) = 6/5$. For large time delays the asymptotic expansion gives

$$\lim_{\tau \to \infty} \overline{\langle \varphi(t+\tau)\varphi(t)\rangle}\Big|_{\theta,\phi,\delta v} = \frac{1}{\tau^2}(1.303 R k^2 C_n^2 \kappa_0^{-\frac{11}{3}} \sigma_v^{-2}). \qquad (6.56)$$

We can calculate the power spectrum from (6.34) by introducing the general expression (6.54) for the temporal covariance:

$$\overline{W_\varphi(\omega)}\Big|_{\theta,\phi,\delta v} = 4\pi^2 Rk^2 \int_{-\infty}^{\infty} d\tau \, \exp(i\omega\tau) \int_0^{\infty} d\kappa \, \kappa \Phi_n(\kappa) \exp\left(-\frac{\tau^2 \kappa^2 \sigma_v^2}{2}\right)$$

$$= \sqrt{2\pi} 4\pi^2 Rk^2 \sigma_v^{-1} \int_0^{\infty} d\kappa \, \Phi_n(\kappa) \exp\left(-\frac{\omega^2}{2\kappa^2 \sigma_v^2}\right)$$

For the von Karman model this can be expressed as follows:

$$\overline{W_\varphi(\omega)}\Big|_{\theta,\phi,\delta v} = (1.458 Rk^2 C_n^2 \kappa_0^{-\frac{8}{3}} \sigma_v^{-1}) U(\tfrac{4}{3}, \tfrac{1}{2}, \omega^2/(2\kappa_0^2 \sigma_v^2)) \qquad (6.57)$$

With this concise result one can compute the power spectrum for a variety of conditions using numerical routines for the second Kummer function.

The high-frequency spectrum emerges on using the asymptotic expansion for the Kummer function:

$$\lim_{\omega \to \infty} \overline{W_\varphi(\omega)}\Big|_{\theta,\phi,\delta v} = \omega^{-\frac{8}{3}} \left(3.674 Rk^2 C_n^2 \sigma_v^{\frac{5}{3}}\right) \qquad (6.58)$$

This is the same as the constant-wind result (6.40) except that the transverse speed $v$ has been replaced by the velocity dispersion $\sigma_v$. The numerical constants are different because one speed is constant whereas the other is a random variable in three dimensions. The spectrum approaches a constant value for low frequencies:

$$\lim_{\omega \to 0} \overline{W_\varphi(\omega)}\Big|_{\theta,\phi,\delta v} = 1.499 Rk^2 C_n^2 \kappa_0^{-\frac{8}{3}} \sigma_v^{-1} \qquad (6.59)$$

This is similar to the result (6.39) but with $v_y$ replaced by $\sigma_v$. The sample-length factor in (6.45) which multiplies the power spectrum will suppress this low-frequency region.

In addition, we should note that the turbulent structure undergoes fundamental changes over time, which is important here. The structure today is rarely the same as it was yesterday. Actual observations show that the turbulent field changes in 10–30 min. This means that frequencies less than $10^{-3}$ Hz may reflect basic changes in the eddy structure rather than low-frequency oscillations. This observation brings us to the next subject.

### *6.2.5 The Influence of Phase Trends*

It is apparent from the microwave record reproduced in Figure 4.1 that phase trends are often encountered. These distort both autocorrelation and power-spectrum measurements. A typical situation is illustrated in Figure 4.7, which shows a stationary phase record superimposed on a linear trend:

$$\psi(t) = \varphi(t) + \eta t \tag{6.60}$$

The rate of change $\eta$ is usually small and the trend is represented by low frequencies in the spectrum. Comparisons of theory and experiments were primarily successful because they relied on the high-frequency portion of the spectrum.

There are several techniques for dealing with phase drift. One approach is to *detrend* the raw data by fitting straight lines or polynomials to a running mean value. The corrected phase records are often stationary when these fitted models are subtracted from the data stream. Alternatively, one can filter the data stream to remove low-frequency components since they relate primarily to phase trends. There is an inherent uncertainty in both processes. One must wonder whether one has removed a secular trend or omitted low-frequency components of the stationary process.

This problem was solved for spaced receivers by focusing attention on the structure function. That method worked well because trends are usually the same for two nearby paths. The same approach does not work when one uses the temporal structure function. Combining (6.31) and (6.60) shows that the temporal structure function increases quadratically with time:

$$\mathcal{D}_\varphi(\tau) = \langle [\varphi(t + \tau) - \varphi(t)]^2 \rangle + \eta^2 \tau^2 \tag{6.61}$$

A continuous increase with time was noted in measurements made as part of developing an infrared interferometer for astronomical observations [37]. A helium–neon laser signal was transmitted over a reflected path between two telescopes 5.5 m apart to measure the instantaneous electric distance in order to compensate for baseline variation. Distance fluctuations were measured to within an accuracy of 0.04 μm. The temporal structure function computed from this data is reproduced in Figure 6.11. For small time delays it rises with a slope somewhat less than 5/3. This is consistent with the prediction for $\mathcal{D}_\varphi$ plotted in Figure 5.8 and represents the influence of the outer scale length. For delays greater than a few seconds the measured structure function does not approach an asymptotic value. This continuous upward drift is probably caused by phase trends, although the slope is considerably

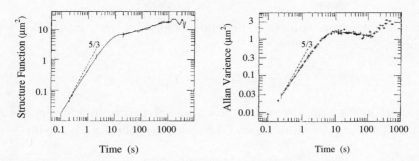

Figure 6.11: The temporal structure function and Allan variance of phase fluctuations measured at 11 μm with an infrared interferometer by Bester, Danchi, Degiacomi, Greenhill and Townes [37].

less than the prediction (6.61). This characteristic has led to the search for other methods to describe such data.

### 6.2.6 The Allan Variance

The *Allan variance* was introduced to deal with such problems [38]. It is defined by the following combination of time-shifted phase terms:

$$\mathbb{A}(\tau) = \langle [\, \varphi(t+\tau) - 2\varphi(t) + \varphi(t-\tau)]^2 \rangle \tag{6.62}$$

This measure is often used to describe random processes because a linear phase trend cancels out. Notice that a trend that is quadratic in time survives.

The Allan variance for single-path phase fluctuations is related to the temporal structure function. To see this we rewrite the basic definition as follows:

$$\mathbb{A}(\tau) = \langle \{ [\varphi(t+\tau) - \varphi(t)][\varphi(t) - \varphi(t-\tau)] \}^2 \rangle$$

which allows the result to be expressed in the following form:

$$\mathbb{A}(\tau) = \mathcal{D}_\varphi(\tau) + \mathcal{D}_\varphi(-\tau) - 2\langle [\varphi(t+\tau) - \varphi(t)][\varphi(t) - \varphi(t-\tau)] \rangle$$

The last term can be expressed in terms of $\mathcal{D}_\varphi(2\tau)$ and the first two terms by using the algebraic identity (5.57):

$$\mathbb{A}(\tau) = 2\mathcal{D}_\varphi(\tau) + 2\mathcal{D}_\varphi(-\tau) - \mathcal{D}_\varphi(2\tau)$$

When the structure function depends solely on the magnitude of its argument,

$$\mathbb{A}(\tau) = 4\mathcal{D}_\varphi(\tau) - \mathcal{D}_\varphi(2\tau). \qquad (6.63)$$

For short paths near the surface the phase structure function is described by the 5/3 scaling law and we would expect the following behavior:

$$\mathbb{A}(\tau) = \text{constant} \times \tau^{\frac{5}{3}} \qquad (6.64)$$

This scaling law is compared with laser distance measurements in Figure 6.11. Their slope is somewhat less than 5/3, which is very probably due to the influence of the outer scale length on the structure function suggested in Figures 5.4, 5.5 and 5.6. The measured Allan variance reached an asymptotic value after approximately 10 s, indicating that phase drift had been removed.

The Allan variance has become a measurement standard in the last decade, but was not used to describe early microwave phase data. It is now being exploited to characterize GPS path-length variations. Single-path range residuals for GPS signals were measured at four widely separated stations in 1994 [39]. The Allan variance was extracted from this data with time delays between 500 and 1200 s. For a typical wind speed $v = 10$ m s$^{-1}$ the displacements would have been 4–120 km. These distances are greater than the height of tropospheric irregularities and we would expect that the 2/3 scaling law (5.51) would describe the phase structure functions. Data taken on a single day was fitted by the following empirical scaling law, which is reasonably close to the prediction:

$$\mathbb{A}(\tau) = \text{constant} \times \tau^{0.50\pm0.10} \qquad (6.65)$$

## 6.3 The Phase Difference for Spherical Waves

Many of the microwave experiments identified in Table 6.1 measured the phase difference between adjacent receivers. They did so for several reasons. First, it is a good deal easier to measure the phase difference than it is to measure single-path phase fluctuations. At microwave frequencies one need only run a waveguide or coaxial line between the receivers to measure this difference. For optical interferometers, the two received signals are usually transmitted in vacuum and combined at a common point to count fringes. The second reason that these measurements are preferred is because phase trends on nearby ray paths are often the same and cancel out on differencing the signals. In addition, these programs were sponsored to measure angle-of-arrival errors for microwave tracking systems. We will learn in Chapter 7 that the angular error is proportional to the phase difference taken between closely spaced receivers.

Table 6.1 indicates that a good deal of experimental attention has been focused on the temporal variability of the phase difference. These programs used a single microwave source monitored by two or more receivers. The transmitters acted like point sources and their signals at the distance of the receiver were well characterized by spherical waves. Several optical experiments also used spherical waves. To interpret their data we must calculate the phase-difference power spectrum for a diverging reference signal.

### 6.3.1 The Autocorrelation of the Phase Difference

We begin by considering the covariance of the phase difference measured between the two receivers. If we use the representation

$$\Delta\varphi(t) = \varphi_1(t) - \varphi_2(t) \tag{6.66}$$

then

$$\langle \Delta\varphi(t+\tau)\,\Delta\varphi(t)\rangle = \langle\varphi_1(t)\varphi_1(t+\tau)\rangle + \langle\varphi_2(t)\varphi_2(t+\tau)\rangle$$
$$-\langle\varphi_1(t)\varphi_2(t+\tau)\rangle - \langle\varphi_1(t+\tau)\varphi_2(t)\rangle. \tag{6.67}$$

The first and second terms are described by the temporal covariance for a single receiver because the normal wind is almost perpendicular to both ray paths and their angular separation is very small in most experiments. The challenge is to calculate the cross correlation between the phase shifts measured at different receivers and different times.

#### 6.3.1.1 Paths Close to the Surface

When the signals travel close to the surface one ignores variations of $C_n^2$ along the propagation paths and assumes that the refractive-index irregularities are isotropic. The general expression (5.5) then describes the covariance of phase fluctuations in terms of the distances $s_1$ and $s_2$ measured along the split rays:

$$\langle \varphi\varphi'\rangle = 4\pi k^2 \int_0^\infty d\kappa\,\kappa\Phi_n(\kappa) \int_0^R ds_1 \int_0^R ds_2\,\frac{\sin(\kappa D)}{D}$$

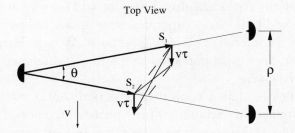

Figure 6.12: The propagation geometry for analyzing the temporal variability of the phase difference between two receivers that monitor a spherical wave.

The separation distances for the third and fourth terms in (6.67) are taken between a point on one ray and a time-displaced point on the other ray. These distances are identified as the cross connections in Figure 6.12. They can be described in terms of the ray coordinates and the angle $\theta$ subtended by the receivers:

$$\langle\varphi_1(t)\varphi_2(t+\tau)\rangle: \quad D^2 = s_1^2 + s_2^2 - 2s_1s_2\cos\theta$$
$$+v^2\tau^2 + 2(s_1 + s_2)v\tau\sin(\theta/2)$$
$$\langle\varphi_1(t+\tau)\varphi_2(t)\rangle: \quad D^2 = s_1^2 + s_2^2 - 2s_1s_2\cos\theta$$
$$+v^2\tau^2 - 2(s_1 + s_2)v\tau\sin(\theta/2)$$

These expressions reduce to the ray-displacement expression used in Section 5.1.1 when $\tau = 0$. Because the two separation distances are symmetrical in $s_1$ and $s_2$ one can recast the double ray-path integration in the following way:

$$\langle\varphi\varphi'\rangle = 8\pi k^2 \int_0^\infty d\kappa\, \kappa\Phi_n(\kappa) \int_0^R ds_1 \int_0^{s_1} ds_2\, \frac{\sin(\kappa D)}{D}$$

On changing the integration variable from $s_2$ to $D$ and dropping the subscript on $s_1$, the covariance becomes

$$\langle\varphi\varphi'\rangle = 8\pi k^2 \int_0^\infty d\kappa\, \kappa\Phi_n(\kappa) \int_0^R ds \int_{D_{\min}}^{D_{\max}} dD\, \frac{\sin(\kappa D)}{(D^2 - \aleph^2)^{\frac{1}{2}}}$$

where the limits on the separation distance are given by

$$D_{\min} = 2s\sin(\theta/2) \pm v\tau$$
$$D_{\max} = [s^2 + v^2\tau^2 - 2sv\tau\sin(\theta/2)]^{\frac{1}{2}}$$

and the parameter $\aleph$ is identified as

$$\aleph = \cos(\theta/2)\,[2s\sin(\theta/2) \pm v\tau].$$

The angle subtended by the receivers $\theta$ is usually quite small and one can make the following approximation:

$$\aleph = D_{\min} = s\theta \pm v\tau$$

The upper limit $D_{\max}$ is large relative to the eddy sizes for propagation near the surface so that

$$\langle\varphi\varphi'\rangle = 8\pi k^2 \int_0^\infty d\kappa\,\kappa\Phi_n(\kappa) \int_0^R ds \int_\aleph^\infty dD\,\frac{\sin(\kappa D)}{(D^2 - \aleph^2)^{\frac{1}{2}}}$$

and the integral on $D$ gives a zeroth-order Bessel function:

$$\langle\varphi\varphi'\rangle = 4\pi^2 k^2 \int_0^\infty d\kappa\,\kappa\Phi_n(\kappa) \int_0^R ds\,J_0(\kappa|s\theta \pm v\tau|) \qquad (6.68)$$

The covariance of the phase difference can be expressed by combining the above results. Setting $s = Ru$ and recalling that $\rho = \theta R$, we have

$$\text{h small:} \quad \langle\Delta\varphi(t+\tau)\,\Delta\varphi(t)\rangle = 4\pi^2 R k^2 \int_0^\infty d\kappa\,\kappa\Phi_n(\kappa)\mathcal{G}(\kappa,\rho,v\tau) \qquad (6.69)$$

where

$$\mathcal{G}(\kappa,\rho,v\tau) = 2J_0(\kappa v\tau) - \int_0^1 du\,J_0(\kappa|\rho u + v\tau|)$$

$$- \int_0^1 du\,J_0(\kappa|\rho u - v\tau|). \qquad (6.70)$$

The meaning of these terms is apparent from Figure 6.12. The first represents the single-path autocorrelation experienced along the separate paths. The integral expressions have the form of spatial correlations for spherical waves. They are characterized by a *sliding value* of the separation between the ray paths as one moves from the transmitter to the receivers. These ray-path displacements are supplemented by a temporal translation $v\tau$ of the entire medium. The term containing $\rho u - v\tau$ corresponds to eddies that are approaching the rays and the term containing $\rho u + v\tau$ represents eddies that are receding from the ray paths.

When the von Karman model is introduced into (6.69) each wavenumber integration has the same form and the result can be identified using (6.26). The autocorrelation for the phase difference is

$$C_{\Delta\varphi}(\tau,\rho) = \frac{1}{\mathcal{N}}(\kappa_0 v\tau)^{\frac{5}{6}} K_{\frac{5}{6}}(\kappa_0 v\tau)$$

$$-\frac{1}{2}\int_0^1 du\,(\kappa_0|\rho u + v\tau|)^{\frac{5}{6}} K_{\frac{5}{6}}(\kappa_0|\rho u + v\tau|)$$

$$-\frac{1}{2}\int_0^1 du\,(\kappa_0|\rho u - v\tau|)^{\frac{5}{6}} K_{\frac{5}{6}}(\kappa_0|\rho u - v\tau|) \qquad (6.71)$$

where the normalization is the same expression evaluated for zero time delay:

$$\mathcal{N} = \frac{1}{2^{\frac{1}{6}}}\Gamma(\tfrac{5}{6}) - \int_0^1 du\,(\kappa_0\rho u)^{\frac{5}{6}} K_{\frac{5}{6}}(\kappa_0\rho u).$$

The outer scale length is easily exceeded by the time delay term $v\tau$ near the surface and the integrals must be done numerically. Most optical and millimeter-wave experiments present their data in terms of the power spectrum and we will return to that description soon.

### 6.3.1.2 Microwave Autocorrelation Measurements on Elevated Paths

The autocorrelation of the phase difference was measured in a comprehensive series of experiments at 9.4 GHz [15]. The inclined path on Maui is described in Figure 4.9. Eight baselines ranging from 2.2 to 4847 ft were employed. Typical measurements for the longest and shortest baselines are reproduced in Figure 6.13. The short baseline simulated a monopulse tracker and the autocorrelation falls to zero in less than 2 s. By contrast, the largest baseline exhibits considerable correlation out to 50 s and beyond. The challenge is that of explaining this data.

One can make some useful observations without constructing a detailed analytical model. We notice that the wavenumber-weighting function (6.70) depends both on the inter-receiver spacing and on the *drift distance* $v\tau$. This suggests a trade-off between $\rho$ and $v\tau$. Their influences should be measured against the outer scale length and the *sample distance* $vT$. The autocorrelation is controlled primarily by the time delay when $\rho$ is small compared with $v\tau$. That is the situation represented by the rapid fall-off in the left-hand panel of Figure 6.13. The terms involving $\rho$ are dominant when the separation between receivers is large and the effect of the time delay is felt only when $v\tau > \rho$.

One can extend this qualitative explanation by building on the description (6.71) developed for paths near the surface. For an elevated path one must

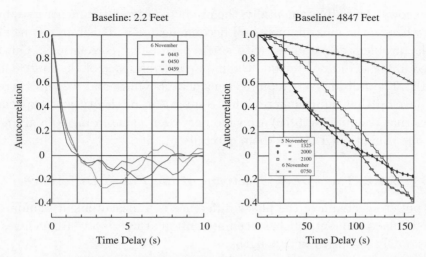

Figure 6.13: The autocorrelation of the phase difference for two inter-receiver separations on Maui measured at 9.4 GHz by Norton, Herbstreit, Janes, Hornberg, Peterson, Barghausen, Johnson, Wells, Thompson, Vetter and Kirkpatrick [15].

recognize the significant anisotropy.[5] This can be represented by using the small values for $\kappa_0$ that correspond to the large horizontal outer scales. Since the outer scale length is large relative both to the inter-receiver spacings and to the time delay term $v\tau$, one can replace the MacDonald functions by their small-argument expansions and perform the parametric integrations. The result depends solely on the ratio $v\tau/\rho$ and is given by

$$ C_{\Delta\varphi}\left(\frac{v\tau}{\rho}\right) = \frac{1}{2}\left|1 + \frac{v\tau}{\rho}\right|^{\frac{8}{3}} \pm \frac{1}{2}\left|1 - \frac{v\tau}{\rho}\right|^{\frac{8}{3}} - \frac{8}{3}\left|\frac{v\tau}{\rho}\right|^{\frac{5}{3}} \tag{6.72} $$

where the plus sign is used for $v\tau < \rho$ and the minus sign for $v\tau > \rho$. The autocorrelation begins at unity and goes slowly to zero as $v\tau/\rho$ increases:

$$ \lim_{x\to\infty}[C_{\Delta\varphi}(x)] = \frac{40}{81}x^{-\frac{1}{3}} $$

Let us compare this description with the data in Figure 6.13. Consider first the small-separation case. For reasonable values of the wind speed $v\tau/\rho$ is large unless the time delay is less than a few seconds. That agrees with the measurements in the left-hand panel. When $\rho = 1482$ m the same velocity predicts that the correlation should be substantial for time delays out to 3000 s, which is greater than the sample length. This is not what the

---

[5] One might question the assumption $\kappa D_{\max} \gg 1$ made in deriving (6.68), but we found in Section 4.2.1 that this is not a problem for elevated paths.

data shows. More fundamentally, the predicted correlation never goes below zero whereas the measurements all become negative. This reminds us of the similar problem encountered with single-path phase correlations. Our simplified analysis did not include the important influence of the finite sample length. The full analysis requires considerable effort and we will therefore not develop that subject here. This comparison of theory and experiment shows that results like (6.72) are valid only over a small range of parameters and one is cautioned to use them carefully.

### 6.3.2 The Power Spectrum of the Phase Difference

The power spectrum of the phase difference is a measurement standard for terrestrial experiments. It has been used often at microwave frequencies and occasionally at optical wavelengths.

#### 6.3.2.1 A Basic Expression for Paths Close to the Surface

We focus first on propagation paths that are close to the surface. In this case we can use the expressions (6.69) and (6.70) to represent the temporal correlation of $\Delta\varphi$. It is then a simple matter to compute the power spectrum using (6.34):

$$W_{\Delta\varphi}(\omega) = 4\pi^2 R k^2 \int_0^\infty d\kappa \; \kappa \Phi_n(\kappa) \Sigma(\kappa, \omega) \qquad (6.73)$$

where

$$\Sigma(\kappa, \omega) = 2 \int_{-\infty}^\infty d\tau \exp(i\omega\tau) \, J_0(\kappa v \tau)$$

$$- \int_0^1 du \int_{-\infty}^\infty d\tau \exp(i\omega\tau) \, J_0(\kappa \, |\rho u - v\tau|)$$

$$- \int_0^1 du \int_{-\infty}^\infty d\tau \exp(i\omega\tau) \, J_0(\kappa \, |\rho u + v\tau|). \qquad (6.74)$$

The Fourier transforms are evaluated by shifting the integration variables:

$$\int_{-\infty}^\infty d\tau \, e^{i\omega\tau} J_0(\kappa|\rho u \pm v\tau|) = 2\exp(\pm i\omega\rho u/v) \int_0^\infty d\tau' \cos(\omega\tau') \, J_0(\kappa v \tau')$$

so that

$$W_{\Delta\varphi}(\omega) = \left\{ 8\pi^2 k^2 R \int_0^\infty d\kappa \; \kappa \Phi(\kappa) \int_0^\infty d\tau \cos(\omega\tau) \, J_0(\kappa v \tau) \right\}$$

$$\times 2 \int_0^1 du \left[ 1 - \cos\left(\frac{\rho \, \omega u}{v}\right) \right].$$

The combination of terms in curly brackets is recognized as the single-path phase spectrum (6.36). The following important relationship emerges and is completely independent of the turbulence spectrum:

$$W_{\Delta\varphi}(\omega) = 2\left[1 - \frac{v}{\rho\omega}\sin\left(\frac{\rho\omega}{v}\right)\right]W_\varphi(\omega) \qquad (6.75)$$

To test this prediction one must explore the relative importance of the single-path spectrum and the receiver term in square brackets. To do so we use the von Karman result (6.38):

$$W_{\Delta\varphi}(\omega) = 4.384 R k^2 C_n^2 v^{\frac{5}{3}}\left[1 - \frac{v}{\rho\omega}\sin\left(\frac{\rho\omega}{v}\right)\right]\frac{1}{\left(\omega^2 + \kappa_0^2 v^2\right)^{\frac{4}{3}}} \qquad (6.76)$$

The receiver term begins quadratically and reaches its asymptotic value at the *corner frequency* defined by

$$\omega_c = v/\rho \qquad (6.77)$$

which is the reciprocal of the crossing time. The relative influence of the terms in (6.76) depends on a comparison of the corner frequency and the threshold frequency:

$$v/\rho \qquad \text{versus} \qquad \kappa_0 v.$$

The numerical value of $\kappa_0\rho$ is decisive since $v$ is a common factor. This parameter is close to or considerably less than unity both for optical and for microwave experiments near the surface.

The high-frequency spectrum is dominated by the single-path result:[6]

$$\omega > \omega_c: \quad W_{\Delta\varphi}(\omega) = 4.384 R k^2 C_n^2 v^{\frac{5}{3}}\left[1 - \frac{v}{\rho\omega}\sin\left(\frac{\rho\omega}{v}\right)\right]\omega^{-\frac{8}{3}} \qquad (6.78)$$

The receiver term modulates the single-path spectrum, although its effect decreases as the frequency increases. Its influence is often smeared out by the wind variability discussed previously.

In the low-frequency regime $v/\rho$ is greater than the threshold frequency $\kappa_0 v$ and the receiver term can be expanded:

$$1 - \frac{v}{\rho\omega}\sin\left(\frac{\rho\omega}{v}\right) \approx \frac{1}{6}\left(\frac{\rho v}{\omega}\right)^2$$

---

[6] This expression for the phase-difference spectrum was first derived by Clifford [40] using diffraction theory. Our result is identical to his expression for $\rho > \sqrt{\lambda R}$, which corresponds to the geometrical optics limit.

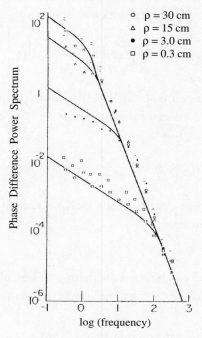

Figure 6.14: The phase-difference power spectrum measured with a diverging laser signal on a 70-m path by Clifford, Bouricius, Ochs and Ackley [41]. The solid curves are the geometrical optics prediction from (6.78) using values of wind speed and $C_n^2$ measured independently at the site.

The rising importance of this term competes with the falling nature of the single-path spectrum. The two effects combine to give a softer frequency scaling:

$$\omega < \omega_c: \qquad W_{\Delta\varphi}(\omega) = (0.730 R k^2 C_n^2 v^{-\frac{1}{3}} \rho^2)\omega^{-\frac{2}{3}} \qquad (6.79)$$

This suggests that spectra measured simultaneously with different baselines should be spread out at small frequencies and this behavior is indeed observed. In most situations the corner frequency represents the transition between the low- and high-frequency forms of the power spectrum.

### 6.3.2.2 Experimental Confirmations

These basic predictions have been confirmed both at optical and at microwave frequencies. In a pioneering experiment a diverging wave was generated with a helium–neon laser. This signal was transmitted over a 70-m path 1.6 m above the surface [41]. Four receivers with spacings ranging from 0.3 to 30 cm provided laterally displaced signals that were reflected to a common

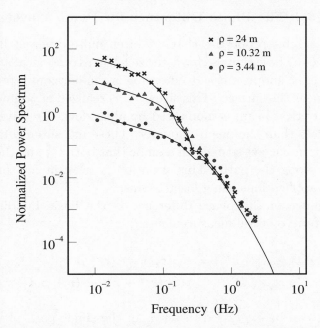

Figure 6.15: The phase-difference power spectrum for three inter-receiver separations measured at 35 GHz on a 28-km path by Mandics, Lee and Waterman [33]. The solid curves correspond to the prediction using the measured values for a cross-path wind speed $v = 3.8\,\mathrm{m\,s^{-1}}$ and $C_n^2 = 3.1 \times 10^{-7}\,\mathrm{m^{-\frac{1}{3}}}$.

point and compared in terms of phase with the signal reaching that point directly. This reference signal passed through a retardation plate to replicate the additional path length suffered by the displaced signals in reaching the common point. Spectra for a typical run are reproduced in Figure 6.14. The solid curves correspond to the prediction (6.75) using measured wind speeds and $C_n^2$ values. This experiment was duplicated in the USSR [42] and later done with even greater accuracy in Japan using a phase-locked interferometer [43]. The agreement between theory and experiment was good in every test.

A comparable experiment was conducted at 35 GHz along a 28-km path near Stanford [33]. This signal traveled higher in the convective boundary layer than the optical link described above. The signal was well approximated by a spherical wave at this range. Three receivers were electrically connected to generate phase-difference time series in real time. Local values of wind speed and $C_n^2$ were established by independent measurements. Phase-difference power spectra from this experiment are reproduced in Figure 6.15 and compared with the prediction (6.76). Again the agreement is remarkably good.

## 6.4 The Phase Difference for Plane Waves

Plane waves exhibit properties that are often quite different from those of spherical waves. These signals characterize both astronomical and terrestrial applications. The simplest case to describe is the horizontal propagation of plane waves near the surface. This situation is realized experimentally when a collimated optical beam is monitored by phase-locked receivers separated laterally by less than the beam diameter. These measurements are usually made close to the surface so one can assume that both $C_n^2$ and the wind speed are constant along the path. This situation is relevant for target-location and imaging applications using laser beams.

The covariance of the phase difference is the basic building block for describing interferometric measurements:

$$\langle \Delta\varphi(t)\, \Delta\varphi(t+\tau) \rangle = \langle [\varphi(\mathbf{r},t) - \varphi(\mathbf{r}+\boldsymbol{\rho},t)]$$
$$\times [\varphi(\mathbf{r},t+\tau) - \varphi(\mathbf{r}+\boldsymbol{\rho},t+\tau)] \rangle \qquad (6.80)$$

This product can be expressed in terms of the single-path phase structure function by using the algebraic identity (5.57). The separation between receivers and time-delay displacement are completely equivalent when Taylor's frozen-medium hypothesis is valid:

$$2\langle \Delta\varphi(t)\, \Delta\varphi(t+\tau) \rangle = \mathcal{D}_\varphi(|\boldsymbol{\rho} - \mathbf{v}\tau|) + \mathcal{D}_\varphi(|\boldsymbol{\rho} + \mathbf{v}\tau|) - 2\mathcal{D}_\varphi(|\mathbf{v}\tau|) \quad (6.81)$$

Since the baseline vector and the influential cross-path component of the wind velocity are parallel in these experiments, the covariance depends solely on scalar arguments:

$$2\langle \Delta\varphi(t)\, \Delta\varphi(t+\tau) \rangle = \mathcal{D}_\varphi(|\rho - v\tau|) + \mathcal{D}_\varphi(|\rho + v\tau|) - 2\mathcal{D}_\varphi(v\tau) \qquad (6.82)$$

### 6.4.1 The Autocorrelation of the Phase Difference

The autocorrelation emerges when the covariance of the phase difference (6.82) is divided by the same result evaluated for zero time delay:

$$C_{\Delta\varphi}(\tau) = \frac{\mathcal{D}_\varphi(|\rho - v\tau|) + \mathcal{D}_\varphi(|\rho + v\tau|) - 2\mathcal{D}_\varphi(v\tau)}{2\mathcal{D}_\varphi(\rho)} \qquad (6.83)$$

For links near the surface the outer scale length is often comparable to the baseline distance $\rho$ and the time-delay displacement $v\tau$. In these situations

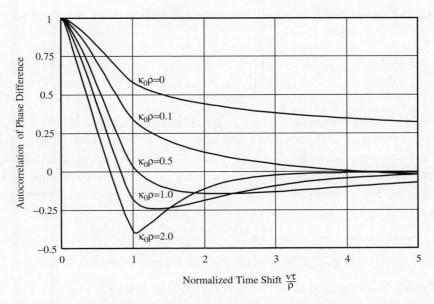

Figure 6.16: The autocorrelation of the phase difference for a plane wave traveling close to the surface plotted as a function of the normalized time shift for various values of the receiver baseline divided by the outer scale length. These predictions are based on (6.84).

one must use the full expression for the phase structure function given by (5.25) in terms of MacDonald functions:

$$C_{\Delta\varphi}(\tau) = \frac{1}{\mathcal{N}}[2(\kappa_0 v\tau)^{\frac{5}{6}}K_{\frac{5}{6}}(\kappa_0 v\tau) - (\kappa_0|\rho + v\tau|)^{\frac{5}{6}}K_{\frac{5}{6}}(\kappa_0|\rho + v\tau|)$$

$$- (\kappa_0|\rho - v\tau|)^{\frac{5}{6}}K_{\frac{5}{6}}(\kappa_0|\rho - v\tau|)] \tag{6.84}$$

where

$$\mathcal{N} = 2\left(\frac{\Gamma\left(\frac{5}{6}\right)}{2^{\frac{1}{6}}} - (\kappa_0\rho)^{\frac{5}{6}}K_{\frac{5}{6}}(\kappa_0\rho)\right) \tag{6.85}$$

This result is plotted in Figure 6.16 as a function of $v\tau/\rho$ for various values of the parameter $\kappa_0\rho$. These computations indicate the strong role played by large eddies when the outer scale length is comparable to the separation.

The inter-receiver separation is usually less than the outer scale length in collimated-beam applications. One can then use the 5/3 approximation for the structure functions provided that $v\tau$ is also less than $L_0$. This condition limits the time delay to small fractions of a second for reasonable wind speeds. In that narrow range

$$
\begin{aligned}
\rho \ll L_0: \\
v\tau \ll L_0:
\end{aligned}
\qquad
C_{\Delta\varphi}(\tau) = \frac{1}{2}\left[\left|1+\frac{v\tau}{\rho}\right|^{\frac{5}{3}} + \left|1-\frac{v\tau}{\rho}\right|^{\frac{5}{3}} - 2\left(\frac{v\tau}{\rho}\right)^{\frac{5}{3}}\right] \qquad (6.86)
$$

Equation (6.86) corresponds to the curve for $\kappa_0\rho = 0$ in Figure 6.16 and is evidently quite different than the other results. It never goes negative and falls to zero very slowly. This reminds us that approximations based on the 5/3 rule are valid only over a small range of separations in time and space. For most applications it is safer to use numerical values for the autocorrelation function unless the outer scale length is very large.[7]

### 6.4.2 Phase-difference Power Spectra

The preceding description of the autocorrelation function provides the basis for estimating the power spectrum of phase-difference measurements made near the surface. Combining (6.34) with the expression (6.82) yields

$$
\begin{aligned}
W_{\Delta\varphi}(\omega) = \frac{1}{2}\int_{-\infty}^{\infty} d\tau \exp(i\omega\tau)\,[\,\mathcal{D}_\varphi(|\rho - v\tau|) \\
+ \mathcal{D}_\varphi(|\rho + v\tau|) - 2\mathcal{D}_\varphi(v\tau)]
\end{aligned}
\qquad (6.87)
$$

The structure function for plane waves was expressed as a weighted integral of the turbulence spectrum in (5.19). We reverse the time and wavenumber integrations after combining these expressions:

$$
\begin{aligned}
W_{\Delta\varphi}(\omega) = 4\pi^2 R k^2 \int_0^{\infty} d\kappa\,\kappa\Phi_n(\kappa)\Bigg( 2\int_{-\infty}^{\infty} d\tau \exp(i\omega\tau)\,J_0(\kappa v\tau) \\
- \int_{-\infty}^{\infty} d\tau \exp(i\omega\tau)\,J_0(\kappa|\rho - v\tau|) \\
- \int_{-\infty}^{\infty} d\tau \exp(i\omega\tau)\,J_0(\kappa|\rho + v\tau|) \Bigg)
\end{aligned}
$$

We can establish a simple relation between the phase difference and single-path phase spectra by shifting the time integrations:

$$
W_{\Delta\varphi}(\omega) = 2\left[1 - \cos\left(\frac{\omega\rho}{v}\right)\right]W_\varphi(\omega) \qquad (6.88)
$$

---

[7] Tatarskii successfully used (6.86) to explain the spatial correlation of time-differenced microwave phase measurements when the signals traveled along a 16-km path [20][22].

This is the plane-wave analog of the relationship (6.75) for propagation of spherical waves near the surface. The interferometer receiver terms which multiply the single-path spectrum in the two cases,

$$2\left[1 - \frac{v}{\omega\rho}\sin\left(\frac{\omega\rho}{v}\right)\right] \quad \text{and} \quad 2\left[1 - \cos\left(\frac{\omega\rho}{v}\right)\right]$$

are quite different in the high-frequency region. The spherical-wave factor approaches unity when $\omega$ exceeds the corner frequency $v/\rho$. By contrast, the plane-wave receiver term oscillates between zero and four for all frequencies. This persistent harmonic modulation is a unique feature of plane waves.

## 6.5 Astronomical Interferometry

Microwave interferometry has been used to measure the positions and structures of distant galaxies for more than fifty years. Optical interferometers are being developed to provide similar information in the visible and infrared bands. Refractive-index irregularities in the troposphere distort the wave-front of arriving microwave and optical signals. Amplitude fluctuations generated by diffraction are not important for interferometric measurements [44]. By contrast, the accuracy of astronomical measurements depends almost entirely on phase errors that are imposed on the ray paths that reach the separated receivers of an array. Phase fluctuations are correlated across the downcoming wave-front, which leads to errors in angular position and limits the resolution of images.

The temporal variability of the phase difference is also an important ingredient of such measurements. The spectrum of phase variations is needed in order to design wave-front correctors for large telescopes and fringe trackers used with interferometers. Appropriate processing of data for the phase difference depends on the way in which the signals change with time. The choices for the sample length, signal filtering and detrending approaches depend on temporal characteristics of the arriving signals.

A more fundamental reason for studying the temporal variability is the powerful observational opportunity that it provides. Taylor's hypothesis describes the atmospheric variability as the horizontal transport of a frozen random medium past the line of sight. The graphical description of this approximation in Figure 6.1 indicates that one can substitute the time delay for inter-receiver separations when Taylor's approximation is a valid description. Although it is expensive to add receivers to an array, one can simulate additional ray paths through the random medium by making time-delayed phase-difference measurements. This inexpensive procedure is a powerful tool for interferometric astronomy when Taylor's hypothesis is valid.

### 6.5.1 The Time-shifted Phase Structure Function

The phase structure function for transmission through the atmosphere is the basic ingredient for describing the performance of an interferometer. It is accurately described by (5.44) when the refractive-index irregularities obey the 2/3 law. However, that expression did not include the moving medium which causes the temporal variability of the phase difference. We now need to enlarge that description by including the additional displacements caused by the wind vector. The wind speed and its direction can change significantly with altitude, time of day and season – as aircraft pilots know well.

We simplify the problem by considering only signals that arrive normal to the receiver baseline. This is the preferred orientation for interferometry, as noted in Chapter 5. On the other hand, it cannot always be used because it is difficult to reconfigure the array rapidly as the source moves across the sky. The ideal propagation geometry is illustrated in Figure 6.17, in which

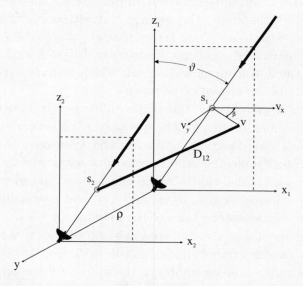

Figure 6.17: The geometry for describing plane waves that arrive at two receivers from a distant source. The receivers are assumed to lie along a baseline that is perpendicular to the plane of propagation. The horizontal wind blows the random medium past the signal paths. The distance between a point on one ray and the time-displaced point on the other ray path is identified by $D_{12}$.

the horizontal wind vector is also identified. Notice that the wind is allowed to vary with height, as is found using balloon-sounding data.

The temporal displacements $\tau v_x$ and $\tau v_y$ are equivalent to additional inter-receiver separations according to Taylor's hypothesis. With that equivalence one can repeat the analysis that began with (5.36) and establish the following expression for the time-shifted phase structure function:

$$
\mathcal{D}_\varphi\left(\rho, \frac{\pi}{2}, \vartheta, \tau, \mathbf{v}\right) = k^2 \int_0^\infty dr\, C_n^2(r) \int_{-r}^r du\, \mathcal{U}(u, \rho, \vartheta, \tau, \mathbf{v}) \qquad (6.89)
$$

where

$$
\mathcal{U}(u, \rho, \vartheta, \tau, \mathbf{v}) = \left| u^2 \mathsf{p}^2 + 2u \frac{v_x \tau \sin \vartheta}{a^2} + \frac{\rho^2 - 2\rho\, v_y \tau + v^2 \tau^2}{a^2} \right|^{\frac{1}{3}} - \left| u^2 \mathsf{p}^2 \right|^{\frac{1}{3}}
$$

$$(6.90)$$

and

$$
v^2 = v_x^2 + v_y^2.
$$

Equation (6.89) corresponds to $\mathcal{D}_\varphi(|\boldsymbol{\rho} - \tau\mathbf{v}|)$ in the covariance expression (6.81). It describes an inclined signal path with zenith angle $\vartheta$ and recognizes elongated irregularities. It also accommodates different height profiles for the wind velocity and $C_n^2$.

When the source is above $5°$ elevation one can replace $\mathsf{p}$ by $\cos\vartheta/c$ and it is then helpful to change from slant range to altitude using $u = \zeta \sec\vartheta$ and $r = z \sec\vartheta$:

$$
\mathcal{D}_\varphi\left(\rho, \frac{\pi}{2}, \vartheta, \tau, \mathbf{v}\right) = k^2 \sec^2 \vartheta\, a^{-\frac{2}{3}} \int_0^\infty dz\, C_n^2(z) \int_{-z}^z d\zeta \left[ -\left| \frac{\zeta a}{c} \right|^{\frac{2}{3}} \right.
$$

$$
\left. + \left| \left(\frac{\zeta a}{c}\right)^2 + 2\zeta v_x \tau \sin \vartheta + \rho^2 - 2\rho v_y \tau + v^2 \tau^2 \right|^{\frac{1}{3}} \right]
$$

$$(6.91)$$

The term $\mathcal{D}_\varphi(|\boldsymbol{\rho} + \tau\mathbf{v}|)$ can be obtained by reversing the sign of $v_y$ since the temporal displacement stretches the baseline in that case. The height integrations must be evaluated numerically for general profiles of turbulent activity and wind velocity.

### 6.5.2 The Autocorrelation of the Phase Difference

With the expressions for the time-shifted structure function given above, one can describe the covariance of the phase difference using (6.81):

$$\langle \Delta\varphi(t)\,\Delta\varphi(t+\tau)\rangle = \frac{1}{2}k^2\sec^2\vartheta\,a^{-\frac{2}{3}}\int_0^\infty dz\,C_n^2(z)\int_{-z}^z d\zeta\,\mathcal{Y}(\zeta,\rho,\tau,\vartheta,\mathbf{v})$$

$$(6.92)$$

where

$$\mathcal{Y}(\zeta,\rho,\tau,\vartheta,\mathbf{v}) = \left|\zeta^2 + 2\zeta\rho v_x\tau\sin\vartheta + \rho^2 - 2\rho v_y\tau + v^2\tau^2\right|^{\frac{1}{3}}$$
$$+ \left|\zeta^2 + 2\zeta\rho v_x\tau\sin\vartheta + \rho^2 + 2\rho v_y\tau + v^2\tau^2\right|^{\frac{1}{3}}$$
$$- 2\left|\zeta^2 + v^2\tau^2\right|^{\frac{1}{3}}.$$

$$(6.93)$$

The time-lagged covariance was measured at 2.7 and 5.0 GHz in early radio-astronomy experiments at Cambridge [44]. Baselines of 750 and 1500 m were used. Data samples were selected for periods during which the rms phase difference changed slowly. The covariance of $\Delta\varphi$ was characterized by well-defined peaks with widths ranging from 100 to 200 s. Comparison of these covariances measured with different baselines left little doubt that a pattern of refractive-index irregularities was being carried across the array on prevailing winds. The speed of this bulk motion was evident in the phase data and was compared with wind profiles taken nearby. The effective wind speed was often several times the surface value, suggesting that the irregularities responsible were at heights of 1–2 km.

There has been a recent revival of interest in analyzing time-lagged covariance data. Discrete refractive-index irregularities in the convective boundary layer are of interest both to astronomers and to meteorologists. Thermals and buoyant plumes of enhanced humidity occur close to the surface. These plumes are sometimes called columns of precipitable water vapor. Such transient structures influence the ray paths differently for widely spaced receivers and can be identified in microwave covariance measurements. An initial test was done using 15-GHz signals from a galactic source and the VLA at the NRAO [45]. Powerful narrow-band signals from communication satellites provide another way to do so with existing arrays [46].

### 6.5.3 The Allan Variance for Interferometric Data

In describing phase-difference measurements for closely spaced receivers, we noted that phase trends are often common to the two paths and cancel

Table 6.2: Microwave measurements of the frequency stability, Allan variance and power spectrum for the phase difference between adjacent receivers of interferometers for signals that travel through the atmosphere; $N$ is the number of receivers

| Instrument and location | $f$ (GHz) | $N$ | Spacings | $\delta f/f$ | $\mathbb{A}(\tau,\rho)$ | $W(\omega,\rho)$ | Ref. |
|---|---|---|---|---|---|---|---|
| NMA, Nobeyama | 22 | 5 | 27–540 m | | x | | 47 |
| BIMA, | 86 | 3 | 12–300 m | x | | | 49 |
| Hat Creek | 86 | 6 | 6–600 m | | x | x | 50 |
| PdBI, | 86 | 3 | 24–288 m | x | | x | 51 |
| Plateau de Bure | | | | | | | |
| VLBI, | 100 | 3 | 400–1200 km | x | | | 49 |
| USA | 89 | 3 | 400–1200 km | | x | | 52 |

Table 6.3: Optical measurements of the temporal structure function, Allan variance and power spectrum for the phase difference between adjacent receivers of interferometers for signals that travel through the atmosphere; $N$ is the number of receivers

| Instrument and location | $\lambda$ ($\mu$m) | $N$ | Spacings (m) | $\mathcal{D}_\varphi(\tau)$ | $\mathbb{A}(\tau,\rho)$ | $W(\omega,\rho)$ | Ref. |
|---|---|---|---|---|---|---|---|
| LPO, La Palma | 0.7 | 2 | 0.15–2.2 | | | x | 53 |
| ISI, | 11 | 2 | 4 | x | x | x | 37 |
| Mount Wilson | | | | | | | |
| Mark III, | 0.7 | 2 | 12 | | | x | 54 |
| Mount Wilson | 0.8 | 5 | 3–31.5 | | | x | 18 |

out on taking their difference. One cannot be confident that this is valid for microwave interferometers. Conditions along the ray paths to widely separated receivers are sometimes quite different. In most VLA and VLBI experiments one prefers to use the Allan variance from which linear phase trends disappear. One can describe these measurements by replacing $\varphi(t)$ by $\Delta\varphi(t)$ in the basic definition (6.62) for the Allan variance:

$$\mathbb{A}(\tau,\rho) = \langle[\Delta\varphi(t+\tau) - 2\,\Delta\varphi(t) + \Delta\varphi(t-\tau)]^2\rangle \qquad (6.94)$$

This gives a convenient measure of the coherence time for an interferometer and sets a limit to the coherent integration time which can be used to avoid reducing the amplitude in averaging data.

The Allan variance is directly related to the power spectrum of phase-difference fluctuations as indicated in Problem 5. It has an advantage over the traditional description in which the low-frequency portion is dominated by the averaging time. By contrast, the Allan variance should be relatively insensitive to the sample length and this is confirmed experimentally [47][48]. Guides to measurements of the Allan variance are provided in Tables 6.2 and Table 6.3.

### 6.5.3.1 An Analytical Description for $\mathbb{A}(\tau, \rho)$

One can describe these experiments by relating the Allan variance of the phase difference to the phase structure function for atmospheric transmission. To accomplish that goal we rewrite the basic expression (6.94) in terms of the time-shifted single-path phase fluctuations measured at the two receivers:

$$\mathbb{A}(\tau, \rho) = \langle \{ [\varphi_1(t+\tau) - 2\varphi_1(t) + \varphi_1(t-\tau)]$$

$$- [\varphi_2(t+\tau) - 2\varphi_2(t) + \varphi_2(t-\tau)] \}^2 \rangle \qquad (6.95)$$

We can express $\mathbb{A}(\tau, \rho)$ in terms of $\mathcal{D}_\varphi(\rho)$ by using the algebraic identity (5.57) and the previous result (6.63) for the single-path Allan variance. If the basic structure function is symmetrical about the origin,

$$\mathbb{A}(\tau, \rho) = 8\mathcal{D}_\varphi(v\tau) + 6\mathcal{D}_\varphi(\rho) - 4\mathcal{D}_\varphi(|\boldsymbol{\rho} - \mathbf{v}\tau|) - 4\mathcal{D}_\varphi(|\boldsymbol{\rho} + \mathbf{v}\tau|)$$

$$- 2\mathcal{D}_\varphi(2v\tau) + \mathcal{D}_\varphi(|\boldsymbol{\rho} - 2\mathbf{v}\tau|) + \mathcal{D}_\varphi(|\boldsymbol{\rho} + 2\mathbf{v}\tau|). \qquad (6.96)$$

The wind vector is represented here by a single value, although we know that it varies with altitude. This is merely a *symbolic* convention since the description for the transmission structure function $\mathcal{D}_\varphi$ that we will use includes the wind vector profile.

One can use this result to make some useful statements about $\mathbb{A}(\tau, \rho)$ that do not depend on the specific structure function. Note first that the Allan variance should be independent of the separation when the inter-receiver spacing is much larger than the temporal displacement:

$$v\tau \ll \rho: \qquad \mathbb{A}(\tau, \rho) = 8\mathcal{D}_\varphi(v\tau) - 2\mathcal{D}_\varphi(2v\tau) \qquad (6.97)$$

The opposite situation occurs for delay-time displacements that are large relative to the spacing:

$$v\tau \gg \rho: \qquad \mathbb{A}(\tau, \rho) = 6\mathcal{D}_\varphi(\rho) \qquad (6.98)$$

This suggests that $\mathbb{A}(\tau, \rho)$ should approach an asymptotic value for large time delays, which should increase with the separation between receivers.

These predictions can be made explicit by using the description (6.91) of the transmission structure function. That is a complicated expression and we will make two assumptions to simplify it. The first has to do with the wind direction. Notice that the wind vector appears in (6.91) with components that are *in* the plane of propagation and *normal* to it. For terrestrial links we found that the wind component along the line of sight has no effect on phase variability. That selection is not apparent in the transmission expression for $\mathcal{D}_\varphi$ because we have delayed the ray-path integration until the last step. We suspect that the linear velocity term in (6.91) is relatively unimportant for low-elevation sources because they are similar to terrestrial paths:

$$\zeta\, v_x \tau \sin\vartheta \simeq 0 \qquad (6.99)$$

The same term vanishes for sources near the zenith because $\vartheta = 0$. In both cases one can replace the vector arguments in (6.96) by their scalar counterparts since $\rho$ and $\mathbf{v}$ are effectively parallel. The same observation allows one to simplify the transmission structure function (6.91) as follows:

$$\mathcal{D}_\varphi\!\left(\rho, \frac{\pi}{2}, \vartheta, \tau, \mathbf{v}\right) = 2k^2 \sec^2\vartheta\, a^{-\frac{2}{3}} \int_0^\infty dz\, C_n^2(z) \int_0^z d\zeta$$
$$\times \{|\zeta^2 + [\rho \pm \tau v(z)]^2|^{\frac{1}{3}} - |\zeta^2|^{\frac{1}{3}}\} \qquad (6.100)$$

which is identical to (5.46) with $\rho$ replaced by $\rho \pm \tau v(z)$. To make further progress one must know how the wind speed and $C_n^2$ vary with height.

This brings us to a second assumption frequently used in estimating the Allan variance. Contrary to meteorological measurements, we assume that the wind vector does not vary with height. At first sight this seems quite reckless. On the other hand, we notice that $C_n^2(z)$ weights different height levels in a commanding way. Only winds at the height range for which $C_n^2$ is a maximum play an important role. The selection of a dominant height range also depends on the inter-receiver separation. We learned in Chapter 5 that a comparatively small height range is influential for interferometer measurements. The wind vector can be relatively constant in this narrow region for the limited periods typically used to make astronomical observations. In this case the transmission structure function (6.100) can be replaced by (5.50) and (5.51) when one is estimating the Allan variance.

The 5/3 scaling-law function is appropriate when the combined displacement is less than the effective height of the troposphere. Using the shorthand notation introduced in (5.62),

$$\rho + v\tau < \frac{a}{c} H : \qquad \mathcal{D}_\varphi(\rho) = \mathsf{Q}\rho^{\frac{5}{3}}$$

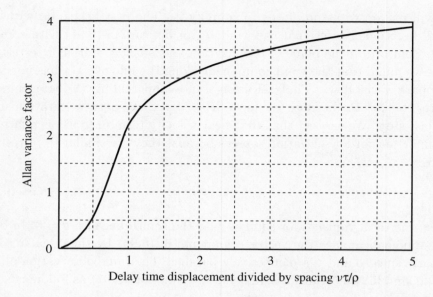

Figure 6.18: Numerical values of the function identified by square brackets in (6.101) plotted as a function of the delay-time displacement divided by the inter-receiver separation. This prediction assumes that the combined displacement $\rho + v\tau$ is less than the effective height of the turbulent atmosphere.

we find the following relationship for the Allan variances:

$$\mathbb{A}(\tau,\rho) = Q\rho^{\frac{5}{3}}\left[6 + 8\left(\frac{v\tau}{\rho}\right)^{\frac{5}{3}} - 4\left|1 + \frac{v\tau}{\rho}\right|^{\frac{5}{3}} - 4\left|1 - \frac{v\tau}{\rho}\right|^{\frac{5}{3}}\right.$$

$$\left. -2\left(2\frac{v\tau}{\rho}\right)^{\frac{5}{3}} + \left|1 + 2\frac{v\tau}{\rho}\right|^{\frac{5}{3}} + \left|1 - 2\frac{v\tau}{\rho}\right|^{\frac{5}{3}}\right] \qquad (6.101)$$

The function in square brackets which multiplies the basic structure function depends solely on the ratio $v\tau/\rho$. It is plotted in Figure 6.18 and exhibits an initial rise followed by a very slow approach to the asymptotic value of 6.

One must be more cautious in using the 2/3 scaling law (5.66) to estimate the Allan variance. The terms in (6.96) containing $\rho$ can certainly be replaced in that way. There is no assurance, however, that $\mathcal{D}_\varphi(v\tau)$ and $\mathcal{D}_\varphi(2v\tau)$ have arguments whose displacements exceed the effective height of the atmosphere. Even when $v\tau$ is comparable to the separation, the terms with difference arguments will not necessarily be represented properly by the 2/3 scaling law.

### 6.5.3.2 Measurements of $\mathbb{A}(\tau, \rho)$

In a comprehensive experiment Wright measured the Allan deviation[8] at 86 GHz using the BIMA array at Hat Creek, California [50]. Simultaneous data was taken with fifteen different inter-receiver spacings for time intervals ranging from 20 to 6000 s. Results for four spacings selected from this large data set are reproduced in Figure 6.19.

Our predictions provide only a very modest explanation for this data. From concurrent measurements of phase structure function it was noted that the effective height of the turbulent atmosphere seemed to cause a break for spacings between 200 and 300 m at this site [50]. The time intervals and receiver spacings employed require different forms of the structure function in the different regions of Figure 6.19.

A reasonable explanation for spacings less than 160 m can be constructed for this data. In this range one can justify using the prediction (6.101) provided that the time interval is not so large that $\rho + v\tau > H$. The curves for spacings of 80 and 160 m exhibit two features. The first is that they reach asymptotic values at approximately $\tau = 100$ s and those asymptotic values increase with the inter-receiver spacing. Both features are consistent with (6.101). The wind speed on this day was measured as $v = 2.2$ m s$^{-1}$, which leads one to expect saturation at $\tau_c \simeq \rho/v \simeq 100$ s, in good agreement with the experiment. The asymptotic values should scale as $\rho^{\frac{5}{3}}$ and this too is consistent.

The second feature of the data was its initial behavior. For very small time intervals the theoretical expression (6.101) suggests that

$$H > \rho \gg v\tau: \quad \mathbb{A}(\tau, \rho) \simeq 0.206 Q (v\tau)^{\frac{5}{3}} \tag{6.102}$$

This behavior is not evident in Figure 6.19, possibly because the time delay begins at $\tau = 20$ s and this range was missed. For larger time delays Figure 6.19 indicates that the structure-function factor is proportional to $v\tau/\rho$, which would predict

$$\mathbb{A}(\tau, \rho) = \text{constant} \times \tau \rho^{\frac{2}{3}}. \tag{6.103}$$

Wright [50] fitted the experimental data in this range to the following scaling law:

$$\mathbb{A}(\tau, \rho) \simeq \text{constant} \times \tau^{0.86 \pm 0.04} \tag{6.104}$$

---

[8] The Allan deviation $\sigma_A$ is simply the square root of the Allan variance $\mathbb{A}(\tau, \rho)$.

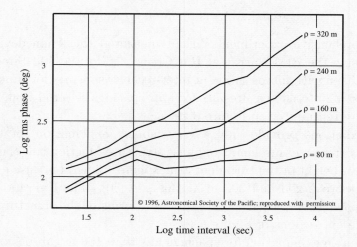

Figure 6.19: The Allan deviation of the phase difference measured as a function of the delay time for various inter-receiver separations at 86 GHz by Wright [50]. These curves were selected from a larger and more complicated data set taken on 21 December 1994 using the BIMA array at Hat Creek.

which is reasonably close to the prediction. The result (6.103) indicates that the larger separations should have greater values in this region and that feature was indeed observed.

Our simplified analysis does not provide a satisfactory explanation for data in the range 160 m $< \rho <$ 600 m. Here we expect to see a transition to the 2/3 scaling law for the structure function for many of the $(\rho, \tau)$ combinations. Our previous reservations regarding the use of the 2/3 relation in estimating (6.96) are thus important. Note that the experimental data continues to rise over the entire interval for these large separations. The pyramid of assumptions we used to simplify the structure function has undoubtedly compromised our description. It is clear that additional work is needed to explain interferometer data in terms of atmospheric transmission models.

### 6.5.4 The Frequency Stability of Arriving Signals

Some of the millimeter-wave signals that arrive from galactic sources are remarkably stable in frequency. This is because they originate in molecular vibrations whose frequency is precisely known and because their red shift is constant. The 86.243-GHz signals often used for such observations result from maser emissions by the SiO molecule in distant quasars. The variability of the frequency of these signals is caused by time-varying phase fluctuations produced in the turbulent atmosphere. Their measurements are identified in Table 6.2.

The frequency stability of arriving signals is defined by the variance of frequency fluctuations [38]:

$$\sigma_f^2 = \tfrac{1}{2}\langle [f(t+\tau) - f(t)]^2 \rangle \qquad (6.105)$$

This is closely related to the Allan variance, as one can see by rewriting the basic definition in the following form:

$$\mathbb{A}(\tau,\rho) = \tau^2 \left\langle \left( \frac{\Delta\varphi(t+\tau) - \Delta\varphi(t)}{\tau} - \frac{\Delta\varphi(t) - \Delta\varphi(t-\tau)}{\tau} \right)^2 \right\rangle$$

The instantaneous electromagnetic frequency is the rate of change of phase with time and the time-differenced terms in this expression approximate this derivative for small time delays:

$$\mathbb{A}(\tau,\rho) = (2\pi\tau)^2 \langle [f(t+\tau) - f(t)]^2 \rangle$$

An important connection between the frequency stability and the Allan deviation is thus established:

$$\sigma_f = \frac{1}{\sqrt{2}\,\pi f} \frac{\sigma_{\mathrm{A}}(\tau,\rho)}{\tau} \qquad (6.106)$$

The frequency fluctuation should decline with increasing separation in time since the Allan variance saturates for large time delays. Measurements of frequency variability were made with the PdBI array in France [51]. The scaling law for $\sigma_f$ versus time was measured for delays ranging from 4 s to 1 min and separations that varied between 24 and 241 m. This suggests that $v\tau > \rho$ for typical wind speeds and one would assume that the Allan deviation was relatively constant for each baseline. This means that the rms frequency shift should vary inversely with the delay time. The measurements gave a slope of $-1.0 \pm 0.2$ in good agreement with that prediction. Similar data was obtained using the BIMA array at Hat Creek [49], where the slope was measured as $-0.83 \pm 0.05$ for baselines ranging from 12 to 300 m.

The frequency variability has also been measured for VLBI configurations. In a collaboration involving the Hat Creek, Owens Valley and Kitt

Peak observatories, signals at 100 GHz were compared against stable frequency sources and later compared with each other [49]. The frequency deviation $\sigma_f$ was measured in the time range 10 s $< \tau <$ 210 s for baselines of 400, 800 and 1200 km. These separations are much greater than the temporal displacement $v\tau$ and one would expect (6.97) to describe the experimental results. The structure-function arguments in that expression should straddle the break created by the effective height of the atmosphere for a typical wind speed of $v = 10$ m s$^{-1}$:

$$\sigma_f = \begin{cases} \text{constant} \times \tau^{-\frac{1}{6}} & v\tau < H \\ \text{constant} \times \tau^{-\frac{2}{3}} & v\tau > H \end{cases} \tag{6.107}$$

The measured data was characterized by the following scaling laws:

$$\sigma_f = \begin{cases} \text{constant} \times \tau^{-0.4\pm0.1} & 10 \text{ s} < \tau < 40 \text{ s} \\ \text{constant} \times \tau^{-0.6\pm0.1} & 40 \text{ s} < \tau < 210 \text{ s} \end{cases} \tag{6.108}$$

The second prediction of (6.107) is thus verified but the first is not. It is possible that the $\tau^{-\frac{1}{6}}$ behavior would have been observed for time delays less than 10 s and that the transition from 5/3 to 2/3 was under way in the range 10 s $< \tau <$ 40 s. In any case, it is clear that actual measurements are the best guide to frequency coherence for VLBI observations.

### 6.5.5 Interferometric Power-spectrum Measurements

The power spectrum can be computed from the covariance of the phase difference by replacing $\varphi$ by $\Delta\varphi$ in the basic relationship (6.34):

$$W_{\Delta\varphi}(\omega, \rho) = \int_{-\infty}^{\infty} d\tau \exp(i\omega\tau) \langle \Delta\varphi(t+\tau)\,\Delta\varphi(t) \rangle \tag{6.109}$$

One can substitute the general expression for the temporal covariance (6.81) and carry out the time-delay integrations when the structure function is known. That is a difficult calculation if the baseline and wind vectors are not perpendicular to the plane of propagation.

In this discussion we shall simplify matters by making the assumption (6.99) that was used to estimate the Allan variance.[9] This ignores the velocity component which lies in the plane of propagation identified in

---

[9] In Problem 7 we show how to calculate $W_{\Delta\varphi}(\omega, \rho)$ without this assumption.

Figure 6.17. This approximation is also supported by astronomical data taken at 86 GHz. Wright found that the structure-function scaling laws exhibited "*little correlation with wind direction*" [50]. Since the power spectrum is derived from $\mathcal{D}_\varphi$, this suggests that $W_{\Delta\varphi}$ may be relatively insensitive to the wind direction. That conjecture will be checked by comparing the results obtained here with interferometric spectra. With this lack of dependence on the wind direction, the plane-wave covariance expression (6.82) depends solely on scalar arguments. Because the power spectrum is directly related to the covariance, we change its components from the structure functions to the phase correlations:

$$W_{\Delta\varphi}(\omega,\rho) = \int_{-\infty}^{\infty} d\tau \exp(i\omega\tau)\langle\varphi^2\rangle[\,2\mu(v\tau) - \mu(|\rho - v\tau|) - \mu(|\rho + v\tau|)]$$

(6.110)

We shift the time variable in the second and third terms to establish a simple relationship between the phase difference and single-path power spectra:

$$\mathbf{v} \parallel \boldsymbol{\rho} \qquad W_{\Delta\varphi}(\omega,\rho) = 4\sin^2\!\left(\frac{\omega\rho}{2v}\right) W_\varphi(\omega) \qquad (6.111)$$

The factor $\sin^2[\omega\rho/(2v)]$ describes the interference pattern of the receivers. It depends only on the inter-receiver separation and the wind speed. The single-path phase power spectrum $W_\varphi(\omega)$ carries the entire burden of describing the anisotropy and altitude limitation of tropospheric irregularities. This result is formally the same as the description (6.88) for propagation of plane waves near the surface. The two outcomes are differentiated by the single-path phase spectrum appropriate to the type of propagation. For terrestrial links $W_\varphi(\omega)$ should be understood as the expression (6.37). We have no such result for atmospheric transmission and will derive one after addressing another important issue.

### 6.5.5.1 The Influence of the Sample Length

Microwave-interferometric data are usually averaged over substantial time intervals, as noted in Table 5.2. The sample length influences experimental data importantly. We explored this role in Section 5.2.3 by using Taylor's hypothesis to characterize atmospheric variability. That description can be simplified now that we have introduced the phase-difference power spectrum. Such measurements necessarily exclude frequencies that are less than the reciprocal sample length. We follow the course explained above in (6.42) to find that the sample-averaged spectrum can be written as the product of three terms [55]. If the sample-averaged mean value is subtracted from the data stream, the variance of the phase difference is described by

$$\overline{\langle \Delta\varphi^2 \rangle} = 4 \int_{-\infty}^{\infty} d\omega \, W_\varphi(\omega) \sin^2\!\left(\frac{\omega\rho}{2v}\right)\!\left(1 - \frac{\sin^2(\omega T/2)}{(\omega T/2)^2}\right) \qquad (6.112)$$

where $T$ is the sample length. The finite data sample therefore suppresses low-frequency components in the single-path spectrum. The receiver term $\sin^2[\omega\rho/(2v)]$ has the same effect and dominates the sample-length term so long as the crossing time is greater than the separation between receivers. These observations are usually made with $\rho < 2vT$ as noted in Table 5.2.

Some authors prefer to estimate the variance of the phase difference without subtracting the mean value [54][55]:

$$\overline{\langle M^2 \rangle} = 4 \int_{-\infty}^{\infty} d\omega \, W_\varphi(\omega) \sin^2\!\left(\frac{\omega\rho}{2v}\right) \frac{\sin^2(\omega T/2)}{(\omega T/2)^2} \qquad (6.113)$$

The single-path phase spectrum is evidently the key to both descriptions.

### 6.5.5.2 The Single-path Phase Power Spectrum for Atmospheric Transmission

One could try to establish the power spectrum from the structure function (6.91) which is based on the 2/3 law for refractive irregularities. That expression assumes that the outer scale length is infinite. One encounters problems in taking its Fourier transform when $\rho$ is replaced by $v\tau$. We believe that the horizontal outer scale length is large but bounded in the lower atmosphere. By contrast, the vertical scale length is quite small. It is important to retain a finite outer scale length so as to identify the influence of the dissimilar horizontal and vertical properties. When the computation is complete, one can allow the horizontal scale length to become large.

To accomplish this ambitious program we return to the description (5.38) for the structure function expressed in terms of the turbulence spectrum. Since the wind vector is assumed to be parallel to the baseline and normal to the downcoming rays, we can set $\beta = \pi/2$ and replace $\rho$ by $v\tau$ in that description. The corresponding covariance is given by the following expression:

$$\langle \varphi(t+\tau)\varphi(t) \rangle = 4\pi k^2 \int_0^\infty ds_1 \int_0^\infty ds_2 \, C_n^2\!\left(\frac{1}{2}(s_1+s_2)\right)$$

$$\times \int_0^\infty dq \, q\Omega(q) \, \frac{\sin\!\left(q\sqrt{\mathsf{p}^2(s_1-s_2)^2 + v^2\tau^2/a^2}\right)}{\sqrt{\mathsf{p}^2(s_1-s_2)^2 + v^2\tau^2/a^2}}$$

$$(6.114)$$

This result includes the influence of anisotropic irregularities through the oblique transmission factor $\mathsf{p}$ defined by (4.37), the stretched wavenumber $q$ and the scaling parameter $a$.

In Section 4.2.1 we found that the transmission parameter can be approximated by $\mathsf{p} \simeq \cos\vartheta/c$ unless the source is very close to the horizon. That situation is usually avoided for astronomical observations and monitoring satellite navigation signals. The sum and difference coordinates defined by (4.36) are used to transform the ray-path integrations. We then make the flat-earth approximation and shift to vertical coordinates $r = z\sec\vartheta$ and $u = \zeta\sec\vartheta$:

$$\langle \varphi(t+\tau)\varphi(t) \rangle = 8\pi k^2 \sec^2\vartheta \, a \int_0^\infty dz \, C_n^2(z) \int_0^z \frac{d\zeta}{v} \int_0^\infty dq \, q\Omega(q)$$

$$\times \left[\tau^2 + \left(\frac{a\zeta}{cv}\right)^2\right]^{-\frac{1}{2}} \sin\!\left(\frac{qv}{a}\sqrt{\tau^2 + \left(\frac{a\zeta}{cv}\right)^2}\right)$$

$$(6.115)$$

The single-path phase power spectrum is calculated by combining this expression for the temporal covariance with the definition (6.34). We let the Fourier integration on the delay time play past the height and wavenumber integrals to operate on the last term:

$$W_\varphi(\omega) = 8\pi k^2 \sec^2\vartheta \, a \int_0^\infty dz \, C_n^2(z) \int_0^z \frac{d\zeta}{v} \int_0^\infty dq \, q\Omega(q) \mathcal{J}(q,\omega,\zeta)$$

where

$$\mathcal{J}(q,\omega,\zeta) = 2 \int_0^\infty d\tau \cos(\omega\tau) \left[\tau^2 + \left(\frac{a\zeta}{cv}\right)^2\right]^{-\frac{1}{2}} \sin\!\left(\frac{qv}{a}\sqrt{\tau^2 + \left(\frac{a\zeta}{cv}\right)^2}\right).$$

This integral can be found in Appendix B:

$$W_\varphi(\omega) \,=\, 8\pi^2 k^2 \sec^2 \vartheta \, a \int_0^\infty dz \, C_n^2(z) \int_0^z \frac{d\zeta}{v}$$

$$\times \int_{\omega a/v}^\infty dq \, q\Omega(q) J_0\!\left(\frac{\zeta}{c}\sqrt{q^2 - \left(\frac{\omega a}{v}\right)^2}\right) \tag{6.116}$$

The level of turbulent activity $C_n^2(z)$ is an explicit function of height. Notice that variations of wind speed with altitude are also accommodated.

At this stage we have made no assumption about the turbulence spectrum except that its anisotropic properties are described by the ellipsoidal model. To proceed further we select the von Karman model. The terrestrial result (6.38) then provides a useful comparison for the more complicated situation we are now considering. The spectrum for atmospheric transmission becomes

$$W_\varphi(\omega) = 0.033 \times 8\pi^2 k^2 \sec^2 \vartheta \int_0^\infty dz \, C_n^2(z) \int_0^z \frac{d\zeta}{v} \, \mathcal{Z}(\zeta, \omega) \tag{6.117}$$

where

$$\mathcal{Z}(\zeta, \omega) = \int_{\omega a/v}^\infty \frac{dq \, q}{(q^2 + \kappa_0^2)^{\frac{11}{6}}} J_0\!\left(\frac{\zeta}{c}\sqrt{q^2 - \left(\frac{\omega a}{v}\right)^2}\right).$$

The wavenumber integration can be done by setting $q^2 = x^2 + (\omega a/v)^2$ and using a result from Appendix D:

$$\mathcal{Z}(\zeta, \omega) = \frac{1}{\Gamma(\frac{11}{6})} \left\{ \frac{\zeta v}{2ac}\left[\omega^2 + \left(\frac{v\kappa_0}{a}\right)^2\right]^{-\frac{1}{2}} \right\}^{\frac{5}{6}} K_{\frac{5}{6}}\!\left(\frac{\zeta a}{cv}\sqrt{\omega^2 + \left(\frac{v\kappa_0}{a}\right)^2}\right) \tag{6.118}$$

The last step is to complete the vertical integrations in (6.117). To do so one must know accurately how $C_n^2$ and $v$ vary with height.

In Chapter 5 we found that interferometric data are influenced primarily by the convective boundary layer. This region is often approximated by a slab model in which one assumes that $C_n^2$ is constant up to a fixed height and vanishes above it. If the wind speed can also be considered constant we can rescale the height integrations and the spectrum becomes

$$W_\varphi(\omega) = \frac{0.033 \times 8\pi^2}{2^{\frac{1}{6}}\Gamma(\frac{11}{6})} \frac{H \sec^2 \vartheta \, k^2 C_n^2 c a^{-\frac{5}{3}} v^{\frac{5}{3}}}{\left[\omega^2 + \left(\frac{\kappa_0 v}{a}\right)^2\right]^{\frac{4}{3}}} \int_0^L dx \left(1 - \frac{x}{L}\right) x^{\frac{5}{6}} K_{\frac{5}{6}}(x) \tag{6.119}$$

where the upper limit is defined by the dimensionless combination

$$L = \frac{a}{c}\frac{H}{v}\sqrt{\omega^2 + \left(\frac{v\kappa_0}{a}\right)^2}.$$

At this point we need to estimate $L$ for typical atmospheric conditions. The effective height of the tropospheric irregularities $H$ is about 1 km and the wind speed is typically $v = 10$ m s$^{-1}$. The elongation ratio $a/c$ is roughly 100 in the troposphere and the power spectrum is seldom measured below $10^{-3}$ Hz. These combine to make $L$ greater than 100 and we therefore let it go to infinity. The surviving integral is a special case of a result to be found in Appendix D and the following expression for the power spectrum emerges:

$$\text{Slab Model:} \quad W_\varphi(\omega) = \frac{2.192H \sec^2\vartheta\, k^2 C_n^2 ca^{-\frac{5}{3}}v^{\frac{5}{3}}}{\left[\omega^2 + \left(\frac{\kappa_0 v}{a}\right)^2\right]^{\frac{4}{3}}} \qquad (6.120)$$
$$v = \text{constant}$$

This result is surprisingly close to the terrestrial-link expression (6.38) but differs from it in several ways that are driven by the medium's anisotropy. The multiplicative factor

$$ca^{-\frac{5}{3}}$$

reduces the absolute level of the power spectrum relative to the terrestrial case. More fundamentally, the frequency dependence of (6.120) is determined by the large value of the elongation ratio $a/c$. This scaling with frequency corresponds to the 5/3 scaling law for the phase structure function as discussed in Problem 8 and is a consequence of the relationships between the effective height of tropospheric irregularities and the other parameters. It is a good description for optical interferometers that employ both small baselines and short averaging times.

A different frequency dependence would result in the region described by the 2/3 scaling law (5.51). This is probably important for microwave arrays that use large separations and long sampling times. One cannot then be confident that (6.120) provides a good description. In addition, it is important to remember that our basic result assumes that $C_n^2$ and $v$ are constant. This is only a crude approximation to the real atmosphere and we need to compare it against interferometer observations before adopting it.

### 6.5.5.3 Interferometric Measurements of Power Spectra

The power spectrum of $\Delta\varphi$ is a common measurement both for microwave and for optical interferometers. One can describe it by combining (6.111) and (6.120) to generate the following scaling law:

$$W_{\Delta\varphi}(\omega, \rho) = \text{constant} \times \frac{\sin^2[\omega\rho/(2v)]}{\left[\omega^2 + \left(\dfrac{\kappa_0 v}{a}\right)^2\right]^{\frac{4}{3}}} \tag{6.121}$$

We need to estimate the threshold frequency for atmospheric transmission before using this expression:

$$\omega_0 = \frac{\kappa_0 v}{a} = 2\pi \frac{v}{aL_0}$$

Here $aL_0$ is the horizontal outer scale length, which is several kilometers along most of the path traveled by the downcoming signal. This means that $\omega_0$ is less than $10^{-2}$ Hz for wind speeds near 10 m s$^{-1}$. The second term in the denominator of (6.121) is often ignored and astronomical measurements are typically compared with the expression

$$W_{\Delta\varphi}(\omega, \rho) = \text{constant} \times \sin^2[\omega\rho/(2v)]\,\omega^{-\frac{8}{3}}. \tag{6.122}$$

This suggests that the spectrum should behave as $\omega^{-\frac{2}{3}}$ below the corner frequency $\omega_c = v/\rho$. Above that value one would expect the spectrum to scale as $\omega^{-\frac{8}{3}}$ but with a continuing harmonic modulation.

Phase-difference power spectra were measured with the Mark III optical interferometer on Mount Wilson [19]. Three fixed and two movable siderostats were used to generate spectra for a variety of baselines, although simultaneous data for different baselines was not obtained. A spectrum measured with a separation of 31.5 m is reproduced in Figure 6.20. This data includes frequencies from $10^{-3}$ to 10 Hz and covers nine orders of magnitude for the power spectrum. The dotted line corresponds to the prediction (6.122). Harmonic modulation caused by the basic interference pattern is evident across the entire spectrum. A transition from the $-2/3$ scaling law to $-8/3$ occurs at a frequency of approximately 0.1 Hz, which agrees with the estimated corner frequency for a wind speed of 10 m s$^{-1}$. The slope of the high-frequency region was measured on different nights with various baselines. The experimental slope values clustered around $-2.55$ and were close to the predicted value $-2.67$.

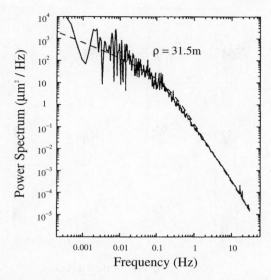

Figure 6.20: The power spectrum of the phase difference measured on Mount Wilson with the Mark III optical interferometer by Buscher, Armstrong, Hummel, Quirrenbach, Mozurkewich, Johnston, Denison, Colavita and Shao [19]. The dotted curve corresponds to (6.122), which implies an infinite outer scale length. This data was taken on 2 July 1989 with an inter-telescope separation of 31.5 m.

Analysis of the Mark III data showed that the spectra often became level below $10^{-2}$ Hz [19]. That could be due to either the finite integration time or the horizontal outer scale length. By comparing these measurements with (6.121), the observed effect was identified with outer-scale flattening. Scale lengths comparable to the spacing between receivers seemed to best explain the data. Including this correction improved the agreement between measured slopes in the high-frequency region and the $-8/3$ prediction. Notice that $L_0 = 30$ m is substantially smaller than the estimates 1 km $< L_0 < 10$ km found elsewhere for the horizontal outer scale length in the anisotropic troposphere.

The wind speeds evident in this data were substantially greater than the measured surface winds, as had been noticed in earlier microwave experiments [44]. This means that our slab model with constant $C_n^2$ and $v$ is too crude. More general profile descriptions were explored by Lindegren [56].

We would like to understand how the phase-difference spectrum depends on the length and orientation of the baseline. Our prediction (6.111) suggests that the separation dependence is uncoupled from the properties of

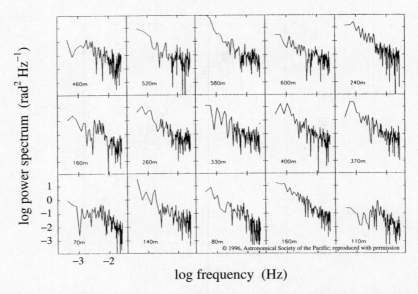

log power spectrum (rad² Hz⁻¹)

log frequency (Hz)

Figure 6.21: Phase-difference power spectra measured simultaneously for fifteen projected baselines by Wright [50]. This data was taken on 21 December 1994 at Hat Creek, California with the BIMA facility using 86-GHz signals.

the tropospheric irregularities and concentrated in a single term:

$$\mathbf{v} \parallel \boldsymbol{\rho} \qquad W_{\Delta\varphi}(\omega, \rho) = \text{constant} \times \sin^2[\omega\rho/(2v)] \qquad (6.123)$$

It is important to remember that this result depends on two assumptions: (1) the wind vector is parallel to the receiver baseline, and (2) both are perpendicular to the plane of propagation. We examine the more general case in which $\mathbf{v}$ and $\boldsymbol{\rho}$ are not parallel in Problem 7. We learn there that the baseline dependence cannot be extracted from the height integrations, as it has been in establishing this simple expression.

To test these predictions requires simultaneous measurements of phase-difference spectra for a number of inter-receiver spacings. Such data was taken in experiments using 86-GHz signals and fifteen inter-receiver separations ranging from 70 to 600 m at Hat Creek [50]. A typical data set is reproduced in Figure 6.21 in which the indicated separations are projected baselines orthogonal to the position of the astronomical source. Our simple relationships suggest that a scaling-law transition should occur at the corner frequency:

$$\omega_c = \frac{v}{2\pi\rho} \text{ Hz}$$

It is significant that the BIMA receivers are not collinear and it is impossible to identify a unique wind speed connecting them. For the shorter baselines of 70 m $< \rho <$ 160 m, one would expect a change in slope beyond approximately $10^{-2}$ Hz if the wind speed were 10 m s$^{-1}$. Unfortunately that is about where the data stops. The relatively flat behavior below this point corresponds to the $\omega^{-\frac{2}{3}}$ prediction, for which effects of the outer scale and sample lengths may also be important. The longer-baseline measurements with 370 m $< \rho <$ 600 m should exhibit a transition at frequencies between $3 \times 10^{-3}$ and $4 \times 10^{-3}$ Hz for the same wind speed. This change in scaling laws is not apparent in Figure 6.21. Measured slopes above these corner frequencies ranged from $-1.2$ to $-2.1$ and are quite different than the expected value $-8/3$. On the other hand, the data does not extend very far above $\omega_c$ and one might be seeing only the predicted transition from $-2/3$ to $-8/3$. It would be desirable to extend such measurements to the larger frequency range shown in Figure 6.20.

### 6.5.5.4 The Applicability of Taylor's Hypothesis

In presenting Taylor's hypothesis for describing wind-borne irregularities we noted that it is usually a good approximation for short-baseline optical interferometers. On the other hand, we found in (6.10) that its conditions need not be satisfied for large microwave arrays. There are several problems. Large irregularities have the greatest influence because of their relative strength and large horizontal size, which is often comparable to the inter-receiver separation. Moreover, their speed is the greatest of all the eddies in the turbulence hierarchy and one cannot be sure that they remain frozen as the mean wind field transports the random structure from one receiver to another. This is compounded for many arrays by the problem that not all the antennas are collinear and one must consider different projections of the wind for different pairs of receivers. To make matters worse, the direction and speed of the wind may not be the same at all the receivers in a large array. Such questions are even more important for VLBI observations.

In view of these problems, it is surprising that so much microwave data has been analyzed by assuming that a frozen medium is being borne past the array on a constant wind. One must look to the results of measurement programs to see whether this approximation is justified. In cross-correlation measurements in the C band using the Cambridge array, Hinder found that the majority of his data *could be* explained by a bulk motion of the atmosphere moving past the telescopes at a constant speed [44]. In the Nobeyama experiments at 22 GHz [47] the spectra of the Allan variance were found to be similar to the wind spectra, although low-frequency components would have been removed in estimating $\mathbb{A}(\tau, \rho)$. Olmi and Downes compared the slopes of phase-difference spectra taken in two different ways at 86 GHz using the PdBI array [51]. They found that Taylor's hypothesis was obeyed in some cases but counter examples were also observed. Measurements reported from the VLA facility at the NRAO [45] have consistently been more skeptical, probably owing to its large size. They have questioned whether there is a unique wind velocity that can connect spatially and temporally displaced phase measurements and note that averaging of data always plays an important role. Wright has also been cautious regarding Taylor's hypothesis on the basis of his BIMA observations at 86 GHz [50]. He notes that turbulence is neither frozen nor carried at a constant speed over the baseline distances employed there. He observes, as we do, that big eddies move faster than do small ones and that their speed is often comparable to the mean wind speed. Wright assumes that the distance–time connection is true only in a statistical sense: $\langle \rho^2 \rangle = \langle v^2 \tau^2 \rangle$ rather than $\rho = v\tau$. Although there is conflicting evidence, the observations unfortunately *do not support* the widespread use of Taylor's hypothesis that has been made in analyzing data taken with microwave arrays.

## 6.6  Problems

### Problem 1

The relation (6.25) is an integral equation connecting the phase covariance and the spectrum of irregularities. Using the orthogonality of the Bessel functions, show that it can be inverted to yield an exact solution:

$$\Phi_n(\kappa) = \frac{1}{4\pi^2 k^2 R} \int_0^\infty \tau J_0(\kappa v \tau) \langle \varphi(t) \varphi(t + \tau) \rangle \, d\tau$$

One could calculate the turbulence spectrum from this relationship if one could measure the correlation accurately for a wide range of time delays. By conducting an error analysis of this relationship, examine how that attractive possibility is frustrated by drift in the experimental data and phase noise in the receivers.

### Problem 2

Regard the connection (6.37) between the power spectrum of phase fluctuations and the distribution of turbulent irregularities as an integral equation.

Show that this relation can be inverted to provide the following equivalent solutions:

$$\Phi_n(\kappa) = \frac{-v^2}{\pi^3 R \, k^2} \frac{d}{d\kappa} \left( \kappa \int_{\kappa U}^{\infty} \frac{W_\varphi(\omega)}{\omega \sqrt{\omega^2 - \kappa^2 v^2}} \, d\omega \right)$$

$$= \frac{-v}{\pi^2 R \, k^2} \frac{d}{d\kappa} \int_0^{\pi/2} W_\varphi(\kappa v \sec z) \, dz$$

What is the physical meaning of these results? How are they likely to be limited by noise in the phase data?

### Problem 3

Consider the role that a finite sample length plays in autocorrelation measurements. Represent the random phase history by its Fourier transform as suggested in (6.42). Use the ensemble average of Fourier frequency components presented in (6.43) to show that the phase covariance is expressed as the following weighted integration of the power spectrum:

$$\langle \varphi(t+\tau)\varphi(t) \rangle = \frac{1}{\pi} \int_0^{\infty} W_\varphi(\omega) \, d\omega \left[ \cos(\omega\tau) + \left( \frac{\sin(\omega T/2)}{\omega T/2} \right)^2 \right.$$

$$\left. -2\cos\left( \frac{\omega\tau}{2} \right) \left( \frac{\sin(\omega T/2)}{\omega T/2} \right) \left( \frac{\sin(\omega(T-\tau)/2)}{\omega(T-\tau)/2} \right) \right]$$

Notice that this reduces to expression (6.45) for the phase variance.

### Problem 4

Calculate the Allan variance for single-path phase variations and express it in terms of a weighted wavenumber integral of the turbulence spectrum. Compute it as a function of $\kappa_0 v \tau$ for the von Karman model.

### Problem 5

Establish the following connection between the Allan variance and the phase-difference power spectrum:

$$\mathbb{A}(\tau,\rho) = \frac{1}{2\pi} \int_{-\infty}^{\infty} d\omega \, W_{\Delta\varphi}(\omega,\rho)[6 - 8\cos(\omega\tau) + 2\exp(-2i\omega\tau)]$$

## Problem 6

Describe the interferometric observations of an overhead source. Assume that the horizontal wind does not vary with altitude and makes an angle $\beta$ with the receiver baseline. Assume that the random medium is isotropic and show that the covariance is

$$2\langle \Delta\varphi(t)\,\Delta\varphi(t+\tau)\rangle = \mathcal{D}_\varphi(|\boldsymbol{\rho} - \mathbf{v}\tau|) + \mathcal{D}_\varphi(|\boldsymbol{\rho} + \mathbf{v}\tau|) - 2\mathcal{D}_\varphi(|\mathbf{v}\tau|)$$

where the vector displacements are given by

$$|\boldsymbol{\rho} \pm \mathbf{v}\tau| = \sqrt{\rho^2 + v^2\tau^2 \pm 2\rho v\tau \cos\beta}.$$

For separations that are less than the effective height of the random atmosphere one can use the 5/3 scaling law for the structure functions. In that case show that the autocorrelations of $\Delta\varphi$ for the two orthogonal directions are given by the following expressions:

$$\mathbf{v} \parallel \boldsymbol{\rho}: \qquad C_{\Delta\varphi}(\tau, \rho) = \frac{1}{2}\left[\left|1 + \frac{v\tau}{\rho}\right|^{\frac{5}{6}} + \left|1 - \frac{v\tau}{\rho}\right|^{\frac{5}{6}} - 2\left(\frac{v\tau}{\rho}\right)^{\frac{5}{3}}\right]$$

$$\mathbf{v} \perp \boldsymbol{\rho}: \qquad C_{\Delta\varphi}(\tau, \rho) = \left[1 + \left(\frac{v\tau}{\rho}\right)^2\right]^{\frac{5}{6}} - \left(\frac{v\tau}{\rho}\right)$$

Compute and plot these functions to explore their dependences on the relative direction of $\mathbf{v}$ and $\boldsymbol{\rho}$.

Use the covariance expression above to describe the phase-difference power spectrum for these conditions. For an arbitrary isotropic spectrum show that

$$W_{\Delta\varphi}(\omega) = \frac{4\pi^2 k^2}{v} \int_0^\infty dz \int_0^\infty d\zeta\, \Phi_n\!\left(\sqrt{\zeta^2 + \omega^2/v^2}, z\right)$$
$$\times \left[1 - \cos\left(\zeta\rho\sin\beta\right)\cos\left(\frac{\rho\omega}{v}\cos\beta\right)\right]$$

as Yura was the first to find [57]. This can be evaluated in terms of a MacDonald function for the von Karman model with parameters $C_n^2$ and $\kappa_0$ that vary with height:

$$W_{\Delta\varphi}(\omega) = 4.390\frac{k^2}{v} \int_0^\infty dz\, C_n^2(z)\frac{1}{\left(\kappa_0^2 + \omega^2/v^2\right)^{\frac{4}{3}}}\mathcal{V}(\omega)$$

where

$$\mathcal{V}(\omega) = 1 - \frac{\cos\left(\dfrac{\rho\omega}{v}\cos\beta\right)}{2^{\frac{1}{3}}\Gamma\left(\frac{4}{3}\right)}\left(\rho\sin\beta\ \sqrt{\kappa_0^2 + \omega^2/v^2}\right)^{\frac{4}{3}}$$
$$\times K_{\frac{4}{3}}\left(\rho\sin\beta\sqrt{\kappa_0^2 + \omega^2/v^2}\right)$$

How does anisotropy influence these results?

## *Problem 7*

Establish a transmission expression for the phase-difference power spectrum that does not restrict the horizontal wind velocity to being parallel to the receiver baseline. Assume that the receiver baseline is perpendicular to the plane of propagation and that the elevation of the source is greater than 5°. Use the ellipsoidal model (2.87) to describe the anisotropic nature of the irregularities and the von Karman spectrum to characterize their wavenumber decomposition. Employ the $(v, \beta)$ description of the wind velocity identified in Figure 6.17 to establish the following expression:

$$W_{\Delta\varphi}(\omega, \rho) = \frac{0.033 \times 8\pi^2 k^2 \sec^2\vartheta\, a}{\Gamma(\frac{11}{6})} \int_0^\infty dz\, C_n^2(z) \int_0^z \frac{d\zeta}{v}$$
$$\times \cos\left(\frac{\omega}{v}\cos\beta\tan\vartheta\right) \mathcal{Z}'(\zeta, \omega)$$

where $\omega_0 = v\kappa_0/a$ and

$$\mathcal{Z}'(\zeta, \omega) = 2\left(\varrho_1\frac{v^2}{2a^2}\left(\omega^2 + \omega_0^2\right)^{-\frac{1}{2}}\right)^{\frac{5}{6}} K_{\frac{5}{6}}\left(\varrho_1\sqrt{\omega^2 + \omega_0^2}\right)$$
$$- \left(\varrho_2\frac{v^2}{2a^2}\left(\omega^2 + \omega_0^2\right)^{-\frac{1}{2}}\right)^{\frac{5}{6}} K_{\frac{5}{6}}\left(\varrho_2\sqrt{\omega^2 + \omega_0^2}\right)\exp\left(+i\frac{\omega\rho}{v}\sin\beta\right)$$
$$- \left(\varrho_3\frac{v^2}{2a^2}\left(\omega^2 + \omega_0^2\right)^{-\frac{1}{2}}\right)^{\frac{5}{6}} K_{\frac{5}{6}}\left(\varrho_3\sqrt{\omega^2 + \omega_0^2}\right)\exp\left(-i\frac{\omega\rho}{v}\sin\beta\right)$$

where the auxiliary functions are defined by the following expressions:

$$\varrho_1^2 v^2 = \zeta^2\left(\frac{a^2}{c^2} - \cos^2\beta\right)$$

$$\varrho_2^2 v^2 = \zeta^2\left(\frac{a^2}{c^2} - \cos^2\beta\tan^2\vartheta\right) + 2\zeta\rho\cos\beta\sin\beta\tan\vartheta + \rho^2\cos^2\beta$$

$$\varrho_3^2 v^2 = \zeta^2\left(\frac{a^2}{c^2} - \cos^2\beta\tan^2\vartheta\right) - 2\zeta\rho\cos\beta\sin\beta\tan\vartheta + \rho^2\cos^2\beta$$

Convince yourself that this result reduces to (6.111) for $\beta = \pi/2$, where the single-path power spectrum is defined by (6.115) and (6.117). Now that you have incorporated this complication, can you accommodate receiver baselines that are not perpendicular to the plane of propagation?

### Problem 8

Show that the single-path phase power spectrum and phase structure functions are related to one another by the following fundamental expression:

$$\mathbf{v} \parallel \boldsymbol{\rho}\colon \qquad \mathcal{D}(\rho) = \frac{2}{\pi}\int_{-\infty}^{\infty} d\omega\,\sin^2\!\left(\frac{\omega\rho}{2v}\right)W_\varphi(\omega)$$

Demonstrate that a frequency scaling law $\omega^{-\frac{8}{3}}$ corresponds to the structure function

$$\mathcal{D}(\rho) = \text{constant} \times \rho^{\frac{5}{3}}.$$

Can you connect the more general descriptions of these benchmark quantities using the general expressions for atmospheric transmission developed in this chapter and Chapter 5?

### References

[1] "The Global Observing System of the World Weather Watch," World Meteorological Organization, WMO No. 872 (1998). (This document can be obtained from World Weather Watch Department, WMO/OMM, Case Postale No 2300, CH-1211 Genève 2, Switzerland.)

[2] G. I. Taylor, "The Spectrum of Turbulence," *Proceedings of the Royal Society*, **164**, No. A 919, 476–490 (18 February 1938).

[3] V. I. Tatarskii, *The Effects of the Turbulent Atmosphere on Wave Propagation* (translated from the Russian and issued by the National Technical Information Office, U.S. Department of Commerce, Springfield, VA 22161, 1971), 88–91.

[4] L. R. Tsvang, "Measurements of the Spectrum of Temperature Fluctuations in the Free Atmosphere," *Izvestiya Akademii Nauk SSSR, Seriya*

*Geofizicheskaya (Bulletin of the Academy of Sciences of the USSR, Geophysical Series)*, No. 1, 1117–1120 (January 1960).

[5] E. E. Gossard, "Power Spectra of Temperature, Humidity and Refractive Index from Aircraft and Tethered Balloon Measurements," *IRE Transactions on Antennas and Propagation*, **AP-8**, No. 2, 186–201 (March, 1960).

[6] L. R. Tsvang, "Some Characteristics of the Spectra of Temperature Pulsations in the Boundary Layer of the Atmosphere," *Izvestiya Akademii Nauk SSSR, Seriya Geofizicheskaya (Bulletin of the Academy of Sciences of the USSR, Geophysical Series)*, **10**, 961–965 (October 1963).

[7] See [3], pages 127–130.

[8] See [3], pages 42 and 61.

[9] A. Ishimaru, *Wave Propagation and Scattering in Random Media*, vol. 2 (Academic Press, San Diego, CA, 1978), 341.

[10] G. V. Azizyan, A. S. Gurvich and M. Z. Kholmyanskiy, "Influence of Transport Velocity Variations on the Cross Spectra of the Temperature Fluctuations," *Izvestiya Akademii Nauk, Fizika Atmosfery i Okeana (Bulletin of the Academy of Sciences of the USSR, Atmospheric and Oceanic Physics)*, **16**, No. 3, 154–159 (1980).

[11] H. G. Nalbandyan, "Locally 'Frozen' Hypothesis in a Random Velocity Field," *Izvestiya Akademii Nauk, Fizika Atmosfery i Okeana, (Bulletin of the Academy of Sciences of the USSR, Atmospheric and Oceanic Physics)*, **34**, No. 3, 268–271 (1998).

[12] V. P. Lukin, V. L. Mironov, V. V. Pokasov and S. S. Khmelevtsov, "Phase Fluctuations of Optical Waves Propagating in a Turbulent Atmosphere," *Radiotekhnika i Electronika (Radio Engineering and Electronic Physics)*, **20**, No. 6, 28–34 (June 1975).

[13] V. P. Lukin, V. V. Pokasov and S. S. Khmelevtsov, "Investigation of the Time Characteristics of Fluctuations of the Phases of Optical Waves Propagating in the Bottom Layer of the Atmosphere," *Izvestiya Vysshikh Uchebnykh Zavedenii, Radiofizika (Soviet Radiophysics)*, **15**, No. 12, 1426–1430 (December 1972).

[14] V. P. Lukin, *Atmospheric Adaptive Optics* (SPIE Optical Engineering Press, Bellingham, Washington, 1995; originally published in Russian in 1986), 85–90.

[15] K. A. Norton, J. W. Herbstreit, H. B. Janes, K. O. Hornberg, C. F. Peterson, A. F Barghausen, W. E. Johnson, P. I. Wells, M. C. Thompson, M. J. Vetter and A. W. Kirkpatrick, *An Experimental Study of Phase Variations in Line-of-Sight Microwave Transmissions* (National Bureau of Standards Monograph 33, U.S. Government Printing Office, Washington, 1 November 1961).

[16] M. C. Thompson and H. B. Janes, "Measurements of Phase-Front Distortion on an Elevated Line-of-Sight Path," *IEEE Transactions on Aerospace and Electronic Systems*, **AES-6**, No. 5, 645–656 (September 1970).

[17] G. M. B. Bouricius and S. F. Clifford, "Experimental Study of Atmospherically Induced Phase Fluctuations in an Optical Signal," *Journal of the Optical Society of America*, **60**, No. 11, 1484–1489 (November 1970).

[18] See [3], pages 259–271.

[19] D. F. Buscher, J. T. Armstrong, C. A. Hummel, A. Quirrenbach, D. Mozurkewich, K. J. Johnston, C. S. Denison, M. M. Colavita and M. Shao, "Interferometric Seeing Measurements on Mt. Wilson: Power

Spectra and Outer Scales," *Applied Optics*, **34**, No. 6, 1081–1096 (20 February 1995).

[20] See [3], pages 320–324.

[21] See [3], pages 7–8.

[22] J. W. Herbstreit and M. C. Thompson, "Measurements of the Phase of Radio Waves over Transmission Paths with Electrical Lengths Varying as a Result of Atmospheric Turbulence," *Proceedings of the IRE*, **43**, No. 10, 1391–1401 (October 1955).

[23] A. P. Deam and B. M. Fannin, "Phase Difference Variations in 9,350 Megacycle Radio Signals Arriving at Spaced Antennas" *Proceedings of the IRE*, **43**, No. 10, 1402–1404 (October 1955).

[24] M. C. Thompson and H. B. Janes, "Measurements of Phase Stability Over a Low-Level Tropospheric Path," *Journal of Research of the NBS – D. Radio Propagation*, **63D**, No. 1, 45–51 (July–August 1959).

[25] H. B. Janes and M. C. Thompson, "Errors Induced by the Atmosphere in Microwave Range Measurements," *Radio Science, Journal of Research NBS/USNC-URSI*, **68D**, No. 11, 1229–1235 (November 1964).

[26] M. C. Thompson, H. B. Janes and A. W. Kirkpatrick, "An Analysis of Time Variations in Tropospheric Refractive Index and Apparent Radio Path Length," *Journal of Geophysical Research*, **65**, No. 1, 193–201 (January 1960).

[27] M. C. Thompson and H. B. Janes, "Measurements of Phase-Front Distortion on an Elevated Line-of-Sight Path," *IEEE Transactions on Aerospace and Electronic Systems*, **AES-6**, No. 5, 645–656 (September 1970).

[28] H. B. Janes, M. C. Thompson, D. Smith and A. W. Kirkpatrick, "Comparison of Simultaneous Line-of-Sight Signals at 9.6 and 34.5 GHz," *IEEE Transactions on Antennas and Propagation*, **AP-18**, No. 4, 447–451 (July 1970).

[29] H. B. Janes, M. C. Thompson and D. Smith, "Tropospheric Noise in Microwave Range-Difference Measurements," *IEEE Transactions on Antennas and Propagation*, **AP-21**, No. 2, 256–260 (March 1973).

[30] M. C. Thompson, L. E. Wood, H. B. Janes and D. Smith, "Phase and Amplitude Scintillations in the 10 to 40 GHz Band," *IEEE Transactions on Antennas and Propagation*, **AP-23**, No. 6, 792–797 (November 1975).

[31] M. C. Thompson and H. B. Janes, "Effects of Sea Reflections on Phase of Arrival of Line-of-Sight Signals," *IEEE Transactions on Antennas and Propagation*, **AP-19**, No. 1, 105–108 (January 1971).

[32] M. C. Thompson, H. B. Janes, L. E. Wood and D. Smith, "Phase and Amplitude Scintillations at 9.6 GHz on an Elevated Path," *IEEE Transactions on Antennas and Propagation*, **AP-23**, No. 6, 850–854 (November 1975).

[33] P. A. Mandics, R. W. Lee and A. T. Waterman, "Spectra of Short-Term Fluctuations of Line-of-Sight Signals: Electromagnetic and Acoustic," *Radio Science*, **8**, No. 3, 185–201 (March 1973).

[34] A. T. Waterman, "Atmospheric Effects: Some Theoretical Relations and Sample Measurements," in *Atmospheric Effects on Radar Target Identification and Imaging*, edited by H. Jeske (Reidel, Dordrecht, 1976).

[35] M. C. Thompson and H. B. Janes, "Antenna Aperture Size Effect on Tropospheric Phase Noise," *IEEE Transactions on Antennas and Propagation*, **AP-14**, No. 6, 800–802 (November 1966).

[36] W. D. Otto, R. J. Hill, A. D. Sarma, J. J. Wilson, E. L. Andreas, J. R. Gosz and D. I. Moore, "Results of the Millimeter-Wave Instrument Operated at Sevilleta, New Mexico," NOAA Technical Memorandum ERL ETL-262, Boulder, Colorado (February 1996).

[37] M. Bester, W. C. Danchi, C. G. Degiacomi, L. J. Greenhill and C. H. Townes, "Atmospheric Fluctuations: Empirical Structure Functions and Projected Performance of Future Instruments," *The Astrophysical Journal*, **392**, No. 2, 357–374 (10 June 1992).

[38] J. A. Barnes, A. R. Chi, L. S. Cutler, D. J. Healey, D. B. Leeson, T. E. McGunigal, J. A. Mullen, W. L. Smith, R. L. Sydnor, R. F. C. Vessot and G. M. R. Winkler, "Characterization of Frequency Stability," *IEEE Transactions on Instrumentation and Measurement*, **IM-20**, No. 2, 105–120 (May 1971).

[39] C. J. Naudet, "Estimation of Tropospheric Fluctuations using GPS Data," TDA Progress Report 42-126, Jet Propulsion Laboratory, Pasadena, CA (15 April 1996).

[40] S. F. Clifford, "Temporal-Frequency Spectra for a Spherical Wave Propagating through Atmospheric Turbulence," *Journal of the Optical Society of America*, **61**, No. 10, 1285–1292 (October 1971).

[41] S. F. Clifford, G. M. B. Bouricius, G. R. Ochs and M. H. Ackley, "Phase Variations in Atmospheric Optical Propagation," *Journal of the Optical Society of America*, **61**, No. 10, 1279–1284 (October 1971).

[42] V. P. Lukin and V. V. Pokasov, "Optical Wave Phase Fluctuations," *Applied Optics*, **20**, No. 1, 121–135 (1 January 1981).

[43] H. Matsumoto and K. Tsukahara, "Effects of the Atmospheric Phase Fluctuation on Long-Distance Measurement," *Applied Optics*, **23**, No. 19, 3388–3394 (1 October 1984).

[44] R. Hinder, "Fluctuations of Water Vapour Content in the Troposphere as Derived from Interferometric Observations of Celestial Radio Sources," *Journal of Atmospheric and Terrestrial Physics*, **34**, No. 7, 1171–1186 (July 1972).

[45] A. R. Jacobson and R. Sramek, "A Method for Improved Microwave-Interferometer Remote Sensing of Convective Boundary Layer Turbulence using Water Vapor as a Passive Tracer," *Radio Science*, **32**, No. 5, 1851–1860 (September-October 1997).

[46] M. Ishiguro, T. Kanzawa and T. Kasuga, "Monitoring of Atmospheric Phase Fluctuations Using Geostationary Satellite Signals," in *Radioastronomical Seeing*, edited by J. E. Baldwin and W. Shouguan (Pergamon Press, London, 1990), 60–63.

[47] T. Kasuga, M. Ishiguro and R. Kawabe, "Interferometric Measurement of Tropospheric Phase Fluctuations at 22 GHz on Antenna Spacings of 27 to 540 m," *IEEE Transactions on Antennas and Propagation*, **AP-34**, No. 6, 797–803 (June 1986).

[48] R. A. Sramek, "VLA Phase Stability at 22 GHz on Baselines of 100 m to 3 km," VLA Test Memo 143, NRAO (20 October 1983). See especially Table II on page 11.

[49] M. C. H. Wright and W. J. Welch, "Interferometer Measurements of Atmospheric Phase Noise at 3 mm," in *Radioastronomical Seeing*, edited by J. E. Baldwin and W. Shouguan (Pergamon Press, London, 1990), 71–74.

[50] M. C. H. Wright, "Atmospheric Phase Noise and Aperture-Synthesis Imaging at Millimeter Wavelengths," *Publications of the Astronomical Society of the Pacific*, **108**, No. 724, 520–534 (June 1996).

[51] L. Olmi and D. Downes, "Interferometric Measurement of Tropospheric Phase Fluctuations at 86 GHz on Antenna Spacings of 24 m to 288 m," *Astronomy and Astrophysics*, **262**, No. 2, 634–643 (September 1992).

[52] A. E. E. Rogers, A. T. Moffet, D. C. Backer and J. M. Moran, "Coherence Limits in VLBI Observations at 3-Millimeter Wavelength," *Radio Science*, **19**, No. 6, 1552–1560 (November–December 1984).

[53] N. S. Nightingale and D. F. Buscher, "Interferometric Seeing Measurements at the La Palma Observatory," *Monthly Notices of the Royal Astronomical Society*, **251**, 155–166 (1991).

[54] M. M. Colavita, M. Shao and D. H. Staelin, "Atmospheric Phase Measurements with the Mark III Stellar Interferometer," *Applied Optics*, **26**, No. 19, 4106–4112 (1 October 1987).

[55] A. F. Dravskikh and A. M. Finkelstein, "Tropospheric Limitations in Phase and Frequency Coordinate Measurements in Astronomy," *Astrophysics and Space Science*, **60**, No. 2, 251–265 (February 1979).

[56] L. Lindegren, "Atmospheric Limitations of Narrow-field Optical Astrometry," *Astronomy and Astrophysics*, **89**, No. 1/2, 41–47 (September 1980).

[57] H. Yura, private communication, 23 July 1998.

# 7

# Angle-of-arrival Fluctuations

Degradation of stellar images is the most familiar example of propagation through random media and is visible to the naked eye. When a star is viewed through a telescope this degradation manifests itself in three ways: (a) as a variation of the image intensity, (b) as image broadening and (c) as wandering of the centroid of the image. This chapter is devoted to the third effect, which has also been called quivering, dancing and jitter. Image wandering is influenced primarily by large irregularities in the lower atmosphere for which ray theory is a good description. Image motion and angle-of-arrival fluctuations are different manifestations of the same *random ray bending* by atmospheric irregularities.

Image motion is readily observed in photographic plates placed at the focal plane of a stationary telescope. If there were no atmosphere, the stellar source would trace a smooth star trail on the plate as the earth and telescope turn together. Actual star trails exhibit random angular fluctuations about this nominal trajectory of 1 or 2 arc seconds as indicated in Figure 7.1. This random motion is observed in all astronomical measurements, although the magnitude varies with time, altitude and location. The error ranges from 0.5 to 2.0 arc seconds at sea level. It decreases with altitude and is usually 0.5 arc seconds on Mauna Kea (14 000 ft) but is sometimes as small as 0.25 arc seconds. Geometrical optics provides a valid description for astronomical quivering over a wide range of applications [1][2][3][4]. We now know that this approach is valid well into the strong-scattering regime.

The development of coherent light sources made it possible to concentrate optical signals at considerable distances. With laser sources one can form collimated beams and focused waves with simple optical trains. This technology gave rise to important applications. Optical communication links with unusually wide bandwidths became possible. Laser range finding and target designation provide unique military applications. With high-power

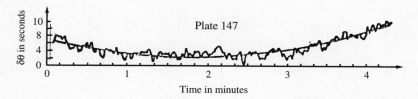

Figure 7.1: A star trail recorded on a photographic plate at the focal plane of a stationary telescope by Kolchinskii [5].

lasers it became possible to deposit large amounts of focused energy onto distant targets, which generated proposals for directed-energy weapons. These applications are sensitive to beam wandering since the signal must remain on the target for a considerable time. That requirement gave rise to vigorous experimental and theoretical studies of beam wandering. It was found that geometrical optics provides a valid description for beam waves in many situations of practical interest [6][7].

## 7.1 Measurement Considerations

A description of angular errors based on the instantaneous direction of the ray tangent was established in Chapter 3. We found there that angular fluctuations are proportional to lateral gradients of the dielectric variations integrated along the nominal ray path. The angular error has two components. One is perpendicular to the plane of the nominal ray path and describes angular wandering out of that plane. The second component lies in the propagation plane and is normal to the ray path. Before developing the consequences of those expressions, it is important to expand and clarify our description of angular errors.

In doing so we must distinguish among several definitions for angle-of-arrival fluctuations. The measured quantities depend strongly on the method and equipment used to sense them. To make these distinctions, consider the *crinkled wave-front* shown arriving at a telescope opening illustrated in Figure 7.2. This irregular wave-front has various components that are naturally described by orthonormal functions closely related to Zernike polynomials [8][9][10][11]. The composite wave-front is described by a weighted series of these functions. The most prominent feature of the disturbed phase front is usually the overall tilt, which is described by the second term in this expansion. Higher-order terms describe refocusing, astigmatism and coma.

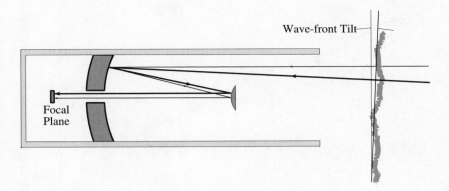

Figure 7.2: An irregular wave-front falling on a Cassegrain telescope, indicating the important role played by wave-front tilt.

We concentrate on *wave-front tilt* since it is the dominant aberration for large telescopes. One can measure this quantity with wave-front sensors mounted at the telescope opening or with a centroid tracker at the focal plane. In adaptive-optics systems this information is used to drive steering mirrors that compensate for the tilt of arriving signals in real time. These techniques can stabilize the image and improve the mean irradiance by almost a factor of ten [12]. The specific performance depends on the type of wave being received and on the tilt sensor used.

### 7.1.1  Interferometers

A direct way to measure wave-front tilt is to use an interferometer at the entry plane of a telescope. Three small receivers are deployed in a plane perpendicular to the direction of propagation, as illustrated in Figure 7.3. The pairs of receivers are connected in order to measure the phase difference. One can express the angular error normal to the ray plane in terms of the individual phase terms:

$$\delta\theta_y = \frac{\lambda}{2\pi\,\Delta y}[\varphi(\Delta y, 0) - \varphi(0, 0)] \tag{7.1}$$

In the limit of small separations this leads to Chandrasekhar's expression (3.35) for the bearing error:

$$\delta\theta_y = \frac{1}{k}\frac{\partial\varphi}{\partial y}$$

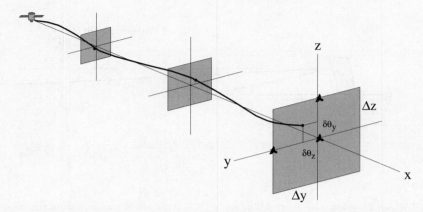

Figure 7.3: The geometry for analyzing the two angle-of-arrival errors which are orthogonal to the direction of propogation with two interferometers.

The angle of arrival described by the ray tangent is therefore identical to the wave-front tilt measured by an interferometer.

The wave-front tilt is proportional to the phase difference measured between adjacent receivers and its variance is therefore proportional to the phase structure function:

$$\langle \delta\theta_y^2 \rangle = \frac{1}{k^2} \frac{\mathcal{D}_\varphi(\Delta y, 0)}{(\Delta y)^2} \tag{7.2}$$

For small separations between receivers one can expand the structure function in a Taylor series to find the component normal to the ray plane.

$$\langle \delta\theta_y^2 \rangle = \frac{1}{2k^2} \left.\frac{\partial^2}{\partial y^2}\mathcal{D}_\varphi(y, z)\right|_{y=z=0} \tag{7.3}$$

In an analogous way the component in the ray-plane error is described by

$$\langle \delta\theta_z^2 \rangle = \frac{1}{2k^2} \left.\frac{\partial^2}{\partial z^2}\mathcal{D}_\varphi(y, z)\right|_{y=z=0} \tag{7.4}$$

and the cross correlation of the two components becomes

$$\langle \delta\theta_y \, \delta\theta_z \rangle = \frac{1}{2k^2} \left.\frac{\partial^2}{\partial y \, \partial z}\mathcal{D}_\varphi(y, z)\right|_{y=z=0} \tag{7.5}$$

These expressions do not assume that the random medium is isotropic.

It is sometimes possible to treat the atmosphere as if it were isotropic. In this case the phase structure function depends only on the scalar distance $\rho$ between the two receivers:

$$\mathcal{D}_\varphi(y, z) = \mathcal{D}_\varphi\left(\sqrt{y^2 + z^2}\right) = \mathcal{D}_\varphi(\rho)$$

The two angular-error components should now have the same variance since there is no preferred direction. One can show this by changing the partial derivatives from rectangular to cylindrical coordinates:

$$\frac{\partial^2 \mathcal{D}_\varphi}{\partial y^2} = \frac{\partial}{\partial y}\left(\frac{\partial \rho}{\partial y}\frac{\partial \mathcal{D}_\varphi}{\partial \rho}\right)$$

$$= \frac{1}{\rho}\left(1 - \frac{y^2}{\rho^2}\right)\frac{\partial \mathcal{D}_\varphi}{\partial \rho} + \frac{y^2}{\rho^2}\frac{\partial^2 \mathcal{D}_\varphi}{\partial \rho^2}$$

A similar expression results for the other component:

$$\frac{\partial^2 \mathcal{D}_\varphi}{\partial z^2} = \frac{1}{\rho}\left(1 - \frac{z^2}{\rho^2}\right)\frac{\partial \mathcal{D}_\varphi}{\partial \rho} + \frac{z^2}{\rho^2}\frac{\partial^2 \mathcal{D}_\varphi}{\partial \rho^2}$$

They have a common form when we take the limits $y = 0$ and $z = 0$ in the correct order:

$$\langle \delta\theta_y^2 \rangle = \langle \delta\theta_z^2 \rangle = \frac{1}{2k^2}\frac{\partial^2}{\partial \rho^2}\mathcal{D}_\varphi(\rho)|_{\rho=0} \qquad (7.6)$$

In a similar manner one can show that the cross correlation vanishes for an isotropic medium:

$$\langle \delta\theta_y \, \delta\theta_z \rangle = 0$$

### 7.1.2 Centroid Trackers

One can also sense wave-front tilt by noting the position of an image in the focal plane of a telescope. The star trail reproduced in Figure 7.1 is an example of this image motion recorded on film. If one replaces the film by a mosaic of photo diodes, the instantaneous position of the image can be measured with considerable precision. In practice, one programs the electronics to measure the location of the *centroid* of the image because the image is broadened by diffraction effects in the atmosphere. The output of these centroid trackers provides a good estimate for wave-front tilt and a running history measured in this way with a solar telescope is reproduced in Figure 7.4. This technique is widely used to remove wave-front tilt in modern optical systems [14][15].

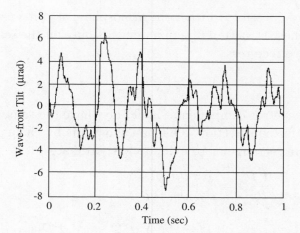

Figure 7.4: Wave-front tilt measured on Sacramento Peak by Acton, Sharbaugh, Roehrig and Tiszauer, using solar pores as sources [13].

The estimate of tilt made by a centroid tracker is sometimes different than the true wave-front tilt. This difference becomes more important for large-diameter telescopes and shorter wavelengths and is also exacerbated by point-ahead requirements and amplitude scintillation. This *centroid aniso-planatism* is due to coma distortions induced by the random medium [16].

## 7.2  Terrestrial Links

Precise measurements of angular wandering have been made on paths near the surface using a controlled transmitter and receivers. They are relatively easy to interpret because the level of turbulence is essentially constant and can be measured by meteorological instruments. The eddies that are primarily responsible for angular errors are nearly isotropic near the surface and the two error components are equal.

### 7.2.1  Plane Waves

Our strategy is to replace the structure function in (7.6) by the wavenumber integral representations developed in Chapter 5. For a plane wave near the surface we use (5.19) to describe the phase structure function as the weighted average of the turbulence spectrum. Since these rays travel in the lower atmosphere we change from the spectrum of dielectric variations to that for refractive-index fluctuations using $\Phi_\varepsilon(\kappa) = 4\Phi_n(\kappa)$:

$$\langle \delta\theta^2 \rangle = \frac{1}{2k^2} \frac{\partial^2}{\partial\rho^2}\Big(8\pi^2 Rk^2 \int_0^\infty d\kappa \ \kappa\Phi_n(\kappa)[1 - J_0(\kappa\rho)]\Big)_{\rho=0} \qquad (7.7)$$

The derivatives operate only on the kernel and one can use the differential equation for the Bessel function to write

$$\langle \delta\theta^2 \rangle = 4\pi^2 R \int_0^\infty d\kappa \; \kappa \Phi_n(\kappa) \left( -\frac{\partial^2}{\partial\rho^2} J_0(\kappa\rho) \right)_{\rho=0}$$

$$= 4\pi^2 R \int_0^\infty d\kappa \; \kappa \Phi_n(\kappa) \kappa^2 \left( J_0(\kappa\rho) - \frac{J_1(\kappa\rho)}{\kappa\rho} \right)_{\rho=0}$$

$$= 4\pi^2 R \int_0^\infty d\kappa \; \kappa^3 \Phi_n(\kappa) \left( \frac{1}{2} \right).$$

The angle-of-arrival variance for closely spaced receivers is thus proportional to the third moment of the turbulence spectrum and should increase linearly with the path length:

$$\text{Plane:} \quad \langle \delta\theta^2 \rangle = 2\pi^2 R \int_0^\infty d\kappa \; \kappa^3 \Phi_n(\kappa) \tag{7.8}$$

The angular error is independent of frequency. On the other hand, we must remember that $C_n^2$ has different values in the microwave and optical bands. This also shows that angle-of-arrival errors are influenced by eddy sizes that are considerably smaller than those which determine the phase variance.

### 7.2.2 Spherical Waves

The path length is relatively short for optical and microwave experiments performed near the surface. In these circumstances, the transmitted field is better described by a spherical wave and we therefore use the result (5.6) for rotated ray paths:

$$\langle \delta\theta^2 \rangle = 4\pi^2 R \int_0^\infty d\kappa \; \kappa \Phi_n(\kappa) \int_0^1 du \left( -\frac{\partial^2}{\partial\rho^2} J_0(\kappa u \rho) \right)_{\rho=0}$$

$$= 4\pi^2 R \int_0^\infty d\kappa \; \kappa \Phi_n(\kappa) \int_0^1 du \, (\kappa u)^2 \left( J_0(\kappa u \rho) - \frac{J_1(\kappa u \rho)}{\kappa u \rho} \right)_{\rho=0}$$

$$= 4\pi^2 R \int_0^\infty d\kappa \; \kappa^3 \Phi_n(\kappa) \int_0^1 du \, u^2 \left( \frac{1}{2} \right)$$

or

$$\text{Spherical:} \quad \langle \delta\theta^2 \rangle = \frac{2}{3}\pi^2 R \int_0^\infty d\kappa \; \kappa^3 \Phi_n(\kappa) \tag{7.9}$$

This result is three times smaller than the corresponding expression for a plane wave. The physical reason is apparent from Figure 5.1. The separation of two parallel rays is everywhere constant and equal to the spacing between receivers. By contrast, the separation between rotated paths varies from zero to the full inter-receiver spacing. This means that the phase fluctuations are more nearly the same on the rotated paths and their difference is smaller.

The angular variance is proportional to the third moment of the turbulence spectrum both in the spherical case and in the plane-wave case. This means that the contributions of large eddies are suppressed and there is no need to characterize the small-wavenumber portion of the spectrum – as we had to do in describing phase variations. Angular variations are specified primarily by the inertial range of the spectrum and one can use the Kolmogorov model with wavenumber cutoffs to describe the spectrum:

$$\langle \delta\theta^2 \rangle = \frac{2}{3}\pi^2 R \int_{\kappa_0}^{\kappa_m} d\kappa \; \kappa^3 \frac{0.033\, C_n^2}{\kappa^{\frac{11}{3}}}$$

The outer scale length is much greater than the inner scale length and a simple result emerges:

$$\text{Spherical:} \quad \langle \delta\theta^2 \rangle = 0.651 R C_n^2 (\kappa_m)^{\frac{1}{3}} \qquad (7.10)$$

The path length is known with great precision and the two meteorological parameters can be measured independently. This formula gives estimates of the angular error that are significantly larger than the measured values. The problem is that we have ignored aperture averaging.

### 7.2.3 Aperture Averaging

Most direction finders use large collectors to enhance the signal-to-noise ratio for differencing of signals. Single-aperture systems use two or more closely spaced feeds mounted at the focus of a parabolic reflector. The phase difference is extracted from the signals measured by each feed. For most applications it is important to consider aperture averaging.

Imagine first that the receiver is a simple array of photo diodes spread over an area A. The wave-front tilt is measured by combining data from the individual detectors and the result is simply the ray-tangent expressions (3.35) averaged over the array:

$$\overline{\delta\theta}_y = \frac{1}{k\mathsf{A}} \int\!\!\int_\mathsf{A} d^2\sigma \; \frac{\partial\varphi}{\partial y} \qquad (7.11)$$

$$\overline{\delta\theta}_z = \frac{1}{k\mathsf{A}} \int\!\!\int_\mathsf{A} d^2\sigma \; \frac{\partial\varphi}{\partial z} \qquad (7.12)$$

These expressions assume that the receiver is boresighted on the source, which is usually the case with steerable arrays. Tatarskii used diffraction theory to show that these expressions are valid even if the light passes through the lenses and mirrors of a telescope [17].

The angle-of-arrival variance can be calculated from these expressions. The phase derivatives can be expressed as double derivatives of the phase structure function taken between two points on the receiving surface. The $y$ component becomes

$$\langle \overline{\delta\theta_y^2} \rangle = \frac{-1}{2k^2 \mathsf{A}^2} \iint_\mathsf{A} dy_1 \, dz_1 \iint_\mathsf{A} dy_2 \, dz_2 \, \frac{\partial^2}{\partial y_1 \, \partial y_2} \mathcal{D}_\varphi(y_1 - y_2, z_1 - z_2)$$

$$= \frac{1}{2k^2 \mathsf{A}^2} \iint_\mathsf{A} dy_1 \, dz_1 \iint_\mathsf{A} dy_2 \, dz_2 \, \frac{\partial^2}{\partial y_1^2} \mathcal{D}_\varphi(y_1 - y_2, z_1 - z_2)$$

with a similar expression for the $z$ component. The two variances should be equal since there is no preferred direction in an isotropic medium. We can therefore describe the common result as the average of the two component variances:

$$\langle \overline{\delta\theta^2} \rangle = \frac{1}{4k^2 \mathsf{A}^2} \iint_\mathsf{A} dy_1 \, dz_1 \iint_\mathsf{A} dy_2 \, dz_2 \left( \frac{\partial^2}{\partial y_1^2} + \frac{\partial^2}{\partial z_1^2} \right)$$

$$\times \mathcal{D}_\varphi \left( \sqrt{(y_1 - y_2)^2 + (z_1 - z_2)^2} \right) \tag{7.13}$$

It is natural to use the cylindrical coordinates identified in Figure 4.11 to express the aperture averages since the reflector or lens is invariably circular. When the Laplacian is cast in terms of the separation $\rho$ between two surface elements,

$$\langle \overline{\delta\theta^2} \rangle = \frac{1}{4k^2 \pi^2 \mathsf{a}^4} \int_0^{a_r} dr_1 \, r_1 \int_0^{a_r} dr_2 \, r_2 \int_0^{2\pi} d\phi_1$$

$$\times \int_0^{2\pi} d\phi_2 \left( \frac{\partial^2}{\partial \rho^2} + \frac{1}{\rho} \frac{\partial}{\partial \rho} \right) \mathcal{D}_\varphi(\rho)$$

where

$$\rho = \sqrt{r_1^2 + r_2^2 - 2r_1 r_2 \cos(\phi_1 - \phi_2)}.$$

The last step is to introduce the wavenumber-spectrum representation of the structure function. We illustrate the calculation for a spherical wave:

$$\langle \overline{\delta\theta^2} \rangle = \frac{1}{4\pi^2 a_{\mathrm{r}}^4} \int_0^{a_{\mathrm{r}}} dr_1 \, r_1 \int_0^{a_{\mathrm{r}}} dr_2 \, r_2 \int_0^{2\pi} d\phi_1 \int_0^{2\pi} d\phi_2$$

$$\times \, 8\pi^2 R \int_0^\infty d\kappa \, \kappa\Phi_n(\kappa) \int_0^1 dw \left( \frac{\partial^2}{\partial\rho^2} + \frac{1}{\rho}\frac{\partial}{\partial\rho} \right) [1 - J_0(\kappa w\rho)]$$

The differential equation for the Bessel function allows the terms in brackets to be replaced by $-\kappa^2 w^2 J_0(\kappa w\rho)$:

$$\langle \overline{\delta\theta^2} \rangle = 2R \int_0^\infty d\kappa \, \kappa^3\Phi_n(\kappa) \int_0^1 dw \, w^2 \frac{1}{a_{\mathrm{r}}^4} \int_0^{a_{\mathrm{r}}} dr_1 \, r_1 \int_0^{a_{\mathrm{r}}} dr_2 \, r_2$$

$$\times \int_0^{2\pi} d\phi_1 \int_0^{2\pi} d\phi_2 \, J_0\left[ \kappa w \sqrt{r_1^2 + r_2^2 - 2r_1 r_2 \cos(\phi_1 - \phi_2)} \right]$$

The addition theorem for Bessel functions (4.32) separates all variables in the four-fold surface integral. The angular integrations eliminate each term except the first and the remaining radial integrations can be done using a result from Appendix D:

$$\langle \overline{\delta\theta^2} \rangle = 2\pi^2 R \int_0^\infty d\kappa \, \kappa^3\Phi_n(\kappa) \int_0^1 dw \, w^2 \left( \frac{2J_1(\kappa w a_{\mathrm{r}})}{\kappa w a_{\mathrm{r}}} \right)^2 \qquad (7.14)$$

The large parentheses define the wavenumber-weighting function for aperture averaging. The influence of eddies smaller than the radius of the receiver is suppressed. One can then use the Kolmogorov model to evaluate the angular variance:

$$\langle \overline{\delta\theta^2} \rangle = 0.132 \times 2\pi^2 R C_n^2 \, a_{\mathrm{r}}^{-2} \int_0^1 dw \int_0^\infty \frac{d\kappa}{\kappa^{\frac{8}{3}}} J_1^2(\kappa w a_{\mathrm{r}})$$

The wavenumber integration can be done by letting $\zeta = \kappa w a_{\mathrm{r}}$ and using a result from Appendix D:

$$\text{Spherical:} \qquad \langle \delta\theta_{\mathrm{sph}}^2 \rangle = 0.844 \, R C_n^2 \, a_{\mathrm{r}}^{-\frac{1}{3}} \qquad (7.15)$$

The corresponding expression for a plane wave is

$$\text{Plane:} \qquad \langle \delta\theta_{\mathrm{pl}}^2 \rangle = 2.251 \, R C_n^2 \, a_{\mathrm{r}}^{-\frac{1}{3}} \qquad (7.16)$$

The effect of aperture averaging is to replace the inner scale length by the radius of the receiver. Since the inner scale length is about 5 mm, the angular error for a telescope with $a_r = 50$ cm will be reduced by 0.15 relative to that measured with closely spaced receivers. Notice that the ratio of these expressions is 3/8 rather than the factor 1/3 established for point receivers. This difference reflects the interaction of aperture averaging and the diverging rays. These results are often expressed in terms of the diameter of the receiver:[1]

$$\text{Spherical:} \quad \langle \delta\theta^2_{\text{sph}} \rangle = 1.064\, R\, C_n^2\, D_r^{-\frac{1}{3}} \tag{7.17}$$

$$\text{Plane:} \quad \langle \delta\theta^2_{\text{pl}} \rangle = 2.837\, R\, C_n^2\, D_r^{-\frac{1}{3}} \tag{7.18}$$

These predictions can be tested since both the distance and the opening are known with considerable precision and $C_n^2$ can be measured independently.

Early confirmation of these expressions came from optical experiments done in the USSR. These used a high-pressure mercury lamp as a point source and a small telescope ($D_r = 8$ cm) as the receiver [18]. The received signal was deflected by a rotating mirror placed near the focal plane and passed through a slit to a photomultiplier. The instantaneous phase front of the arriving wave could be measured in this way. The variance and power spectrum of angular bearing errors were measured under weak-scattering conditions at night for five different path lengths ranging from 125 to 2000 m. This experiment confirmed the linearity of the scaling with distance. Values for $C_n^2$ were inferred from gradient measurements of temperature and wind speed. This did not provide a tight test because it depended on theories regarding the generation of turbulence.

A second series of experiments was performed with the same equipment under conditions of relatively strong scattering [19]. The intensity of turbulence was measured directly by noting temperature fluctuations near the path. A comparison of measured values for the angular variance and the predictions based on $C_n^2$ data is reproduced in Figure 7.5. The correlation was good at short distances, ranging from 91% at 175 m to 79% at 500 m. Basic agreement was retained at 1750 m, where strong scattering was consistently encountered. The difference became more significant as the

---

[1] The first studies of this problem [1][4] considered two slits separated by a distance $D_r$ and obtained the following expression for a plane wave:

$$\langle \delta\theta^2_{\text{pl}} \rangle = 2.91\, R\, C_n^2\, D_r^{-\frac{1}{3}}$$

The difference between this result and (7.18) reflects the approximation used.

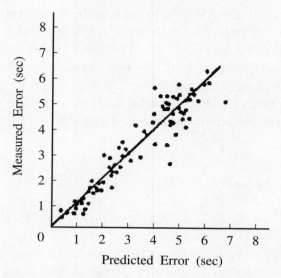

Figure 7.5: A comparison of the geometrical optics prediction with angle-of-arrival errors measured on an optical link by Gurvich and Kallistratova [19].

intensity fluctuations increased. We shall examine angle-of-arrival variations in the context of strong scattering later in these volumes.

### 7.2.4 Beam Waves

Modern terrestrial experiments have usually been performed with collimated or focused beams. One can easily observe spot wander when the transmitter creates a narrow beam. We shall consider this problem in the context of optical communications. To avoid loss of signal the receiving aperture must be large enough to ensure that the beam does not wander beyond its capture area. This means that one should include the influence of aperture averaging in many situations and it is not difficult to do so [6]. Descriptions of beam wander and angle-of-arrival fluctuations were developed from diffraction theory using the Markov approximation [20][21]. It was found later that geometrical optics provides a simpler description for beam waves that is valid for many situations of practical interest [6][7][22]. One can describe movements of the centroid of the spot with this approximation – but not image blurring, which is a result of diffraction.

To describe the deflection of such signals by random irregularities, we must first characterize the unperturbed beam wave. We assume that the

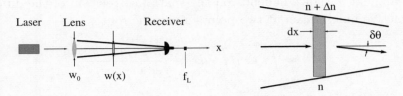

Figure 7.6: An optical communication system using a beam wave focused at a distance $f_L$ beyond the receiver. The figure is expanded on the right-hand side to illustrate a thin slice of the random medium. Beam deflection is caused by there being different values of the refractive index at its upper and lower boundaries.

transmitted signal is focused at a distance $f_L$ and has an initial beam diameter $w_0$ as illustrated in Figure 7.6. In the geometrical optics approximation the width of the beam at a distance $x$ is simply

$$w(x) = w_0 \left| 1 - \frac{x}{f_L} \right| \tag{7.19}$$

This assumes that beam spreading due to diffraction is not significant.

Now consider the enlargement shown on the right-hand side of Figure 7.6. The values of the refractive index at the top and bottom are slightly different because the random medium changes with position:

$$\Delta n(x) = n[x, +\tfrac{1}{2}w(x)] - n[x, -\tfrac{1}{2}w(x)] \tag{7.20}$$

We treat this slice of the medium as a weak prism. The optical thickness of a ray passing through its top is $(n + \Delta n)\,dx$ and must be compared with $n\,dx$ for a ray at the bottom of the prism. The arriving wave is therefore deflected by a small angle

$$d\theta = \frac{\Delta n(x)}{w(x)}\,dx$$

and the received signal has a bearing error that is the sum of all slices along the path:

$$\delta\theta = \int_0^R dx\, \frac{\Delta n(x)}{w(x)} \tag{7.21}$$

In this approximation, we have replaced the gradient in (3.35) by the difference in refractive index at the two boundaries of the beam divided by its width.

The angular variance depends on the spatial correlation of the difference in refractive index taken at two points along the route:

$$\langle \delta\theta^2 \rangle = \int_0^R dx_1 \int_0^R dx_2 \, \frac{\langle \Delta n(x_1)\, \Delta n(x_2) \rangle}{w(x_1)w(x_2)}$$

If we abbreviate by writing $w_1 = w(x_1)$ and $w_2 = w(x_2)$, the correlation is written in the following form using the algebraic identity (5.57):

$$\begin{aligned}
2\langle \Delta n(x_1)\, \Delta n(x_2) \rangle &= \langle [n(x_1, \tfrac{1}{2}w_1) - n(x_2, -\tfrac{1}{2}w_2)]^2 \rangle \\
&+ \langle [n(x_1, -\tfrac{1}{2}w_1) - n(x_2, \tfrac{1}{2}w_2)]^2 \rangle \\
&- \langle [n(x_1, \tfrac{1}{2}w_1) - n(x_2, \tfrac{1}{2}w_2)]^2 \rangle \\
&- \langle [n(x_1, -\tfrac{1}{2}w_1) - n(x_2, -\tfrac{1}{2}w_2)]^2 \rangle
\end{aligned}$$

Each of the averaged differences can be expressed in terms of the structure function for refractive-index variations. One can use the 2/3 law to describe them since we believe that the inertial range plays the largest role:

$$\begin{aligned}
\langle \Delta n(x_1)\, \Delta n(x_2) \rangle &= C_n^2 \{ [(x_1 - x_2)^2 + \tfrac{1}{4}(w_1 + w_2)^2]^{\frac{1}{3}} \\
&- [(x_1 - x_2)^2 + \tfrac{1}{4}(w_1 - w_2)^2]^{\frac{1}{3}} \}
\end{aligned}$$

We convert to sum and difference coordinates,

$$u = x_1 - x_2 \qquad \text{and} \qquad x = \tfrac{1}{2}(x_1 + x_2)$$

and note that the level of turbulence $C_n^2$ depends on the average path position:

$$\langle \delta\theta^2 \rangle = \int_0^R dx \, C_n^2(x) \int_{-z}^{z} du \, \frac{[u^2 + \tfrac{1}{4}(w_1 + w_2)^2]^{\frac{1}{3}} - [u^2 + \tfrac{1}{4}(w_1 - w_2)^2]^{\frac{1}{3}}}{w(x + \tfrac{1}{2}u)w(x - \tfrac{1}{2}u)}$$

$$(7.22)$$

The beam-diameter functions can be expanded about the nominal position:

$$w_1 = w\left(x + \frac{1}{2}u\right) = w(x) + \frac{1}{2}\frac{dw}{dx}$$

$$w_2 = w\left(x - \frac{1}{2}u\right) = w(x) - \frac{1}{2}\frac{dw}{dx}$$

The following replacements are valid since the beam diameter varies slowly with distance:

$$w_1 + w_2 = 2w(x) \qquad \text{and} \qquad w_1 - w_2 = 0$$

Most of the contribution to the first integral in (7.22) comes from very small values of the difference coordinate and one can extend its limits of integration to infinity without serious error:

$$\langle \delta\theta^2 \rangle = \int_0^R dx \, \frac{C_n^2(x)}{w^2(x)} \int_{-\infty}^{\infty} du \, \{[u^2 + w^2(x)]^{\frac{1}{3}} - (u^2)^{\frac{1}{3}}\}$$

The $u$ integration converges slowly but can be evaluated by expressing the difference of terms as a parametric integral:

$$\mathcal{J}(\eta) = \int_{-\infty}^{\infty} du \, [(u^2 + \eta^2)^{\frac{1}{3}} - (u^2)^{\frac{1}{3}}] = \frac{2}{3}\eta^2 \int_0^1 d\zeta \int_0^{\infty} \frac{du}{(u^2 + \zeta\eta^2)^{\frac{2}{3}}}$$

$$= \frac{2}{3}\eta^{\frac{5}{3}} \int_0^1 \frac{d\zeta}{\zeta^{\frac{1}{6}}} \int_0^{\frac{\pi}{2}} d\varphi \, (\cos\varphi)^{-\frac{2}{3}} = \frac{2}{5}\eta^{\frac{5}{3}} \frac{\Gamma(\frac{1}{2})\Gamma(\frac{1}{6})}{\Gamma(\frac{2}{3})} = 2.9143\,\eta^{\frac{5}{3}}$$

The angle-of-arrival variance becomes

$$\langle \delta\theta^2 \rangle = 2.914 \int_0^R dx \, \frac{C_n^2(x)}{w^{\frac{1}{3}}(x)}. \tag{7.23}$$

In most horizontal propagation experiments $C_n^2$ is constant and can be taken outside the integration:

$$\langle \delta\theta^2 \rangle = 2.914 C_n^2 \int_0^R dx \, w^{-\frac{1}{3}}(x)$$

Collimated beams are of considerable practical importance. In this case the beam width $w(x)$ is constant and equal to the diameter of the transmiter, $w_0$:

$$\text{Collimated:} \qquad \langle \delta\theta^2 \rangle = 2.914 \, R \, C_n^2 \, w_0^{-\frac{1}{3}} \tag{7.24}$$

The beam diameter at the transmitter has now replaced the diameter of the receiver which appeared in the expressions for plane and spherical waves. This result can be tested since $R$ and $w_0$ are known with some precision and $C_n^2$ can be measured independently. A careful experiment was performed

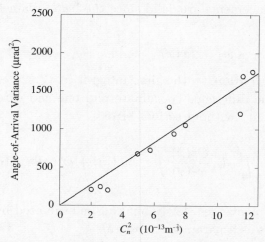

Figure 7.7: A comparison of the angular variance and turbulence level measured by Churnside and Lataitis on a 240-m path near Boulder [6]. The circles represent observed values and the straight line is the prediction for a collimated beam.

in Boulder using a helium–neon laser and an optical train that generated a collimated beam with $w_0 = 12.5$ cm [6]. This signal traveled over ak 240-m path and passed through a reciprocal optical train that focused it onto a 384-by-491-detector array of charge-coupled devices. The array noted the instantaneous position of the wandering beam. $C_n^2$ was measured independently with an optical scintillometer. A favorable comparison of the experimental data and the predictions for a collimated beam is reproduced in Figure 7.7.

Other types of beam waves can be described by combining the analytical description for the beam diameter (7.19) with the general expression (7.23):

$$\langle \delta\theta^2 \rangle = 2.914 \, R C_n^2 \, w_0^{-\frac{1}{3}} \int_0^R dv \left| 1 - \frac{v}{f_{\mathrm{L}}} \right|^{-\frac{1}{3}}$$

For $R \leq f_{\mathrm{L}}$ the integral can be expressed in terms of a hypergeometric function as noted in Appendix H:

$$\langle \delta\theta^2 \rangle = 2.914 \, R C_n^2 \, w_0^{-\frac{1}{3}} \, {}_2F_1\left( \frac{1}{3}, 1, 2 ; \frac{R}{f_{\mathrm{L}}} \right) \qquad (7.25)$$

When the transmitted signal is focused at the receiver:

$$\text{Focused:} \quad \langle \delta\theta^2 \rangle = 4.371 \, R C_n^2 \, w_0^{-\frac{1}{3}} \tag{7.26}$$

The same techniques allow one to calculate the image wander for beam waves that are reflected by mirrors and corner reflectors [6][22].

Some applications emphasize linear displacements of the centroid of the image. The same approximation can be used to describe this spot wandering:

$$p = \int_0^R dx \, \frac{R-x}{w(x)} \, \Delta n(x) \tag{7.27}$$

The variance is calculated using the method outlined above:

$$\langle p^2 \rangle = 0.971 \, R^3 \, C_n^2 \, w_0^{-\frac{1}{3}} \, {}_2F_1\left(\frac{1}{3}, 1, 4 \, ; \, \frac{R}{f_L}\right) \tag{7.28}$$

This result takes the following forms for the two cases of primary interest:

$$\text{Collimated:} \quad \langle p^2 \rangle = 0.971 \, R^3 \, C_n^2 \, w_0^{-\frac{1}{3}} \tag{7.29}$$

$$\text{Focused:} \quad \langle p^2 \rangle = 1.092 \, R^3 \, C_n^2 \, w_0^{-\frac{1}{3}} \tag{7.30}$$

These expressions are quite close to the results of calculations based on the Markov approximation which includes diffraction and strong scattering effects [20][21]. It seems clear that ray theory describes angle-of-arrival fluctuations over a wide range of scattering conditions.

## 7.3 Optical Astronomical Signals

The star trail reproduced in Figure 7.1 illustrates the quivering of a stellar image. When the solar limb is observed by a telescope, eddies in the atmosphere that are large relative to the aperture cause similar tilting of the wave-front and random motion of the solar image. To describe these observations we introduce a more general formulation of the angular errors than we have used thus far. The sources are distant and one can always describe the arriving signal as a plane wave. We analyze the downcoming ray using the coordinate system illustrated in Figure 4.13. The slant range along the ray is denoted by $s$, while $u$ and $v$ locate surface elements on the receiving aperture. We write the aperture-averaged expressions for angular errors (7.11) and (7.12) in terms of the lateral coordinates:

$$\overline{\delta\theta_u} = \frac{1}{k\mathsf{A}} \iint_\mathsf{A} du\, dv\; \frac{\partial}{\partial u}\varphi(u,v) \tag{7.31}$$

$$\overline{\delta\theta_v} = \frac{1}{k\mathsf{A}} \iint_\mathsf{A} du\, dv\; \frac{\partial}{\partial v}\varphi(u,v) \tag{7.32}$$

### 7.3.1 Angular Position Errors

We show how to calculate the angular variances, working first with the $u$ component:

$$\langle\overline{\delta\theta_u^2}\rangle = \frac{1}{k^2\mathsf{A}^2} \iint_\mathsf{A} du_1\, dv_1 \iint_\mathsf{A} du_2\, dv_2\; \frac{\partial^2}{\partial u_1\, \partial u_2}\langle\varphi(u_1,v_1)\varphi(u_2,v_2)\rangle$$

The correlation of phase fluctuations between parallel rays is related to the wavenumber spectrum of refractive-index fluctuations by the basic expression (5.36):

$$\langle\overline{\delta\theta_u^2}\rangle = \frac{1}{\mathsf{A}^2} \int d^3\kappa\, \kappa_u^2$$

$$\times \int_0^\infty ds_1 \int_0^\infty ds_2\, \Phi_n[\boldsymbol{\kappa}, \tfrac{1}{2}(s_1+s_2)] \exp\{i\,[\kappa_s(s_1-s_2)]\}$$

$$\times \iint_\mathsf{A} du_1\, dv_1 \iint_\mathsf{A} du_2\, dv_2\, \exp\{i[\kappa_u(u_1-u_2)+\kappa_v(v_1-v_2)]\}$$

$$\tag{7.33}$$

The partial derivatives have produced a factor $\kappa_u^2$ in this case and would give $\kappa_v^2$ for the other component. The integration variables are separated and one can address the influences of different effects sequentially.

Aperture smoothing is described by the double surface integration:

$$\mathcal{M}(\boldsymbol{\kappa}) = \frac{1}{\mathsf{A}^2} \iint_\mathsf{A} du_1\, dv_1$$

$$\times \iint_\mathsf{A} du_2\, dv_2\, \exp\{i[\kappa_u(u_1-u_2)+\kappa_v(v_1-v_2)]\} \tag{7.34}$$

For a circular opening it is natural to use the cylindrical coordinates identified in Figure 4.11 to describe the surface elements:

$$\mathcal{M}(\boldsymbol{\kappa}) = \frac{1}{\pi^2 a_{\rm r}^4} \int_0^{a_{\rm r}} dr_1 \, r_1 \int_0^{a_{\rm r}} dr_2 \, r_2 \int_0^{2\pi} d\phi_1 \int_0^{2\pi} d\phi_2$$

$$\times \exp\{i[\kappa_u(r_1 \sin\phi_1 - r_2 \sin\phi_2) + \kappa_v(r_1 \cos\phi_1 - r_2 \cos\phi_2)]\}$$

The angular integrations can be done using the following result:

$$\int_0^{2\pi} d\phi \exp[i(p \sin\phi + q \cos\phi)] = 2\pi J_0\left(\sqrt{p^2 + q^2}\right)$$

so that

$$\mathcal{M}(\boldsymbol{\kappa}) = \left(\frac{2}{a_{\rm r}^2} \int_0^{a_{\rm r}} dr \, rJ_0\left(r\sqrt{\kappa_u^2 + \kappa_v^2}\right)\right)^2.$$

The indefinite integral can be done using a result from Appendix D:

$$\mathcal{M}(\boldsymbol{\kappa}) = \left(\frac{2J_1\left(a_{\rm r}\sqrt{\kappa_u^2 + \kappa_v^2}\right)}{a_{\rm r}\sqrt{\kappa_u^2 + \kappa_v^2}}\right)^2 \tag{7.35}$$

The ray-path integrations in (7.33) are dominated by the exponential term which oscillates rapidly. We use the sum and difference coordinates defined by (4.36) to write

$$\langle \overline{\delta\theta_u^2} \rangle = \int d^3\kappa \, \kappa_u^2 \mathcal{M}(\boldsymbol{\kappa}) \int_0^\infty dr \, \Phi_n(\boldsymbol{\kappa}, r) \int_{-r}^r du \exp(i\kappa_s u)$$

$$= 2 \int d^3\kappa \, \kappa_u^2 \mathcal{M}(\boldsymbol{\kappa}) \int_0^\infty dr \, \Phi_n(\boldsymbol{\kappa}, r) \frac{\sin(\kappa_s r)}{\kappa_s}.$$

The last term approximates a delta function because the slant range is large relative to the eddy sizes:

$$\langle \overline{\delta\theta_u^2} \rangle = 2\pi \int_0^\infty dr \int d^3\kappa \, \Phi_n(\boldsymbol{\kappa}, r)\kappa_u^2 \delta(\kappa_s)\left(\frac{2J_1\left(a_{\rm r}\sqrt{\kappa_u^2 + \kappa_v^2}\right)}{a_{\rm r}\sqrt{\kappa_u^2 + \kappa_v^2}}\right)^2 \tag{7.36}$$

The path of an optical signal is determined by temperature fluctuations in the lower atmosphere. As in the case of horizontal propagation, the angular variance for stellar signals on oblique paths is proportional to the third moment of the turbulence spectrum. This means that the result should be determined primarily by the inertial region and be relatively insensitive to the largest eddies. As a first approximation, we assume that the spectrum is isotropic in this regime. We will return to this question in Section 7.3.5 and describe the influence of elongated irregularities. When the spectrum is isotropic it is convenient to express the wavenumber integrations in the spherical coordinates defined by Figure 4.13 and convert the integration over slant range to one over altitude:

$$\langle \overline{\delta\theta_u^2} \rangle = 2\pi \sec\vartheta \int_0^\infty dz \int_0^\infty d\kappa \; \kappa^2 \Phi_n(\kappa, z) \int_0^\pi d\psi \sin\psi \int_0^{2\pi} d\omega$$
$$\times (\kappa \sin\psi \sin\omega)^2 \delta(\kappa \cos\psi) \left( \frac{2J_1(a_r\kappa \sin\psi)}{a_r\kappa \sin\psi} \right)^2$$

The delta function collapses the polar integration and the azimuth integration is simple:

$$\text{Isotropic:} \qquad \langle \overline{\delta\theta_u^2} \rangle = 2\pi^2 \sec\vartheta \int_0^\infty dz \int_0^\infty d\kappa \; \kappa^3 \Phi_n(\kappa, z) \left( \frac{2J_1(a_r\kappa)}{a_r\kappa} \right)^2$$

$$(7.37)$$

The orthogonal error component $\delta\theta_v$ is identical. These results are similar to the description for horizontal propagation except that the path length has been replaced:

$$R \to \sec\vartheta \int_0^\infty dz$$

This similarity allows us to reuse the calculations which led to (7.16) and the angle-of-arrival variance for astronomical sources becomes

$$\langle \overline{\delta\theta^2} \rangle = 2.251 a_r^{-\frac{1}{3}} \sec\vartheta \int_0^\infty dz \; C_n^2(z). \qquad (7.38)$$

This result is often expressed in terms of the diameter of the telescope objective:

$$\langle \overline{\delta\theta^2} \rangle = 2.84 D_r^{-\frac{1}{3}} \sec\vartheta \int_0^\infty dz \; C_n^2(z) \qquad (7.39)$$

These expressions depend on the integrated profile of $C_n^2$. One must modify this formula to recognize the spherical nature of the atmosphere if the source is close to the horizon, as discussed in Problem 2.

This basic prediction (7.39) was confirmed in several ways. To test the scaling with zenith angle, Kolchinskii compared numerous astronomical observations made with a 40-cm telescope at the Goloseyevo Observatory [23]. His data is reproduced in Figure 7.8. Using the general model

$$\delta\theta_{\text{rms}} = \text{constant} \times (\sec\vartheta)^v \qquad (7.40)$$

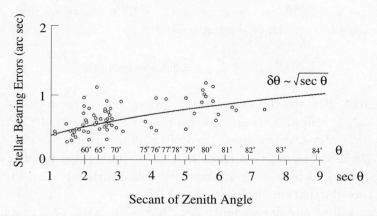

Figure 7.8: The root mean square angle of arrival for stellar images measured as a function of the zenith angle by Kolchinskii using the 40-cm telescope at Goloseyevo [23].

he found that the best fit to all data is

$$v = 0.53 \pm 0.03 \tag{7.41}$$

which is consistent with the theoretical result. One can also test the scaling with the telescope opening:

$$\delta\theta_{\mathrm{rms}} = \mathrm{constant} \times a_{\mathrm{r}}^{-\frac{1}{6}} \tag{7.42}$$

Simultaneous solar-limb observations were made with 10- and 40-cm telescopes [24]. This data showed that the angle-of-arrival variations are highly correlated (96%) and stood in the predicted ratio 1.23. Kolchinskii examined star-trail data for many different telescopes [5] and found that this data is also well approximated by (7.42). The scaling of angle-of-arrival data with the elevation of the source and the telescope opening thus confirms the prediction of geometrical optics.

For practical applications one is usually concerned with the absolute value of the bearing fluctuations. This depends on the level of turbulence integrated over the entire atmosphere. Tropospheric temperature data and amplitude-scintillation spatial correlations indicate that the integrated profile value is approximately

$$\int_0^\infty dz\, C_n^2(z) = 10^{-11}\,\mathrm{m}^{\frac{1}{3}} \tag{7.43}$$

A telescope with a 40-cm opening and a stellar source at high angles of elevation should experience fluctuations of

$$\delta\theta_{\text{rms}} \simeq 7 \text{ µrad} = 1.5 \text{ arc seconds}$$

which agrees with the image quivering measured by telescopes of this size at sea level [5]. Simultaneous observations of $C_n^2$ and angular fluctuations of solar-limb observations made in France [25] further confirm the basic result (7.38). Strong support came from solar-limb experiments performed in the USSR, in which both surface-based and airborne measurements of temperature fluctuations were employed [26].

A great deal of angular-variance data has been taken at many locations. By contrast, there have been very few measurements of the height profile of turbulent activity. The validation of geometrical optics means that one can use stellar quivering to provide reliable estimates for the integrated profile of $C_n^2$ using the relationship

$$\int_0^\infty dz\, C_n^2(z) = 0.444\, a_{\text{r}}^{\frac{1}{3}} \cos\vartheta\, \langle \delta\theta^2 \rangle \qquad (7.44)$$

This approach has been developed in France using both the sun and stars as sources [25][27][28]. An Australian group extended the technique by comparing double stars with nearby single stars [29].

### 7.3.2 The Spatial Correlation of Angular Errors

One can also estimate the correlation between image motions that occur at different points in the focal plane of a telescope. This correlation can be measured by comparing adjacent star trails on a photographic plate. Detector arrays placed in the focal plane now provide a direct and more accurate way to measure this quantity. One must treat the in-plane and out-of-plane errors separately because the baseline vector introduces a preferred direction. Figure 7.9 identifies the two situations. In the example on the left-hand side the two receivers are deployed along an axis perpendicular to the plane of the arriving signals. The separation vector can be described in the earth coordinates $(x, y, z)$ or with the $(s, u, v)$ coordinates centered on the line of sight:

$$\boldsymbol{\rho}_\perp = \rho_u \mathbf{i}_u = \rho_u \mathbf{i}_y \qquad (7.45)$$

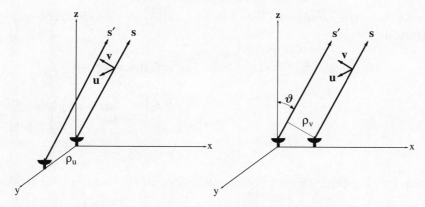

Figure 7.9: The geometry used to describe the spatial correlation for angle-of-arrival fluctuations at adjacent receivers. The case on the left-hand side corresponds to receivers deployed along a baseline perpendicular to the plane of the arriving signals. The case on the right-hand side has the receivers deployed in that plane.

For the case on the right-hand side, the receivers are deployed in the plane of propagation and their separation is described by

$$\boldsymbol{\rho}_\parallel = \rho_v \mathbf{i}_v = \rho_v (-\mathbf{i}_x \cos \vartheta + \mathbf{i}_z \sin \vartheta). \tag{7.46}$$

The spatial correlations measured along these two perpendicular directions will be different, even when the medium is isotropic.

Using the basic expression (7.31), the covariance for the out-of-plane angular wander taken at the two receivers becomes

$$\langle \delta\theta_u(0) \ \delta\theta_u(\rho_u) \rangle$$

$$= \frac{1}{\mathcal{A}^2} \iint_{\mathcal{A}} du_1 \, dv_1 \iint_{\mathcal{A}} du_2 \, dv_2 \int_0^\infty ds_1 \int_0^\infty ds_2$$

$$\times \frac{\partial^2}{\partial u_1 \, \partial u_2} \int d^3\kappa \, \Phi_n[\kappa, \tfrac{1}{2}(s_1 + s_2)]$$

$$\times \exp\{i\,[\kappa_s(s_1 - s_2) + \kappa_u(u_1 - u_2 + \rho_u) + \kappa_v(v_1 - v_2)]\}$$

The integration variables are again separated. The aperture averages are identical to those encountered previously and the result is given by (7.35).

One can use the analysis that led to (7.36) to describe the ray-path integrations:

$$\langle \delta\theta_u(0)\, \delta\theta_u(\rho_u)\rangle = 2\pi \int_0^\infty ds \int d^3\kappa\, \Phi_n(\kappa,s)\kappa_u^2\, \delta(\kappa_s)$$

$$\times \exp(i\kappa_u\rho_u)\left(\frac{2J_1\left(a_\mathrm{r}\sqrt{\kappa_u^2+\kappa_v^2}\right)}{a_\mathrm{r}\sqrt{\kappa_u^2+\kappa_v^2}}\right)^2$$

$$(7.47)$$

We employ spherical wavenumber coordinates to evaluate this expression:

$$\langle \delta\theta_u(0)\, \delta\theta_u(\rho_u)\rangle = 2\pi \int_0^\infty ds \int_0^\infty d\kappa\, \kappa^3 \Phi_n(\kappa,s)\left(\frac{2J_1\left(\kappa a_\mathrm{r}\right)}{\kappa a_\mathrm{r}}\right)^2$$

$$\times \int_0^{2\pi} d\omega \sin^2\omega\, \exp(i\kappa\rho_u \sin\omega)$$

The last integral can be done analytically to give the wavenumber-weighting function for the angle-of-arrival correlation between two sensors deployed *normal to* the plane of propagation:

$$\langle \delta\theta_u(0)\, \delta\theta_u(\rho_u)\rangle = 2\pi \int_0^\infty ds \int_0^\infty d\kappa\, \kappa^3 \Phi_n(\kappa,s)\left(\frac{2J_1\left(\kappa a_\mathrm{r}\right)}{\kappa a_\mathrm{r}}\right)^2$$

$$\times \left(J_0(\kappa\rho_u) - \frac{J_1(\kappa\rho_u)}{\kappa\rho_u}\right)$$

$$(7.48)$$

In a similar way the correlation of bearing angles between two sensors *in* the plane of propagation gives the second wavenumber-weighting function:

$$\langle \delta\theta_v(0)\, \delta\theta_v(\rho_u)\rangle = 2\pi \int_0^\infty ds \int_0^\infty d\kappa\, \kappa^3 \Phi_n(\kappa,s)\left(\frac{2J_1(\kappa a_\mathrm{r})}{\kappa a_\mathrm{r}}\right)^2 \frac{J_1(\kappa\rho_v)}{\kappa\rho_v}$$

$$(7.49)$$

These expressions reduce to the common variance (7.38) as the separation goes to zero. The same techniques can be applied to the situation on the right-hand side of Figure 7.9, for which the receiver baseline lies in the plane of propagation. One finds the same outcomes but with a reversal of direction:

$$\langle \delta\theta_v(0)\, \delta\theta_v(\rho_v)\rangle = \langle \delta\theta_u(0)\, \delta\theta_u(\rho_u)\rangle$$

$$\langle \delta\theta_v(0)\, \delta\theta_v(\rho_u)\rangle = \langle \delta\theta_u(0)\, \delta\theta_u(\rho_v)\rangle$$

The basic expressions (7.48) and (7.49) can be evaluated analytically with the inertial model if one disregards inner-scale effects and aperture averaging. When the outer scale length is large relative to the separation between receivers

$$\langle \delta\theta_v(0)\,\delta\theta_v(\rho_u)\rangle = 1.303\,\sec\vartheta \int_0^\infty dz\, C_n^2(z) \int_0^\infty d\kappa\, \kappa^{-\frac{2}{3}}\frac{J_1(\kappa\rho_u)}{\kappa\rho_u}$$

The wavenumber integral can be done using a result from Appendix D:

$$\langle \delta\theta_v(0)\,\delta\theta_v(\rho_u)\rangle = \rho^{-\frac{1}{3}}\,3.856\,\sec\vartheta \int_0^\infty dz\, C_n^2(z) \qquad (7.50)$$

The other component differs by a numerical factor:

$$\langle \delta\theta_u t(0)\,\delta\theta_u(\rho_u)\rangle = \rho^{-\frac{1}{3}}\,2.570\,\sec\vartheta \int_0^\infty dz\, C_n^2(z) \qquad (7.51)$$

This variation with the inter-receiver separation was confirmed by the French group working with the refracting solar telescope at Nice ($a_r = 20$ cm and $f_L = 10$ m). The sun was viewed through a narrow rectangular slit whose largest dimension was perpendicular to the solar limb [25][27][30]. The illumination was thus proportional to slope fluctuations of the arriving wave-front. The angle of arrival was measured by a linear array of diodes in the plane of the pupil, which could be aligned parallel or perpendicular to the solar limb. Data from this series is reproduced in Figure 7.10 and shows that the correlation of these two components follows the $\rho^{-\frac{1}{3}}$ scaling law for separations between 5.8 and 35 mm.

It is significant that these expressions do not reduce to the variance result for small separations. To correct for this problem we must include aperture averaging using the general formulations of (7.48) and (7.49):

$$\langle \delta\theta_u(0)\,\delta\theta_u(\rho_u)\rangle = 1.303\,\sec\vartheta \int_0^\infty dz\, C_n^2(z) \int_{\kappa_0}^\infty \frac{d\kappa}{\kappa^{\frac{2}{3}}} \left(\frac{2J_1(\kappa a_r)}{\kappa a_r}\right)^2$$

$$\times \left(J_0(\kappa\rho_u) - \frac{J_1(\kappa\rho_u)}{\kappa\rho_u}\right) \qquad (7.52)$$

This expression does reduce to the mean square value of (7.38) when the spacing goes to zero. It is usually important to retain the outer-scale cut-off because $\kappa_0 a_r$ is approximately unity for many telescopes. The height integration operates both on $C_n^2$ and on $\kappa_0$, which vary with altitude in quite different ways. We found before that the variation of $C_n^2$ is far greater

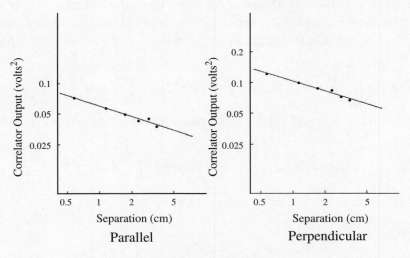

Figure 7.10: The spatial correlation of angular fluctuations in the focal plane of a solar telescope measured parallel and perpendicular to the solar limb by Borgnino and Vernin [27]. The straight lines are the inertial-range predictions described by (7.50) and (7.51).

than that of $\kappa_0$ and the integrations are effectively uncoupled. The normalized spatial correlations computed from (7.52) are plotted in Figure 7.11 as a function of $\rho/a_r$ for several values of $\kappa_0 a_r$. It is evident that the outer scale length *is important* for spatial correlations – in contrast to variance estimates. The orthogonal component gives a similar expression:

$$\langle \delta\theta_v(0)\,\delta\theta_v(\rho_u)\rangle$$

$$= 1.303 \sec\vartheta \int_0^\infty dz\, C_n^2(z) \int_{\kappa_0}^\infty \frac{d\kappa}{\kappa^{\frac{2}{3}}} \left(\frac{2J_1(\kappa a_r)}{\kappa a_r}\right)^2 \frac{J_1(\kappa\rho_u)}{\kappa\rho_u} \qquad (7.53)$$

The solar telescope at Nice ($a_r = 20$ cm) was used to measure the spatial correlation of angular fluctuations in the focal plane [30]. Longitudinal and transverse correlations were measured for separations ranging from 0 to 20 cm with three different diaphragm openings. The data agrees remarkably well with the predictions of Figure 7.11 if the parameter $\kappa_0 a_r \simeq 3$. This suggests that the average outer scale length was about 40 cm. This result is a little surprising until one recalls that this is the telescope diameter. It is quite possible that the opening of the instrument generated eddies of this size above the telescope or that they were formed within the barrel.

Now that we have established analytical descriptions for the spatial correlations of $\delta\theta$, we can return to the simpler problem of horizontal propagation.

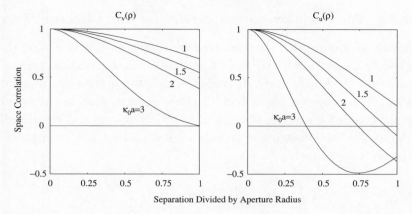

Figure 7.11: The predicted spatial correlation of angular errors orthogonal and parallel to the inter-receiver separation. The correlations are plotted as functions of the separation divided by the aperture radius for several values of $\kappa_0 a_r$.

Spatial correlations measured in this way have been used to measure the inner scale length of atmospheric turbulence. By comparing the wander of adjacent beams on short horizontal paths, one can measure their correlation as a function of beam separation both in the vertical and in the horizontal direction. The difference of the correlations is the important quantity

$$\Upsilon(\rho) = \langle \delta\theta_v(0)\,\delta\theta_v(\rho)\rangle - \langle \delta\theta_u(0)\,\delta\theta_u(\rho)\rangle \tag{7.54}$$

If there is no appreciable aperture averaging we can use (7.48) and (7.49) with a constant $C_n^2$ to write

$$\Upsilon(\rho) = 2\pi R \int_0^\infty d\kappa\ \kappa^3 \Phi_n(\kappa) J_2(\kappa\rho) \tag{7.55}$$

This result depends importantly on large wavenumbers and we use the following expression to describe both the inertial and the dissipation range:

$$\Phi_n(\kappa) = 0.033 C_n^2 \kappa^{-\frac{11}{3}} \mathcal{F}(\kappa\ell_0) \tag{7.56}$$

Here $\ell_0$ is the inner scale length and $\mathcal{F}(0) = 1$. The function $\Upsilon(\rho)$ was evaluated for five different models of $\mathcal{F}(\kappa\ell_0)$ and the results were substantially the same in each case: $\Upsilon(\rho)$ rises from zero at $\rho = 0$, reaches a maximum value for $\rho = \ell_0$, and then falls to zero for large separations [31]. Parallel laser beams were used to measure this quantity with small receivers on a 130-m path in Italy [32]. The experimental data followed the prediction and gave values of $\ell_0$ ranging from 4 to 8 mm. These estimates are confirmed by frequency and spatial correlation measurements of amplitude fluctuations, which will be discussed later in these volumes.

### 7.3.3 *The Angular-error Correlation for Adjacent Stars*

One can also compute the correlation of bearing errors for two stars that are separated by a small angle. We need this result to interpret the correlation of image motion for double stars. It is also required in order to estimate source averaging. The common ingredient in these observations is the covariance of angular errors for rays that arrive from adjacent sources. We identify the two sources by their bearing vectors $\mathbf{s}$ and $\mathbf{s}'$ so that the covariance for a point receiver becomes

$$
\begin{aligned}
\langle \delta\theta_n \, \delta\theta'_n \rangle &= \frac{1}{4} \int_0^\infty ds \int_0^\infty ds' \; \frac{\partial^2}{\partial n \, \partial n'} \langle \Delta\epsilon(\mathbf{s}) \, \Delta\epsilon(\mathbf{s}') \rangle \\
&= \frac{1}{4} \int_0^\infty ds \int_0^\infty ds' \; \frac{\partial^2}{\partial n \, \partial n'} \int d^3\kappa \, \Phi_\varepsilon[\kappa, \tfrac{1}{2}(s+s')] \exp[i\boldsymbol{\kappa} \cdot (\mathbf{s}-\mathbf{s}')] \\
&= \int d^3\kappa \, \kappa_n^2 \int_0^\infty ds \int_0^\infty ds' \; \Phi_n[\kappa, \tfrac{1}{2}(s+s')] \exp[i\boldsymbol{\kappa} \cdot (\mathbf{s}-\mathbf{s}')].
\end{aligned}
$$

$$(7.57)$$

For small angular separations the scalar product $\boldsymbol{\kappa} \cdot (\mathbf{s}-\mathbf{s}')$ can be described in terms of the spherical wavenumber coordinates identified in Figure 7.12:

$$
\boldsymbol{\kappa} \cdot (\mathbf{s}-\mathbf{s}') \simeq \kappa(s-s') \cos\psi + \delta\kappa s \sin\psi \cos\omega
$$

The main contribution to the double ray-path integration comes from small values of the difference $s-s'$ and one can write

$$
\begin{aligned}
\int_0^\infty ds \int_0^\infty ds' \; &\Phi_n[\kappa, \tfrac{1}{2}(s+s')] \exp[i\kappa(s-s')\cos\psi] \\
&\times \exp(i\kappa\delta s \sin\psi \cos\omega) = 2\pi\delta(\kappa\cos\psi) \int_0^\infty ds \, \Phi_n(\kappa, s) \exp(i\delta\kappa s \cos\omega).
\end{aligned}
$$

For angular fluctuations that are in the plane defined by the source

$$
\begin{aligned}
\langle \delta\theta_u(0) \, \delta\theta_u(\delta) \rangle & \\
&= 2\pi \int_0^\infty ds \int_0^\infty d\kappa \, \kappa^3 \Phi_n(\kappa, s) \int_0^{2\pi} d\omega \cos^2\omega \, \exp(i\delta\kappa s \cos\omega) \\
&= 4\pi^2 \int_0^\infty ds \int_0^\infty d\kappa \, \kappa^3 \Phi_n(\kappa, s) \left( J_0(\delta\kappa s) - \frac{J_1(\delta\kappa s)}{\delta\kappa s} \right).
\end{aligned}
$$

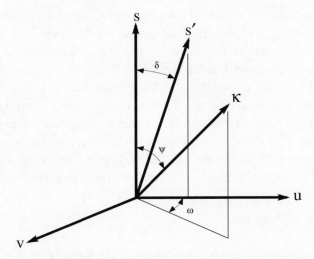

Figure 7.12: The coordinate system used to analyze the correlation of bearing errors for adjacent stars.

Had we included aperture smoothing, the wavenumber-weighting function (7.35) would also have appeared. Changing from slant range to altitude and introducing the inertial spectrum gives

$$\langle \delta\theta_u(0)\,\delta\theta_u(\delta) \rangle = 0.132\,\pi^2 \sec\vartheta \int_0^\infty dz\; C_n^2(z) \int_{\kappa_0}^\infty d\kappa\; \kappa^{-\frac{2}{3}}$$

$$\times \left( \frac{2J_1(\kappa a_{\mathrm{r}})}{\kappa a_{\mathrm{r}}} \right)^2 \left( J_0(\delta\kappa z \sec\vartheta) - \frac{J_1(\delta\kappa z \sec\vartheta)}{\delta\kappa z \sec\vartheta} \right) \quad (7.58)$$

The last term represents the wavenumber-weighting function for adjacent stars and indicates how atmospheric conditions at different levels influence the correlation of angular scintillations. It couples the wavenumber and height integrations in a necessary way because the ray separation changes continuously as the signals pass through the atmosphere. The wavenumber integration is now limited by aperture averaging and by the angular separation. The two cutoff points coincide when the telescope radius is equal to the local separation of the downcoming rays:

$$a_{\mathrm{r}} = \delta z \sec\vartheta$$

Most openings are less than 1 m wide so that aperture averaging is influential only for eddies close to the observatory. On the other hand, the level of turbulence activity is greatest at the surface and small values of $z$ are most important in evaluating (7.58). This means that one must include both the outer-scale cutoff $\kappa_0$ and the aperture weighting function.[2] Near the surface the atmosphere is reasonably isotropic and the outer scale length is approximately 1 m. Kolchinskii used the MIR-12 telescope at the Goloseyevo Observatory ($a_r = 20$ cm and $f_L = 5.5$ m) to measure the correlation of angular fluctuations for adjacent star trails recorded on photographic plates [33]. He compared pairs of stars with angular separations ranging from 5.6 to 61 arc minutes. The measured correlations were 0.7 at the smallest separation and less than 0.08 at the largest. This data is consistent with numerical evaluations of (7.58) if the important contributions occurred in the first 20 m of the troposphere.

### *7.3.4  Angular-error Averaging by Extended Sources*

One can use the expression for the angular covariance to estimate the influence of source averaging [34]. This is important for radio sources in distant galaxies and is sometimes important for observations of planets in our solar system. Similar considerations apply to ground-based imaging of orbiting spacecraft. In this application we regard the two vectors **s** and **s′** introduced in Figure 7.12 as bearing lines to points on a circular source of angular width $\beta$. The angle between these vectors $\delta$ is related to their spherical coordinates measured relative to the vector which defines the center of the source:

$$\cos \delta = \cos \theta_1 \cos \theta_2 + \sin \theta_1 \sin \theta_2 \cos(\phi_1 - \phi_2)$$

The polar angles are small for astronomical sources and one can make the approximation

$$\delta = \sqrt{\theta_1^2 + \theta_2^2 - 2\theta_1 \theta_2 \cos(\phi_1 - \phi_2)}$$

If the radiation coming from different surface elements is not coherent, the source-averaged angle of arrival can be written

$$\overline{\langle \delta\theta_u^2 \rangle} = \frac{1}{(\pi \beta^2)^2} \int_0^\beta d\theta_1 \; \theta_1 \int_0^{2\pi} d\phi_1 \int_0^\beta d\theta_2 \; \theta_2 \int_0^{2\pi} d\phi_2 \langle \delta\theta_u(0) \, \delta\theta_u(\delta) \rangle$$

---

[2] Notice that the angular covariance becomes

$$\langle \delta\theta_u(0) \, \delta\theta_u(\delta) \rangle = \text{constant} \times \delta^{-\frac{1}{3}}$$

if we assume that $\kappa_0 = 0$ and $a_r = 0$. This is the angular version of the simplified spatial covariance expression (7.51) but is *not* consistent with astronomical data [33].

or, on combining expressions,

$$\overline{\langle\delta\theta_u^2\rangle} = 2\pi \int_0^\infty ds \int_0^\infty d\kappa\, \kappa^3 \Phi_n(\kappa,s) \frac{1}{(\pi\beta^2)^2} \int_0^\beta d\theta_1\, \theta_1 \int_0^{2\pi} d\phi_1 \int_0^\beta d\theta_2\, \theta_2$$

$$\times \int_0^{2\pi} d\phi_2 \frac{2J_1\left(\kappa s\sqrt{\theta_1^2 + \theta_2^2 - 2\theta_1\theta_2\cos(\phi_1-\phi_2)}\right)}{\kappa s\sqrt{\theta_1^2 + \theta_2^2 - 2\theta_1\theta_2\cos(\phi_1-\phi_2)}}.$$

The method described in Appendix I is exploited to evaluate the surface integrals using two integrals from Appendix D:

$$\mathcal{J}(\kappa s\beta) = \frac{1}{\pi^2\beta^4} \int_0^\beta d\theta_1\, \theta_1 \int_0^{2\pi} d\phi_1 \int_0^\beta d\theta_2\, \theta_2$$

$$\times \int_0^{2\pi} d\phi_2 \frac{J_1\left(\kappa s\sqrt{\theta_1^2 + \theta_2^2 - 2\theta_1\theta_2\cos(\phi_1-\phi_2)}\right)}{\kappa s\sqrt{\theta_1^2 + \theta_2^2 - 2\theta_1\theta_2\cos(\phi_1-\phi_2)}}$$

$$= \frac{8}{\pi} \int_0^1 dw\, \sqrt{1-w^2} \int_0^{2w} dx\, x\, \frac{J_1(\kappa s\beta x)}{\kappa s\beta x}$$

$$= \frac{8}{\pi(\kappa s\beta)^2} \int_0^{\frac{\pi}{2}} d\phi\, \sin^2\phi\, [1 - J_0(2\kappa s\beta\cos\phi)]$$

$$= \frac{8}{\pi(\kappa s\beta)^2} \Bigg( \int_0^{\frac{\pi}{2}} d\phi\, \sin^2\phi - \frac{1}{2} \int_0^{\frac{\pi}{2}} d\phi\, [1 - \cos(2\phi)]$$

$$\times J_0(2\kappa s\beta\cos\phi) \Bigg)$$

$$= \frac{2}{(\kappa s\beta)^2} \left[ 1 - J_0^2(\kappa s\beta) - J_1^2(\kappa s\beta) \right]$$

The source-averaged variance for a point receiver is thus found:

$$\overline{\langle\delta\theta_u^2\rangle} = 2\pi^2 \int_0^\infty ds \int_0^\infty d\kappa\, \kappa^3 \Phi_n(\kappa,s)$$

$$\times \left\{ \frac{4}{(\beta\kappa s)^2} \left[ 1 - J_0^2(\beta\kappa s) - J_1^2(\beta\kappa s) \right] \right\} \tag{7.59}$$

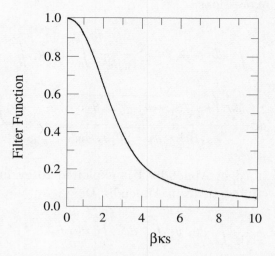

Figure 7.13: The spectral weighting function for source averaging which couples wavenumbers and slant range for a finite source of angular width $\beta$.

The wavenumber-weighting function for source averaging in curly brackets couples the altitude and wavenumber integrations. It is plotted in Figure 7.13 and shows that wavenumbers larger than $4/(s\beta)$ are relatively unimportant.

Measured image degradation is often the result both of source averaging and of receiver averaging. We combine the plane-wave result (7.37) with the source-averaged expression above and convert the path integral into an integration over height:

$$\overline{\langle \delta\theta_u^2 \rangle} = 2\pi^2 \sec\vartheta \int_0^\infty dz \int_0^\infty d\kappa \; \kappa^3 \Phi_n(\kappa, z)\left(\frac{2J_1(\kappa a_\mathrm{r})}{\kappa a_\mathrm{r}}\right)^2$$
$$\times \frac{4}{\beta\kappa z \sec\vartheta}\left[1 - J_0^2(\beta\kappa z \sec\vartheta) - J_1^2(\beta\kappa z \sec\vartheta)\right] \qquad (7.60)$$

Both weighting functions suppress the contributions of large wavenumbers. The second term which represents source averaging depends on altitude and is bounded by the effective height of the atmosphere. This shows that aperture averaging dominates the observation if $a_\mathrm{r} > \beta H \sec\vartheta$, whereas source averaging prevails if $a_\mathrm{r} < \beta H \sec\vartheta$.

It is useful to examine this conclusion in the context of planetary observations. Jupiter is the largest planet in our solar system and subtends an angle of $\beta = 49$ arc seconds from the earth. The effective height of the troposphere is a few kilometers so that $\beta H \approx 0.5$ m. Telescopes with openings this large will experience both source averaging and aperture averaging when observing Jupiter. The other planets in our solar system are smaller than Jupiter and set a lower threshold on the telescope size. One must therefore use both weighting functions to interpret most observations of this type.

### 7.3.5  The Influence of Anisotropy

Our discussion of astronomical bearing errors thus far has assumed that the lower atmosphere is isotropic. From airborne temperature measurements we know that eddies in the free atmosphere are highly elongated in the horizontal direction. This anisotropy is also manifested in fluctuations in the concentration of water vapor and is an important consideration for estimating the phase variance and the phase structure function at microwave frequencies. Yet we have consistently ignored such features in this chapter.

It is helpful to summarize the line of reasoning which led us to the above assumption. Recall first that the angle of arrival is measured by taking the phase difference between two closely spaced receivers and dividing that difference by the separation. This measurement is described analytically by the derivative of phase taken normal to the ray path. The resulting estimate of variance is proportional to the spectrum's third moment, which means that contributions of large eddies are suppressed and the outer scale length does not appear – at least for a Kolmogorov spectrum. The influence of small eddies is eliminated by aperture smoothing. The inertial regime is therefore uniquely important for estimating the angular variance. In this range the decaying eddies are more nearly isotropic than are the large eddies at the input stage. Moreover, the altitude weighting function $C_n^2(z)$ has its largest values near the surface where the eddies are more symmetrical than are those high in the troposphere. This line of argument suggests that anisotropy of the atmosphere does not strongly influence angle-of-arrival measurements.

We need to confirm that conclusion by careful analysis. One can calculate the angular variance for anisotropic media if the ellipsoidal spectrum model (2.87) is a valid description of the troposphere. Using the propagation expression (7.36) for astronomical signals, the variance of the angular component measured normal to the ray plane becomes

$$\langle \overline{\delta\theta_u^2} \rangle = 2\pi \int_0^\infty dr \, C_n^2(r) \int_{-\infty}^\infty d\kappa_x \int_{-\infty}^\infty d\kappa_y \int_{-\infty}^\infty d\kappa_z \, \kappa_u^2 \, \delta(\kappa_s)$$

$$\times \, abc \, \Omega \left( \sqrt{a^2 \kappa_x^2 + b^2 \kappa_y^2 + c^2 \kappa_z^2} \right) \left( \frac{2 J_1 \left( a_r \sqrt{\kappa_u^2 + \kappa_v^2} \right)}{a_r \sqrt{\kappa_u^2 + \kappa_v^2}} \right)^2 . \quad (7.61)$$

This description contains a mixture of wavenumber components referenced to the atmospheric coordinates and to line-of-sight coordinates. The geometry of Figure 4.13 shows how to relate them:

$$\kappa_s = \kappa_x \sin \vartheta + \kappa_z \cos \vartheta$$
$$\kappa_u = \kappa_y$$
$$\kappa_v = -\kappa_x \cos \vartheta + \kappa_z \sin \vartheta$$

With these transformations the delta function collapses the $\kappa_z$ integration:

$$\langle \overline{\delta \theta_u^2} \rangle = 2\pi \sec \vartheta \int_0^\infty dr\, C_n^2(r) \int_{-\infty}^\infty d\kappa_x \int_{-\infty}^\infty d\kappa_y\, \kappa_y^2$$
$$\times\, abc\, \Omega\left( \sqrt{(a^2 + c^2 \tan^2 \vartheta)\kappa_x^2 + b^2 \kappa_y^2} \right)$$
$$\times \left( \frac{2J_1\left(a_{\mathrm{r}}\sqrt{\kappa_y^2 + \kappa_x^2 \sec^2 \vartheta}\right)}{a_{\mathrm{r}}\sqrt{\kappa_y^2 + \kappa_x^2 \sec^2 \vartheta}} \right)^2$$

The following rescaling renders the spectrum isotropic:

$$\ell_x = \kappa_x \sqrt{a^2 + c^2 \tan^2 \vartheta} \qquad \text{and} \qquad \ell_y = \kappa_y b$$

and we then shift to polar coordinates:

$$\langle \overline{\delta \theta_u^2} \rangle = 2\pi \int_0^\infty dr\, C_n^2(r)\left( \frac{1}{b^2 \mathsf{p}} \right) \int_0^\infty d\ell\, \ell^3 \Omega(\ell)$$
$$\times \int_0^{2\pi} d\phi\, \sin^2 \phi \left( \frac{2J_1\left( a_{\mathrm{r}}\ell\sqrt{\dfrac{\sin^2 \phi}{b^2} + \dfrac{\cos^2 \phi}{a^2 c^2 \mathsf{p}^2}} \right)}{a_{\mathrm{r}}\ell\sqrt{\dfrac{\sin^2 \phi}{b^2} + \dfrac{\cos^2 \phi}{a^2 c^2 \mathsf{p}^2}}} \right)^2$$

Here we have used the parameter introduced in (4.37) to describe oblique paths:

$$\mathsf{p} = \sqrt{\frac{\sin^2 \vartheta}{a^2} + \frac{\cos^2 \vartheta}{c^2}}$$

The inertial range dominates the wavenumber integration and we use the Kolmogorov model to describe it:

$$\langle \overline{\delta \theta_u^2} \rangle = 0.033 \times 2\pi \int_0^\infty dr \, C_n^2(r) \left( \frac{1}{b^2 \mathsf{p}} \right) \int_{\kappa_0}^\infty \frac{d\ell}{\ell^{\frac{2}{3}}}$$

$$\times \int_0^{2\pi} d\phi \, \sin^2 \phi \left( \frac{2J_1 \left( a_{\mathrm{r}} \ell \sqrt{\dfrac{\sin^2 \phi}{b^2} + \dfrac{\cos^2 \phi}{a^2 c^2 \mathsf{p}^2}} \right)}{a_{\mathrm{r}} \ell \sqrt{\dfrac{\sin^2 \phi}{b^2} + \dfrac{\cos^2 \phi}{a^2 c^2 \mathsf{p}^2}}} \right)^2$$

One can omit the small-wavenumber cutoff $\kappa_0$ because the $\ell$ integration converges at the origin. The transformation

$$\zeta = \ell a_{\mathrm{r}} \sqrt{\frac{\sin^2 \phi}{b^2} + \frac{\cos^2 \phi}{a^2 c^2 \mathsf{p}^2}}$$

then separates the integrations:

$$\langle \overline{\delta \theta_u^2} \rangle = 0.033 \times 8\pi^2 a_{\mathrm{r}}^{-\frac{1}{3}} \int_0^\infty dr \, C_n^2(r) \left( \frac{1}{b^2 \mathsf{p}} \right) \int_0^\infty \frac{d\zeta}{\zeta^{\frac{8}{3}}} J_1^2(\zeta)$$

$$\times \frac{1}{\pi} \int_0^{2\pi} d\phi \, \sin^2 \phi \left( \frac{\sin^2 \phi}{b^2} + \frac{\cos^2 \phi}{a^2 c^2 \mathsf{p}^2} \right)^{-\frac{1}{6}}$$

The $\zeta$ integral is found in Appendix D and the azimuth integration can be expressed in terms of hypergeometric functions:

$$\langle \overline{\delta \theta_u^2} \rangle = 2.252 a_{\mathrm{r}}^{-\frac{1}{3}} \int_0^\infty dr \, C_n^2(r) \left( \frac{1}{b^2 \mathsf{p}} \right) \mathcal{R}_1(a, b, c, \vartheta) \tag{7.62}$$

where

$$\mathcal{R}_1(a, b, c, \vartheta) = \left( \frac{2a^2 b^2 c^2 \mathsf{p}^2}{b^2 + a^2 c^2 \mathsf{p}^2} \right)^{\frac{1}{6}} \left[ {}_2F_1\left( \frac{1}{12}, \frac{7}{12}, 1; \varpi^2 \right) + \frac{\varpi}{6} \, {}_2F_1\left( \frac{7}{12}, \frac{13}{12}, 2; \varpi^2 \right) \right] \tag{7.63}$$

and we abbreviate

$$\varpi = \frac{b^2 - a^2 c^2 \mathsf{p}^2}{b^2 + a^2 c^2 \mathsf{p}^2} \tag{7.64}$$

The parameter $\varpi$ vanishes for all values of $\vartheta$ if the medium is isotropic and the previous result (7.38) emerges. The in-plane error has a similar description:

$$\langle \overline{\delta\theta_v^2} \rangle = 2.252 a_{\rm r}^{-\frac{1}{3}} \int_0^\infty dr\, C_n^2(r) \left( \frac{\cos^2 \vartheta}{a^2 c^2 {\sf p}^3} \right) \mathcal{R}_2(a,b,c,\vartheta) \qquad (7.65)$$

The function $\mathcal{R}_2(a,b,c,\vartheta)$ is identical to (7.63) with a minus sign for the second term in square brackets. The difference between the two expressions represents the effect of aperture averaging carried out in the two directions normal to the ray path. It is significant that the scaling with the aperture size is the same in both cases.

One can simplify these expressions in many situations of practical interest. The functions $\mathcal{R}_1$ and $\mathcal{R}_2$ do not vary greatly and can often be replaced by unity. The atmospheric random medium usually has the same characteristics in both horizontal directions so that $a = b$. On changing from slant range to altitude integrations, the two bearing errors become

$$\langle \overline{\delta\theta_u^2} \rangle = 2.252 a_{\rm r}^{-\frac{1}{3}} \sec\vartheta \int_0^\infty dz\, C_n^2(z) \left[ \frac{c}{a} (c^2 \sin^2 \vartheta + a^2 \cos^2 \vartheta)^{-\frac{1}{2}} \right] \qquad (7.66)$$

$$\langle \overline{\delta\theta_v^2} \rangle = 2.252 a_{\rm r}^{-\frac{1}{3}} \sec\vartheta \int_0^\infty dz\, C_n^2(z) [ac \cos^2 \vartheta (c^2 \sin^2 \vartheta + a^2 \cos^2 \vartheta)^{-\frac{3}{2}}]$$

$$(7.67)$$

The dimensionless scaling parameters $a$ and $c$ depend on height. Their values at each level are weighted by the local value of $C_n^2$. The level of turbulent activity is greatest at the surface, where they are similar. This approximate symmetry disappears rapidly with altitude but the $C_n^2$ profile is decreasing even faster. To find corrections to the zenith-angle scaling law (7.40) one must use profiles for $a$, $c$ and $C_n^2$ to carry out the integration. Surface values of the scaling parameters are often sufficiently accurate for this purpose and the factors in square brackets can be removed from the height integrations.

## 7.4 Microwave Tracking of Satellites

Microwave tracking of earth satellites and interplanetary spacecraft is widely used to establish their orbital parameters. This takes several forms. Early satellites carried transmitters to communicate the scientific data they were acquiring. It was quickly recognized that direction finding could measure the elevation and azimuth angles of these satellites as they moved across the sky. Some satellites also carried tracking beacons that provided a narrow-band signal at a known frequency to facilitate orbit determination. These microwave sources were tracked with single-aperture radars and with inter-ferometers. Many of the microwave phase-difference experiments identified in Table 5.1 were undertaken in order to predict the accuracy of these pas-sive angular tracking systems.[3] It was found that angular measurements alone could provide precise satellite orbits by using (*a*) tracking data from many passes, (*b*) curve-fitting techniques and (*c*) smoothing of data.

A significant improvement occurred when cooperative transponders were placed on spacecraft. Transponders receive, amplify and retransmit a micro-wave signal sent to the satellite by a ground control station. Round-trip mea-surements provide range and range-rate data. The combination of range and angular data is a superior way to compute the orbits of earth-orbiting satel-lites. Doppler data is uniquely important for determining the trajectories of spacecraft sent to other planets since they are moving away from the earth most of the time.

The more difficult problem is that of how to track silent satellites and launching debris in earth orbit. This is increasingly important because the inventory of such objects is rising steadily. One must use radar skin tracking to monitor this debris. The strength of the reflected signal decreases as the fourth power of the slant range, so large transmitter powers and apertures are required, even if the satellites are in low earth orbit. The same problem occurs for radar tracking of short-range ballistic missiles, for which angular measurements provide a useful adjunct to range and range-rate information.

These tasks are limited by angle-of-arrival fluctuations and we need to understand such errors. The microwave frequencies employed are usually high enough that they are not influenced by ionospheric irregularities. In describing microwave propagation through the troposphere in Section 4.2 we found that anisotropy has an important influence on the way GPS range residuals scale with the zenith angle. We should expect a similar sensitivity in dealing with microwave tracking and will avoid the isotropic expression (7.38) used for optical signals. Instead, we exploit the descriptions derived in Section 7.3.5 for anisotropic media. The expressions (7.66) and (7.67) describe the in-plane and out-of-plane components of the bearing error for

---

[3] The initial interest in such experiments was generated by the American ballistic missile programs which used radio guidance in the first versions of Atlas, Thor and Titan. The same data was later used to estimate the limitations of microwave range instrumentation systems and orbital tracking systems.

arbitrary zenith angles and unrestricted values of the scaling parameters. We learned in Section 2.3.5 that $a/c > 100$ along most of the propagation path, which means that one can make the following approximation:

$$(c^2 \sin^2 \vartheta + a^2 \cos^2 \vartheta)^{\frac{1}{2}} \rightarrow a \cos \vartheta \quad \text{for} \quad E = \pi/2 - \vartheta > c/a$$

The two errors are thus identical and given by the following expression:

$$\langle \delta\theta^2 \rangle = 2.251 \sec^2 \vartheta \; a_r^{-\frac{1}{3}} \int_0^\infty dz \; C_n^2(z) \frac{c}{a^2} \tag{7.68}$$

Measurements of elevation and azimuth errors were made to support the American program to develop ballistic missile defense systems. Transmissions at 7.3 GHz from the IDCSP communication satellite in semi-synchronous orbit were tracked by the Haystack antenna ($a_r = 18.4$ m) using a monopulse feed array [35][36]. The frequency was high enough and the dish large enough that foreground reflections could be ignored above a few degrees elevation. The orbits of these satellites were known very precisely from other means. The bearing angle measured by the radar was compared with ephemeris data to estimate the rms elevation error as a function of the zenith angle to an accuracy of within 0.4 millidegrees. A summary of the data taken in this program is reproduced in Figure 7.14 and exhibits significant diurnal and seasonal variability.

We need to find out whether the prediction (7.68) is consistent with this data. The first task is that of comparing the measured variation with elevation angle with the anisotropic scaling law:

$$\delta\theta_{\text{rms}} = \text{constant} \times \sec \vartheta$$

This expression fits the data in Figure 7.14 remarkably well, even for elevation angles as small as a few degrees. For an isotropic medium it predicts

$$\delta\theta_{\text{rms}} = \text{constant} \times \sqrt{\sec \vartheta}$$

which does not agree with the data. With this confirmation, we can estimate the integrated profile from the tracking data in Figure 7.14. A typical value for the bearing error is 1 millidegree at 10° elevation.

$$\text{Microwave:} \quad \int_0^\infty dz \; C_n^2(z) \frac{c}{a^2} = 6 \times 10^{-11} \, \text{m}^{\frac{1}{3}}$$

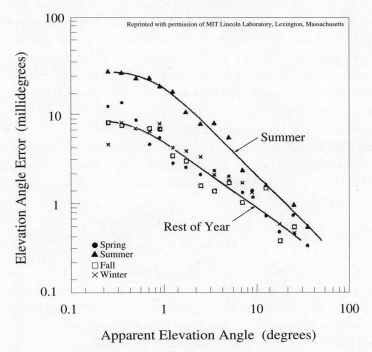

Figure 7.14: Elevation errors measured at 7.3 GHz by Crane [35].

The ratio of scaling parameters is approximately $10^{-2}$, which suggests the following value:

$$\text{Microwave:} \qquad \int_0^\infty dz \, C_n^2(z) \frac{1}{a} \simeq 10^{-8} \, \text{m}^{\frac{1}{3}} \qquad (7.69)$$

Integrated profiles of $C_n^2(z)$ have been measured by other means. The OTS communication satellite in synchronous orbit was tracked at 11.8 GHz by two antennas in England [37]. One antenna was depointed from the source and its gain slope exploited to measure the loss of signal relative to the boresighted antenna. Variations in the angle of arrival were thus converted into amplitude differences that could be precisely measured through their power spectra. The following integrated profiles were established in this way and there is evidently considerable variability:

$$\text{Daytime:} \qquad \int_0^\infty dz \, C_n^2(z) = 6.6 \times 10^{-8} \, \text{m}^{\frac{1}{3}}$$

$$\text{Nighttime:} \qquad \int_0^\infty dz \, C_n^2(z) = 3.4 \times 10^{-11} \, \text{m}^{\frac{1}{3}} \qquad (7.70)$$

There is a second problem, which does not involve random fluctuations. The mean elevation angle is also distorted by the average refractive-index profile [36][38]. This effect depends primarily on the elevation angle and the surface refractivity. Even if these quantities are measured accurately at the tracker, there is an irreducible bias error associated with the uncertainty of the profile. This error is comparable to the fluctuating components. Attempts to separate the effects have been frustrating [36]. Another problem is associated with the temporal variability of $\delta\theta$. The period over which the fluctuating component is averaged determines whether errors are counted as random components or bias. That complicates the identification of sources of error. For these reasons modern satellite-tracking schemes rely largely on range and Doppler data.

## 7.5 Radio Astronomy

Radio emissions from planets, stars and galaxies are routinely observed and provide an important way to study the universe around us. This new type of astronomy has been available since the Second World War when microwave receivers were developed for radar. The first signals were detected at VHF frequencies but this was gradually extended to millimeter wavelengths as sensitive receivers became available. The origin of these signals is of great interest. It is important to determine whether their locations coincide with galaxies that are known from optical astronomy. With these techniques one can often establish the configuration of the source to a precision that optical telescopes cannot provide.

Interferometers are used to locate and define the spatial characteristics of radio sources because phase differences can be reliably measured at microwave frequencies. Large radio telescopes have also been built to enhance the signal-to-noise ratios of these faint signals and provide precise location of sources. Angle-of-arrival errors induced by the atmosphere set the threshold for these position measurements.

Both the troposphere and the ionosphere contribute to the angular error of microwave signals. The frequency scalings of their contributions are quite different. The tropospheric component is independent of frequency whereas the ionospheric component is proportional to the wavelength squared. The influence of the two regions is therefore complementary. Angular errors are determined primarily by the ionosphere below approximately 3000 MHz. Above this frequency the troposphere exerts the greatest influence. The threshold is not well defined and there is considerable overlap of the two effects. The microwave frequencies chosen for observational campaigns are usually well above or far below this arbitrary boundary and provide a convenient way to distinguish the two effects.

### 7.5.1 Ionospheric Influences

Plasma irregularities in the ionosphere and interstellar space exert a strong influence on signals below 3000 MHz. The frequencies employed in these experiments are well above the plasma frequency of the ionosphere and the simple relation between the dielectric constant and electron-density variations given in Section 2.4 is appropriate. The aperture-averaged bearing errors in this case are similar to (7.31) and (7.32). In terms of the line-of-sight coordinates identified in Figure 7.14 we have

$$\delta\theta_u = -\frac{1}{2}r_e\lambda^2 \frac{1}{A}\iint_A du\, dv \int_0^\infty ds\, \frac{\partial}{\partial u}\delta N(s,u,v) \qquad (7.71)$$

$$\delta\theta_v = -\frac{1}{2}r_e\lambda^2 \frac{1}{A}\iint_A du\, dv \int_0^\infty ds\, \frac{\partial}{\partial v}\delta N(s,u,v). \qquad (7.72)$$

The angular variance depends on the spatial correlation of electron-density variations, as we see by writing out the $u$ component:

$$\langle\delta\theta_u^2\rangle = \frac{1}{4}r_e^2\lambda^4 \frac{1}{A^2}\iint_A du\, dv \iint_A du'\, dv' \int_0^\infty ds \int_0^\infty ds'$$

$$\times \frac{\partial^2}{\partial u\, \partial u'}\langle\delta N(s,u,v)\, \delta N(s',u',v')\rangle$$

The wavenumber representation of the electron-density variations introduced in Section 2.4 gives the following expression:

$$\langle\delta\theta_u^2\rangle = \frac{1}{4}\, r_e^2\lambda^4\frac{1}{A^2}\iint_A du\, dv \iint_A du'\, dv' \int d^3\kappa\, \kappa_u^2 \int_0^\infty ds \int_0^\infty ds'$$

$$\times \Psi_N[\boldsymbol{\kappa}, \tfrac{1}{2}(\mathbf{s}-\mathbf{s}')]\exp\{i[\kappa_s(s-s') + \kappa_u(u-u') + \kappa_v(v-v')]\}$$

The $s$ and $s'$ integrations are converted to sum and difference coordinates as defined by (4.36). The difference integration drives $\kappa_s$ to zero because the slant range to ionospheric irregularities is always much greater than their size:

$$\langle\delta\theta_u^2\rangle = \frac{\pi}{2}r_e^2\lambda^4 \int_0^\infty dr \int d^3\kappa\, \Psi_N(\boldsymbol{\kappa}, r)\delta(\kappa_s)\kappa_u^2$$

$$\times \frac{1}{A^2}\iint_A du\, dv \iint_A du'\, dv' \exp\{i[\kappa_u(u-u') + \kappa_v(v-v')]\}$$

The aperture averages were evaluated in (7.35):

$$\langle \delta\theta_u^2 \rangle = \frac{\pi}{2} r_e^2 \lambda^4 \int_0^\infty dr \int d^3\kappa\, \Psi_N(\boldsymbol{\kappa}, r) \delta(\kappa_s) \kappa_u^2 \left( \frac{2J_1(a_r\sqrt{\kappa_u^2 + \kappa_v^2})}{a_r\sqrt{\kappa_u^2 + \kappa_v^2}} \right)^2 \quad (7.73)$$

In spherical wavenumber coordinates this gives the familiar aperture-averaging wavenumber-weighting function:

$$\langle \delta\theta_u^2 \rangle = \frac{\pi}{2} r_e^2 \lambda^4 \int_0^\infty dr \int_0^\infty d\kappa\, \kappa^2 \int_0^\pi d\psi \sin\psi \int_0^{2\pi} d\omega\, \Psi_N(\kappa, \psi, \omega; r)$$

$$\times\, \delta(\kappa\cos\psi)(\kappa\sin\psi\sin\omega)^2 \left( \frac{2J_1(\kappa a\sin\psi)}{\kappa a\sin\psi} \right)^2$$

$$= \frac{\pi}{2} r_e^2 \lambda^4 \int_0^\infty dr \int_0^\infty d\kappa\, \kappa^3$$

$$\times \int_0^{2\pi} d\omega \sin^2\omega\, \Psi_N\left(\kappa, \frac{\pi}{2}, \omega; r\right) \left( \frac{2J_1(\kappa a_r)}{\kappa a_r} \right)^2$$

The spectrum of plasma irregularities depends on all three components of the wavenumber because the ionospheric eddies are elongated in the direction of the magnetic field. We use the power-law model introduced in (2.135) to describe these anisotropic irregularities. The angle of arrival is sensitive to the spectral index $\nu$ and we cannot choose it for analytical convenience, as we did in estimating the phase variance:

$$\Psi_N(\boldsymbol{\kappa}, r) = \frac{\langle \delta N^2 \rangle Q_\nu \kappa_0^{\nu-3}}{2\pi\kappa^\nu [1 + (\mathcal{A}^2 - 1)\cos^2\Theta]^{\frac{\nu}{2}}} \qquad \kappa_0 \le \kappa \le \kappa_m$$

Here $\Theta$ is the angle between the wavenumber and magnetic field vectors. We are free to orient the $u$ and $v$ vectors in any helpful manner. The expression (4.53) that defines $\Theta$ simplifies if we choose the axis so that the terrestrial magnetic field lies in the $(s, v)$ plane,

$$\cos\Theta = \sin\gamma\cos\omega$$

where $\gamma$ is the angle included between $\mathbf{H}$ and $\mathbf{s}$. We combine these results and convert from slant range to altitude:

$$\langle \delta\theta_u^2 \rangle = r_e^2 \lambda^4 Q_\nu \sec\vartheta \int_0^\infty dz\, \langle \delta N^2 \rangle \kappa_0^{\nu-3} \int_{\kappa_0}^{\kappa_m} \frac{d\kappa}{\kappa^{\nu-3}} \left( \frac{2J_1(\kappa a_r)}{\kappa a_r} \right)^2$$

$$\times \int_0^{\pi/2} d\omega\, \frac{\sin^2\omega}{[1 + (\mathcal{A}^2 - 1)\sin^2\gamma\cos^2\omega]^{\frac{\nu}{2}}} \qquad (7.74)$$

We have left the electron-density variance and scale lengths inside the height integration because they probably change with altitude.

The influence of aperture averaging is separated in (7.74) and is expressed by the following wavenumber integral:

$$\mathcal{W}_\nu(a_\mathrm{r}) = \int_{\kappa_0}^{\kappa_\mathrm{m}} \frac{d\kappa}{\kappa^{\nu-3}} \left( \frac{2J_1(\kappa a_\mathrm{r})}{\kappa a_\mathrm{r}} \right)^2 = 4a_\mathrm{r}^{\nu-4} \int_{\kappa_0 a_\mathrm{r}}^{\kappa_\mathrm{m} a_\mathrm{r}} \frac{dx}{x^{\nu-1}} J_1^2(x) \qquad (7.75)$$

We do not know the spectral index $\nu$ from basic physics and must look to experiments to establish its value. In the next volume we will learn that it can be determined from the frequency scaling of amplitude fluctuations. These experiments give values in the range

$$3 < \nu < 4.$$

In this range the upper limit $\kappa_\mathrm{m} a_\mathrm{r}$ can be taken to infinity because radio-astronomy receivers are always larger than the electron gyroradius in the ionosphere.[4] The outer scale length of ionospheric irregularities is large compared with these receivers and the lower limit $\kappa_0 a_\mathrm{r}$ can be ignored. Within the indicated range of $\nu$ values we can therefore use a result from Appendix D to write[5]

$$\mathcal{W}_\nu(a_\mathrm{r}) = a_\mathrm{r}^{\nu-4} \left[ \frac{2^{3-\nu}\Gamma(2-\nu/2)\Gamma(\nu-1)}{\Gamma^2(\nu/2)\Gamma(1+\nu/2)} \right] = \Gamma_\nu a_\mathrm{r}^{\nu-4}. \qquad (7.76)$$

With these results the angular variance can be expressed in terms of the parameters $\nu, \mathcal{A}, \gamma, \kappa_0$ and $\langle \delta N^2 \rangle$ as

$$\langle \delta\theta_u^2 \rangle = \lambda^4 \mathcal{C}_u \int_0^\infty ds \, \langle \delta n^2(s) \rangle \kappa_0^{\nu-3}(s) \qquad (7.77)$$

where the various constants have been combined as follows:

$$\mathcal{C}_u = 2\pi r_\mathrm{e}^2 Q_\nu \Gamma_\nu a_\mathrm{r}^{\nu-4} \int_0^{\pi/2} d\omega \, \frac{\sin^2 \omega}{[1 + (\mathcal{A}^2 - 1)\sin^2 \gamma \cos^2 \omega]^{\frac{\nu}{2}}} \qquad (7.78)$$

The $v$ component is identical to (7.77) except for a change in the azimuth integration:

$$\mathcal{C}_v = 2\pi r_\mathrm{e}^2 Q_\nu \Gamma_\nu a_\mathrm{r}^{\nu-4} \int_0^{\pi/2} d\omega \, \frac{\cos^2 \omega}{[1 + (\mathcal{A}^2 - 1)\sin^2 \gamma \cos^2 \omega]^{\frac{\nu}{2}}} \qquad (7.79)$$

---

[4] This issue is not so clear for the interstellar plasma.

[5] One cannot ignore the lower limit $\kappa_0 a_\mathrm{r}$ if $\nu \geq 4$. The upper limit $\kappa_\mathrm{m} a_\mathrm{r}$ plays an important role if $\nu < 3$. It is only in the narrow range indicated above that one can neglect the inner and outer scale lengths for estimates of the angular variance.

The parameters in these expressions are not known well enough to support confident predictions. On the other hand, their scalings with frequency, aperture size and zenith angle are independent of these parameters.

The frequency dependence of (7.77) reflects the basic relation between the dielectric constant and the electron density:

$$\delta\theta_{\mathrm{rms}} = \mathrm{constant} \times \lambda^2 \qquad (7.80)$$

Early radio-astronomy observations made in England used interferometers to measure the bearing fluctuation of four galactic radio sources [39]. At a wavelength of 8 m the apparent position of the source varied from 2 to 3 minutes of arc under normal conditions. It grew to half a degree during disturbed conditions in the ionosphere. Similar observations at 3.7 and 1.4 m confirmed the predicted wavelength scaling. Later comparison of radio signals taken at 53 and 108 MHz also supported the prediction [40].

The frequency scaling (7.80) is consistent with either ionospheric or interstellar influences since both are weak plasmas. The variation of bearing errors with the elevation angle provides a way to distinguish between the two mechanisms. If the angular error is produced by the ionosphere, one should use $s = z \sec\vartheta$ to relate the slant range along the propagation path to height and the zenith angle. In this case we would expect the measurements to follow

$$\delta\theta_{\mathrm{rms}} = \mathrm{constant} \times \sqrt{\sec\vartheta}. \qquad (7.81)$$

The measurements of angular error ought to be independent of the zenith angle if the very long path through interstellar space is primarily responsible. An early experiment at Boulder tracked the radio source in Cygnus-A. Bearing errors were measured with an interferometer for elevation angles ranging from 10° to 40° [40]. The data is reproduced in Figure 7.15 and is consistent with (7.81). An experiment using satellite transmissions at 150 and 400 MHz was conducted in order to isolate the influence of ionospheric irregularities [41] and the data agreed with (7.81). This suggests that the primary contribution to $\delta\theta$ comes from the earth's atmosphere. That conclusion is supported by the correlation of $\delta\theta$ to conditions in the ionosphere measured by radio soundings.

It is reasonable that nearby irregularities should have the greatest influence on the arriving phase front. Satellite photography is degraded very little by tropospheric turbulence, whereas ground-based astronomical observations are limited in a fundamental way [42][43]. This *bottom-of-the-ocean* effect means that the interstellar plasma plays a negligible role in setting the angular error of downcoming radio waves.

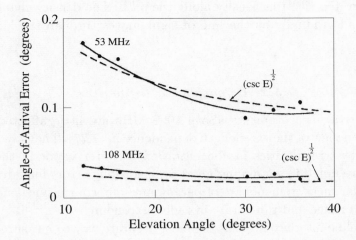

Figure 7.15: Elevation errors for 53- and 108-MHz signals measured as a function of the elevation angle by Lawrence, Jesperson and Lamb [40].

The scaling of angular variance with respect to antenna size is also given by the basic relationship (7.76):

$$\delta\theta_{\mathrm{rms}} = \text{constant} \times (a_{\mathrm{r}})^{\frac{\nu}{2}-2} \tag{7.82}$$

This represents a weak dependence for the range of indices $3 < \nu < 4$. Simultaneous observations with different dishes seem not to have been made.

Variations in the bearing angle of radio sources can tell one a good deal about large-scale irregularities in the ionosphere. Their influence is observed as a gradual wandering of the source bearing. This angular movement is due to refractive bending by horizontal gradients of the average electron density. For a thin layer it is described by

$$\delta\theta(t) = \frac{1}{2}\int ds\,\frac{\partial}{\partial n}\Delta\epsilon(s,n,t) = -\frac{1}{2}r_{\mathrm{e}}\lambda^2\,\Delta H\,\frac{\partial N_0(x,t)}{\partial x}. \tag{7.83}$$

By observing how the apparent direction of the source changes with time one concludes that large wavelike structures exist in the ionosphere [44][45][46]. These traveling ionospheric disturbances are caused by acoustic gravity waves that propagate at heights of several hundred kilometers [47]. They travel from west to east, moving at speeds of several hundred meters per second. The wavelengths of these harmonic structures range from less than 100 to more than 1000 km. They are also observed by monitoring the Faraday rotation of the polarization of satellite signals, which is proportional to the integrated electron density along the path. The density changes both suddenly and gradually as the line of sight moves through the large-scale structures.

### 7.5.2  Tropospheric Influences

Angle-of-arrival measurements above 5000 MHz are little affected by the ionosphere in view of the wavelength dependence in (7.77). They *are* strongly influenced by irregularities in concentration of water vapor in the troposphere, as we noted in examining microwave tracking errors for satellite signals. Angular measurements are of paramount importance since one cannot measure the range and range rate in radio astronomy.

Long-baseline interferometry is the most accurate way to measure angular position. This technique was analyzed in Chapter 5, where we found that the best available accuracy is determined by the receiver baseline, the phase structure function and the zenith angle. If the baseline lies in the plane of propagation

$$\langle \delta\theta^2 \rangle = \sec^2\vartheta \, \frac{\mathcal{D}_\varphi\left(\rho\right)}{k^2\rho^2}. \tag{7.84}$$

If the baseline is normal to the plane the factor $\sec^2\vartheta$ is absent. Experimental phase-difference data – like that plotted in Figure 5.16 – can be used to estimate the angular error for various baselines. The prediction also depends strongly on the averaging time. Stotskii compared such estimates with the angular performances of several early interferometers and found good agreement [48]. Wright recently prepared a summary of microwave and millimeter-wave phase structure functions that allows one to estimate the performance of an interferometer with some confidence [49].

One can use the same approach to estimate the angular errors experienced by large radio telescopes. For a Cassegrain telescope with a primary reflector and subreflector at the focus like that shown in Figure 7.2, the multiple feeds act like elements of an interferometer with a baseline equivalent to the large diameter. Data taken with the Pulkovo Radio Telescope in the USSR gave an error of 1 arc second [48], which agrees with the estimate made from the phase-difference data presented in Figure 5.14 for $\rho = 100$ m.

Large radio telescopes regularly measure angular errors that are much greater both than their inherent accuracy and than the bearing error due to random irregularities. These events are called *anomalous refraction* and create a significant problem for radio-astronomy observations. The 100-m Effelsberg radio telescope operating at 23 GHz found that sources were displaced by 40 arc seconds for periods of up to 30 s [50]. Observations with the 30-m IRAM telescope operating at 86 and 230 GHz in Spain revealed angular displacements as great as 20 arc seconds [50][51]. Similar events at 350 GHz were found with the 15-m James Clerk Maxwell telescope in Hawaii [52]. These events are common during daylight hours but are rare at night. Multifrequency probing leaves little doubt that they are caused by moist air pockets that distort the downcoming beam. These structures can also distort interferometer observations if they fill one path but not the other, although the effect is diminished by the longer baseline term in the denominator of (7.84).

## 7.6 Problems

### *Problem 1*

Develop an expression for $\langle \delta\theta^2 \rangle$ that covers the transition from very small to very large receivers. Assume that the unperturbed field is a plane wave traveling along a horizontal path. Describe the spectrum with a Gaussian expression for the energy-loss region:

$$\Phi_n(\kappa) = 0.033 C_n^2 \kappa^{-\frac{11}{3}} \exp(-\kappa^2/\kappa_{\mathrm{m}}^2)$$

In this expression the parameter $\kappa_{\mathrm{m}}$ is related to the inner scale length $\ell_0$ by

$$\kappa_{\mathrm{m}} = 5.92/\ell_0.$$

Show that your expression reduces to (7.16) when the aperture is considerably larger than the inner scale length. Show that, when the aperture is small relative to the inner scale length,

$$\lim_{a_{\mathrm{r}} \ll \ell_0} \langle \delta\theta^2 \rangle = 3.554 R C_n^2 (\kappa_{\mathrm{m}})^{\frac{1}{3}}.$$

Why is this different than the plane-wave result for closely spaced receivers?

### *Problem 2*

Use the analytical approach developed in Chapter 6 to estimate the auto-correlation and power spectrum for angle-of-arrival fluctuations imposed on a horizontal propagation path. Assume that a wind blows normal to the path with constant speed $v$. Neglect effects of self motion and aperture smoothing to show that the temporal correlations for the two angular components are

$$\langle \overline{\delta\theta_u(t)\,\delta\theta_u(t+\tau)} \rangle = 4\pi^2 R \int_0^\infty d\kappa\;\kappa^3 \Phi_n(\kappa) \left( J_0(\kappa v\tau) - \frac{J_1(\kappa v\tau)}{\kappa v\tau} \right)$$

$$\langle \overline{\delta\theta_v(t)\,\delta\theta_v(t+\tau)} \rangle = 4\pi^2 R \int_0^\infty d\kappa\;\kappa^3 \Phi_n(\kappa) \frac{J_1(\kappa v\tau)}{\kappa v\tau}$$

and that the corresponding power spectra are

$$W_u(f) = \frac{4R\pi^2}{v} \int_{2\pi f/v}^\infty d\kappa\;\kappa^2 \Phi_n(\kappa) \left[ 1 - \left( \frac{2\pi f}{\kappa v} \right)^2 \right]^{\frac{1}{2}}$$

$$W_v(f) = \frac{4R\pi^2}{v} \int_{2\pi f/v}^\infty d\kappa\;\kappa^2 \Phi_n(\kappa) \left\{ 1 - \left[ 1 - \left( \frac{2\pi f}{\kappa v} \right)^2 \right]^{\frac{1}{2}} \right\}.$$

Show that the low-frequency behavior of these power spectra for a Kolmogorov spectrum is given by the following general form:

$$W(f) = \text{constant} \times f^{-\frac{2}{3}}$$

Compare this with the measurements made on optical paths [19][53][54].

## Problem 3

Generalize the expression for the angular error of an optical ray that has passed completely through the spherical atmosphere. Assume that the source is near the horizon and that the profile of $C_n^2$ varies only with the radial distance. Omit gradual bending by the average index of refraction and show that

$$\langle \delta\theta^2 \rangle = 2.251\, a_{\mathrm{r}}^{-\frac{1}{3}} \int_{r_{\mathrm{E}}}^\infty dr\; r \frac{C_n^2(r-R)}{\sqrt{r^2 - r_{\mathrm{E}}^2\,\sin^2\vartheta}}$$

where $r_{\mathrm{E}}$ is the radius of the earth. Confirm that this result reduces to (7.38) when the zenith angle $\vartheta$ is less than $84°$. Can you invert this integral equation to establish the profile if the angular error is measured as a function of the zenith angle?

## Problem 4

Use the interferometer formulation for the angle of arrival to show that the spatial covariance of bearing errors at adjacent receivers separated by a distance $\rho$ in the $u$ direction is given by

$$\langle \delta\theta_u(0)\,\delta\theta_u(\rho)\rangle = -\frac{1}{2k^2}\,\frac{\partial^2 \mathcal{D}_\varphi(\rho)}{\partial\rho^2}$$

and

$$\langle \delta\theta_v(0)\,\delta\theta_v(\rho)\rangle = -\frac{1}{2k^2}\frac{1}{\rho}\,\frac{\partial \mathcal{D}_\varphi(\rho)}{\partial\rho}.$$

Use the similarity approximation for the structure function established in Chapter 5 which depends on Fried's coherence radius:

$$\mathcal{D}_\varphi(\rho) = 2\left(\frac{\rho}{r_0}\right)^{\frac{5}{3}}.$$

Show that the two covariances are

$$\langle \delta\theta_u(0)\,\delta\theta_u(\rho)\rangle = \rho^{-\frac{1}{3}}\frac{10}{9}\frac{1}{k^2}r_0^{-\frac{5}{3}}$$

$$\langle \delta\theta_v(0)\,\delta\theta_v(\rho)\rangle = \rho^{-\frac{1}{3}}\frac{5}{3}\frac{1}{k^2}r_0^{-\frac{5}{3}}.$$

These expressions are the same as the approximations (7.50) and (7.51) which also neglected aperture averaging and the outer-scale region.

## References

[1] V. A. Krasil'nikov, "On Fluctuations of the Angle-of-Arrival in the Phenomenon of Twinkling of Stars," *Doklady Akademii Nauk SSSR (Soviet Physics – Doklady)*, **65**, No. 3, 291–294 (1949). (*Note: publication data refers to the Russian journal; no known English translation is available.*)

[2] S. Chandrasekhar, "A Statistical Basis for the Theory of Stellar Scintillation," *Monthly Notices of the Royal Astronomical Society*, **112**, No. 5, 475–483 (1952).

[3] R. B. Muchmore and A. D. Wheelon, "Line-of-Sight Propagation Phenomenon – I. Ray Treatment," *Proceedings of the IRE*, **43**, No. 10, 1437–1449 (October 1955).

[4] V. I. Tatarskii, *Wave Propagation in a Turbulent Medium*, translated by R. A. Silverman (Dover, New York, 1967), 234–257.

[5] I. G. Kolchinskii, "Optical Instability of the Earth's Atmosphere According to Stellar Observations," originally published in *Naukova Dumka*, Kiev (1967) and translated into English by the Aeronautical Chart and Information Center, St. Louis, Missouri (February 1969).

[6] J. H. Churnside and R. J. Lataitis, "Angle-of-Arrival Fluctuations of a Reflected Beam in Atmospheric Turbulence," *Journal of the Optical Society of America A*, **4**, No. 7, 1264–1272 (July 1987).

[7] J. H. Churnside and R. J. Lataitis, "Wander of an Optical Beam in the Turbulent Atmosphere," *Applied Optics*, **29**, No. 7, 926– 930 (1 March 1990).

[8] M. Born and E. Wolf, *Principles of Optics*, 6th Ed. (Pergamon Press, New York, 1980), 464 *et seq.*

[9] D. L. Fried, "Statistics of a Geometric Representation of Wave-front Distortion," *Journal of the Optical Society of America*, **55**, No. 11, 1427–1435 (November 1965).

[10] C. B. Hogge and R. R. Butts, "Frequency Spectra for the Geometric Representation of Wave-front Distortions Due to Atmospheric Turbulence," *IEEE Transactions on Antennas and Propagation*, **AP-24**, No. 2, 144–154 (March 1976).

[11] R. J. Sasiela, *Electromagnetic Wave Propagation in Turbulence* (Springer-Verlag, Berlin, 1994), 47 *et seq.*

[12] R. L. Fante, "Electromagnetic Beam Propagation in Turbulent Media: An Update," *Proceedings of the IEEE*, **68**, No. 11, 1424–1443 (November 1980).

[13] D. S. Acton, R. J. Sharbaugh, J. R. Roehrig and D. Tiszauer, "Wave-Front Tilt Power Spectral Density from the Image Motion of Solar Pores," *Applied Optics*, **31**, No. 21, 4280–4284 (20 July 1992).

[14] H. T. Yura and M. T. Tavis, "Centroid Anisoplanatism," *Journal of the Optical Society of America A*, **2**, No. 5, 765–773 (May 1985).

[15] J. H. Churnside, M. T. Tavis, H. T. Yura and G. A. Tyler, "Zernike-Polynomial Expansion of Turbulence-Induced Centroid Anisoplanatism," *Optics Letters*, **10**, No. 6, 258–260 (June 1985).

[16] H. T. Yura and M. T. Tavis, "Centroid Anisoplanatism," *Journal of the Optical Society of America A*, **2**, No. 5, 765–773 (May 1985).

[17] V. I. Tatarskii, *The Effects of the Turbulent Atmosphere on Wave Propagation* (translated from the Russian and issued by the National Technical Information Office, U.S. Department of Commerce, Springfield, VA 22161, 1971), 284–289.

[18] V. M. Bovsheverov, A. S. Gurvich and M. A. Kallistratova, "Experimental Investigations of the 'Quivering' of Artificial Light Sources," *Izvestiya Vysshikh Uchebnykh Zavedenii, Radiofizika (Soviet Radiophysics)*, **4**, No. 5, 219–232 (1961). The results of this early experiment are reproduced on pages 300–305 of [15].

[19] A. S. Gurvich and M. A. Kallistratova, "Experimental Study of the Fluctuations in Angle of Incidence of a Light Beam under Conditions of Strong Intensity Fluctuations," *Izvestiya Vysshikh Uchebnykh Zavedenii, Radiofizika (Soviet Radiophysics)*, **11**, No. 1, 37–40 (January 1968).

[20] V. I. Klyatskin and A. I. Kon, "On the Displacement of Spatially-Bounded Light Beams in a Turbulent Medium in the Markovian-Random-Process Approximation," *Izvestiya Vysshikh Uchebnykh Zavedenii, Radiofizika (Soviet Radiophysics)*, **15**, No. 9, 1056–1061 (September 1972).

[21] V. L. Mironov and V. V. Nosov, "On the Theory of Spatially Limited Light Beam Displacements in a Randomly Inhomogeneous Medium," *Journal of the Optical Society of America*, **67**, No. 8, 1073–1080 (August 1977).

[22] J. H. Churnside, "Angle-of-Arrival Fluctuations of Retroreflected Light in the Turbulent Atmosphere," *Journal of the Optical Society of America A*, **6**, No. 2, 275–279 (February 1989).

[23] I. G. Kolchinskii, "Some Results of Observations of the Vibration of Images of Stars at the Main Astronomical Observatory of the Academy of Sciences of the Ukrainian SSR at Goloseyevo," *Astronomicheskii Zhurnal (Soviet Astronomy)*, **1**, No. 4, 624–636 (April 1957).

[24] J. Borgnino, G. Ceppatelli, G. Ricort and A. Righini, "Lower Atmosphere and Solar Seeing: An Experiment of Simultaneous Measurements of Nearby Turbulence by Thermal Radiosondes, by Angle-of-Arrival Statistics and Image Motion Observation," *Astronomy and Astrophysics*, **107**, No. 2, 333–337 (March 1982).

[25] J. Borgnino, J. Vernin, C. Aime and G. Ricort, "A Study of the Degradation of Daytime Astronomical Images due to Turbulence in the Lower Atmosphere by Measurements of the Standard Deviation of the Angle-of-Arrival," *Solar Physics*, **64**, No. 2, 403–415 (December 1979).

[26] M. A. Kallistratova, "Fluctuations in the Direction of Propagation of Light Waves in an Inhomogeneous Turbulent Medium," *Izvestiya Vysshikh Uchebnykh Zavedenii, Radiofizika (Soviet Radiophysics)*, **9**, No. 1, 33–36 (1966).

[27] J. Borgnino and J. Vernin, "Experimental Verification of the Inertial Model of Atmospheric Turbulence from Solar Limb Motion," *Journal of the Optical Society of America*, **68**, No. 8, 1056–1062 (August 1978).

[28] E. Moroder and A. Righini, "The Evaluation of Night Time Seeing from Polar Star Trails," *Astronomy and Astrophysics*, **23**, No. 2, 307–310 (March 1973).

[29] L. Campbell and W. G. Elford, "Measurement of Atmospheric Isoplanatism Using Stellar Scintillation," *Journal of Atmospheric and Terrestrial Physics*, **52**, No. 4, 313–320 (April 1990).

[30] J. Borgnino and F. Martin, "Correlation between Angle-of-Arrival Fluctuations on the Entrance Pupil of a Solar Telescope," *Journal of the Optical Society of America*, **67**, No. 8, 1065–1072 (August 1977).

[31] A. Consortini, "Role of the Inner Scale of Atmospheric Turbulence in Optical Propagation and Methods to Measure It," in *Scattering in Volumes and Surfaces*, edited by M. Nieto-Vesperinas and J. C. Dainty (Elsevier Science, Amsterdam, 1990), 73–90.

[32] A. Consortini, P. Pandolfini, C. Romanelli and R. Vanni, "Turbulence Investigation at Small Scale by Angle-of-Arrival Fluctuations of a Laser Beam," *Optica Acta*, **27**, No. 8, 1221–1228 (August 1980).

[33] See [5], page 164 *et seq.* of the English translation.

[34] M. A. Kallistratova and A. I. Kon, "Fluctuations in the Angle-of-Arrival of Light Waves from an Extended Source in a Turbulent Medium," *Izvestiya Vysshikh Uchebnykh Zavedenii, Radiofizika (Soviet Radiophysics)*, **9**, No. 6, 636–639 (1966).

[35] R. K. Crane, "Low Elevation Angle Measurement Limitations Imposed by the Troposphere: An Analysis of Scintillation Observations Made at Haystack and Millstone," MIT Lincoln Laboratory Technical Report No. 518 (18 May 1976).

[36] R. K. Crane, "Analysis of Tropospheric Effects at Low Elevation Angles," USAF Rome Air Development Center, Technical Report RADC-TR-78-252 (November 1978).

[37] E. Vilar and H. Smith, "A Theoretical and Experimental Study of Angular Scintillations in Earth Space Paths," *IEEE Transactions on Antennas and Propagation*, **AP-34**, No. 1, 2–10 (January 1986).

[38] B. R. Bean and E. J. Dutton, *Radio Meteorology* (National Bureau of Standards Monograph 92, U.S. Government Printing Office, Washington, March 1966), 49 *et seq.*

[39] A. Hewish, "The Diffraction of Galactic Radio Waves as a Method of Investigating the Irregular Structure of the Ionosphere," *Proceedings of the Royal Society*, **214**, No. A 1119, 494–514 (9 October 1952).

[40] R. S. Lawrence, J. L. Jesperson and R. C. Lamb, "Amplitude and Angular Scintillations of the Radio Source Cygnus-A Observed at Boulder, Colorado," *Journal of Research of the NBS-D. Radio Propagation*, **65D**, No. 4, 333–350 (July–August 1961).

[41] R. K. Crane, "Variance and Spectra of Angle-of-Arrival and Doppler Fluctuations Caused by Ionospheric Scintillation," *Journal of Geophysical Research*, **83**, No. A5, 2091–2102 (1 May 1978).

[42] D. L. Fried, "Limiting Resolution Looking Down Through the Atmosphere," *Journal of the Optical Society of America*, **56**, No. 10, 1380–1384 (October 1966).

[43] A. D. Wheelon, "Corona: The First Reconnaissance Satellites," *Physics Today*, **50**, No. 2, 24–30 (February 1997).

[44] J. E. Titheridge, "The Characteristics of Large Ionospheric Irregularities," *Journal of Atmospheric and Terrestrial Physics*, **30**, No. 1, 73–84 (January 1968).

[45] A. R. Jacobson, R. S. Massey and W. C. Erickson, "A Study of Transionospheric Refraction of Radio Waves using the Clark Lake Radio Observatory," *Annales Geophysicae*, **9**, No. 8, 546–552 (August 1991).

[46] T. A. Th. Spoelstra, "Combining TID Observations: NNSS and Radio Interferometry Data," *Journal of Atmospheric and Terrestrial Physics*, **54**, No. 9, 1185–1195 (September 1992).

[47] H. G. Booker, "The Role of Acoustic Gravity Waves in the Generation of Spread-F and Ionospheric Scintillation," *Journal of Atmospheric and Terrestrial Physics*, **41**, No. 5, 501–515 (May 1979).

[48] A. A. Stotskii, "Tropospheric Limitations of the Measurement Accuracy on Coordinates of Cosmic Radio Source," *Izvestiya Vysshikh Uchebnykh Zavedenii, Radiofizika (Soviet Radiophysics)*, **19**, No. 11, 1167–1169 (November 1976).

[49] M. C. H. Wright, "Atmospheric Phase Noise and Aperture-Synthesis Imaging at Millimeter Wavelengths," *Publications of the Astronomical Society of the Pacific*, **108**, No. 724, 520–534 (June 1996).

[50] W. J. Altenhoff, J. W. M. Baars, D. Downes and J. E. Wink, "Observations of Anomalous Refraction at Radio Wavelengths," *Astronomy and Astrophysics*, **184**, Nos 1–2, 381–385 (October 1987).

[51] D. Downes and W. J. Altenhoff, "Anomalous Refraction at Radio Wavelengths," in *Radioastronomical Seeing*, edited by J. E. Baldwin and W. Shouguan (Pergamon Press, London, 1990), 31–40.

[52] S. Church and R. Hills, "Measurements of Daytime Atmospheric 'Seeing' on Mauna Kea Made with the James Clerk Maxwell Telescope," in *Radioastronomical Seeing*, edited by J. E. Baldwin and W. Shouguan (Pergamon Press, London, 1990), 75–80.

[53] See [17], pages 300–304.

[54] A. S. Drofa, "Spectra of Center-of-Gravity Shifts of Light Beams in a Turbulent Atmosphere," *Izvestiya Vysshikh Uchebnykh Zavedenii, Radiofizika (Soviet Radiophysics)*, **21**, No. 8, 840–845 (August 1978).

# 8
# Phase Distributions

The statistical distributions of signals passing through turbulent media are important for several reasons. Engineering applications often require the expected distribution of the level and sometimes the phase of the signal. Scientific applications of these distributions are equally important. The distributions of amplitude and phase are largely independent of the model used to describe the random media and depend primarily on the underlying theory of propagation. Comparison of such predictions with signal-distribution data therefore provides a test of various methods for describing the transmission.

In examining the basis for geometrical optics in Chapter 3 we found that the variance of logarithmic amplitude must be much less than unity where this approximation can be used. By contrast, the phase variance can be quite large when this method is valid. These conditions mean that the field strength is almost completely determined by phase fluctuations in this regime. For that reason we concentrate on the distribution of the phase of the signal.

## 8.1 The Single-path Phase Distribution

Geometrical optics relates the phase fluctuations imposed on electromagnetic signals to dielectric variations encountered along the nominal ray path by the relationship introduced in (3.24):

$$\varphi(t) = \frac{\pi}{\lambda} \int_0^R ds \, \Delta\varepsilon(s,t)$$

### 8.1.1 $\Delta\varepsilon$ *Gaussian*

The dielectric constant can be measured using temperature and humidity sensors for a propagation path close to the surface. This data shows that $\Delta\varepsilon$ often behaves like a Gaussian random variable. When it does, one can be sure that $\varphi(t)$ is also a Gaussian random variable because it is a linear combination of the $\Delta\varepsilon$. Since the average phase fluctuation vanishes,

$$\mathsf{P}(\varphi)\,d\varphi = \frac{d\varphi}{\sigma\sqrt{2\pi}}\exp\left(-\frac{\varphi^2}{2\sigma^2}\right) \tag{8.1}$$

where $\sigma^2$ denotes the phase variance evaluated in Chapter 4:

$$\sigma^2 = \langle\varphi^2\rangle \tag{8.2}$$

Some experiments measure the probability that the phase exceeds a prescribed level. The phase distribution is symmetrical about the origin and the *cumulative probability* is written

$$\mathcal{P}(|\varphi| \geq \varphi_0) = 2\int_{\varphi_0}^{\infty} d\varphi\, W(\varphi). \tag{8.3}$$

The integration can be expressed in terms of the error function when the distribution is Gaussian:

$$\mathcal{P}(|\varphi| \geq \varphi_0) = 1 - \mathrm{erf}\left(\frac{\varphi_0}{\sigma\sqrt{2}}\right) \tag{8.4}$$

Before comparing these results with phase measurements it is important to examine two assumptions that have been made.

### 8.1.2 $\Delta\varepsilon$ *Non-Gaussian*

There are many instances when the measured $\Delta\varepsilon$ do not behave like Gaussian random variables. Temperature data shows that $\Delta T$ sometimes has a log-normal distribution. Fortunately the path integration of these non-Gaussian variations *is* normally distributed if the transmission distance is much greater than the outer scale length. Following a suggestion of Tatarskii, we break the path into discrete segments whose width is a significant number of outer scale lengths. The phase of the received signal is the sum of random phase shifts imposed by various segments:

$$\varphi(t) = \frac{\pi}{\lambda}\left(\int_{\mathrm{seg}\,1} ds\,\Delta\varepsilon + \int_{\mathrm{seg}\,2} ds\,\Delta\varepsilon + \int_{\mathrm{seg}\,3} ds\,\Delta\varepsilon + \cdots\right)$$

The $\Delta\varepsilon$ in different segments are uncorrelated because their widths are large compared with $L_0$, with the exceptions of the border regions which are relatively unimportant. The phase shifts imposed by individual segments are therefore independent. The Central Limit theorem tells us that the sum of a large number of independent quantities is normally distributed. This means that the phase fluctuations should be described by a Gaussian distribution. That reasoning is valid only if the path length is larger than the horizontal outer scale length. One should not apply (8.1) to short paths or to situations for which the effective outer scale length is comparable to the path length.

### 8.1.3 The Effect of Phase Trends

The second problem arises because the $\Delta\varepsilon$ seldom represent a stationary process. The measured single-path phase history reproduced in Figure 4.1 shows that trends are an important feature of such data. The phase of a signal is influenced both by random variations of the dielectric constant and by gradual changes in the ambient value:

$$\varepsilon(r,t) = \varepsilon_0(t) + \Delta\varepsilon(r,t) \tag{8.5}$$

The mean value $\varepsilon_0$ changes during the day and with the seasons. It can also change suddenly when a weather front or a large intermittent structure moves through the path. The simulated phase record of Figure 4.7 illustrates random variations superimposed on a gradual change of the electrical path length. If this trend is ignored and the fluctuations are measured relative to an estimated average value, the phase excursions appear progressively larger as one goes to either end of the data sample. These large positive and negative values distort the probability density function relative to the normal distribution which is expected under stable conditions.

This situation can be analyzed by noting that the probability density function and the characteristic function are related to one another by Fourier transforms:

$$\langle \exp(iq\varphi) \rangle = \int_{-\infty}^{\infty} d\varphi\, \mathsf{P}(\varphi) \exp(iq\varphi) \tag{8.6}$$

One can calculate the probability distribution from the inverse Fourier transform if the characteristic function is known:

$$\mathsf{P}(\varphi) = \frac{1}{2\pi} \int_{-\infty}^{\infty} dq\, \langle \exp(iq\varphi) \rangle \exp(-iq\varphi) \tag{8.7}$$

The phase trend is often a linear function of time during the data sample and the *measured phase* is written

$$\psi(t) = \varphi(t) + \eta t \tag{8.8}$$

where the first term is the random component and the second represents a trend. The characteristic function can be represented as the product of two terms because the trend and random component are independent:

$$\langle \exp[iq\psi(t)] \rangle = \langle \exp[iq\varphi(t)] \rangle \langle \exp(iq\eta t) \rangle$$

The statistical average of the trend term is simply a time average over the sample length:

$$\langle e^{iq\eta t} \rangle = \frac{1}{2T} \int_{-T}^{T} dt \, \exp(iq\eta t) = \frac{\sin(q\eta T)}{q\eta T}$$

The characteristic function of the random phase variations can be described as the Fourier transform of the distribution for the stationary process. We denote this *true* distribution by $P_0(\varphi)$ and write

$$\langle e^{iq\varphi} \rangle = \int_{-\infty}^{\infty} d\varphi \, P_0(\varphi) \exp(iq\varphi).$$

The distribution of measured phase fluctuations emerges when we combine these expressions:

$$\begin{aligned}
P(\psi) &= \frac{1}{2\pi} \int_{-\infty}^{\infty} dq \, e^{-iq\psi} \frac{\sin(q\eta T)}{q\eta T} \int_{-\infty}^{\infty} d\varphi \, P_0(\varphi) e^{iq\varphi} \\
&= \frac{1}{2\pi} \int_{-\infty}^{\infty} d\varphi \, P_0(\varphi) \int_{-\infty}^{\infty} dq \, e^{-iq(\psi-\varphi)} \frac{\sin(q\eta T)}{q\eta T} \\
&= \frac{1}{2\pi\eta T} \int_{-\infty}^{\infty} d\varphi \, P_0(\varphi) \begin{cases} \pi, & \text{for } |\psi - \varphi| < \eta T \\ 0, & \text{for } |\psi - \varphi| > \eta T \end{cases}
\end{aligned}$$

The measured and true distributions are therefore connected by a simple relationship:

$$P(\psi) = \frac{1}{2\eta T} \int_{\psi-\eta T}^{\psi+\eta T} d\varphi \, P_0(\varphi) \tag{8.9}$$

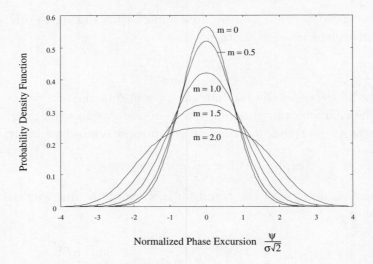

Figure 8.1: Predicted probability density functions for Gaussian phase fluctuations superimposed on a linear trend.

Since we believe that $P_0(\varphi)$ is Gaussian, the measured phase distribution can be expressed in terms of the error function defined in Appendix C:

$$P(\psi)\,d\psi = \frac{d\psi}{\sigma\sqrt{2}}\frac{1}{4m}\left[\mathrm{erf}\left(\frac{\psi}{\sigma\sqrt{2}}+m\right)-\mathrm{erf}\left(\frac{\psi}{\sigma\sqrt{2}}-m\right)\right] \qquad (8.10)$$

The parameter $m$ describes the relative importance of the phase trend and the fluctuating component:

$$m = \frac{\eta T}{\sigma\sqrt{2}} \qquad (8.11)$$

This result is plotted in Figure 8.1 as a function of $\psi/(\sigma\sqrt{2})$ for several values of $m$. Trends enhance the probability of observing large phase excursions, which is indicated by substantial broadening relative to the Gaussian model. Phase trends evidently play an increasingly important role as the sample length increases or the drift becomes more pronounced.

### 8.1.4 Experimental Results

There are surprisingly few measurements of the single-path phase distribution against which our prediction can be tested. The numerous microwave programs identified in Table 4.1 all used considerable averaging, which frustrates this test. Time averages of a non-Gaussian random variable rapidly

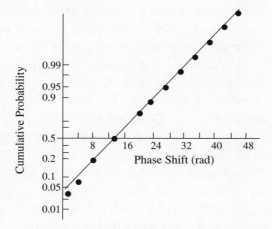

Figure 8.2: The cumulative distribution of single-path phase fluctuations measured with a laser signal and a Michelson interferometer on paths near the surface by Lukin, Pokasov and Khmelevtsov [1]. The straight line represents a Gaussian distribution.

become Gaussian as the smoothing time increases, in the same way that path averaging of the $\Delta\varepsilon$ produces a Gaussian phase distribution.

The laser experiment illustrated in Figure 4.3 provided the first verification of a Gaussian phase distribution in the context of random wave propagation [1]. A laser signal was transmitted along a horizontal path 1.5 m above the surface to distances of 95 and 200 m. The outer scale length was approximately equal to the path height and was therefore far less than the distance. Using the argument presented in Section 8.1.2, it follows that the phase distribution should be Gaussian, even if the $\Delta\varepsilon$ themselves are distributed in a non-Gaussian way. The distribution of instantaneous phase fluctuations is reproduced in Figure 8.2 and evidently gives strong support to the prediction (8.1).

A satellite experiment measured the instantaneous phase fluctuations of microwave signals imposed by the ionosphere. The DNA Wideband Satellite radiated ten mutually coherent signals from a circular orbit 1000 km above the earth. Simultaneous phase fluctuations were measured at 138, 400 and 1290 MHz and drift was removed by detrending [2]. The data was sampled 100 times each second and the measured probability density functions were consistently described by normal distributions. These results confirm the prediction (8.1) for ionospheric phase errors.

Absolute range measurements from GPS satellites promise to provide good tests for the phase errors induced by the troposphere. Satellite transmitters, ground receivers and orbital tracking stations are being steadily improved [3]. Orbit-determination and geodetic reference systems are being refined together with the software that is used to correct for the average atmosphere and zenith angle [4]. This correction capability is approaching the point where range errors induced by irregularities in concentration of water vapor are the dominant residual effect. These will soon set the absolute limit on accuracy of GPS measurements. The range residuals are directly related to phase fluctuations by $\varphi = k\, \delta R$, which means that their distribution should provide a good test of (8.1).

An experimental probability density function for GPS range residuals was extracted from data taken during one day at a single receiving station working with one satellite [5]. The resulting distribution was decidedly *non-Gaussian*, with a sizable shift towards large readings. This is what one would expect from (8.1). The sample length employed was 450 s and the range-residual histories were not detrended.[1] The distributions in Figure 8.1 are therefore appropriate for this comparison.

## 8.2 The Phase-difference Distribution

We are primarily interested in the probability distribution for the difference of phase measured at adjacent receivers:

$$\Delta\psi = \psi_1 - \psi_2 \tag{8.12}$$

On introducing expression (8.8) for the measured phase shifts we see that the phase difference depends only on the random components:

$$\Delta\psi = \varphi_1 - \varphi_2 \tag{8.13}$$

since the trend term is almost the same for the two ray paths. We believe that $\varphi_1$ and $\varphi_2$ are each Gaussian random variables. Since $\Delta\psi$ is a linear combination of Gaussian random variables, we can be sure that it is normally distributed.

### 8.2.1 The Predicted Distribution

To establish the appropriate parameters in the phase-difference distribution one begins with the joint probability distribution for $\varphi_1$ and $\varphi_2$. We can assume that they have the same variance because the random medium is

---

[1] Had a linear phase drift been present in this data it would not have spoiled the agreement between the Allan variance and the same measurements noted in Section 6.2.6.

homogeneous over the small separations employed. The bivariate distribution can be written in terms of the spatial correlation $\mu(\rho)$ as follows:

$$\mathsf{P}(\varphi_1, \varphi_2)\, d\varphi_1\, d\varphi_2 = \frac{d\varphi_1\, d\varphi_2}{2\pi\sigma^2 \sqrt{1 - \mu^2(\rho)}} \exp\left(-\frac{\varphi_1^2 + \varphi_2^2 - 2\mu(\rho)\,\varphi_1\varphi_2}{2\sigma^2(1 - \mu^2(\rho))}\right)$$

(8.14)

We introduce sum and difference variables to establish the distribution of the phase difference:

$$u = \varphi_1 - \varphi_2 \qquad v = \tfrac{1}{2}(\varphi_1 + \varphi_2)$$

and obtain

$$\mathsf{P}(u, v)\, du\, dv = \frac{du\, dv}{2\pi\sigma^2 \sqrt{1 - \mu^2(\rho)}} \exp\left(-\frac{u^2}{4\sigma^2(1 - \mu(\rho))} + \frac{v^2}{\sigma^2(1 + \mu(\rho))}\right)$$

The distribution of the phase difference is found by integrating this expression over all possible values of $v$:

$$\mathsf{P}(\varphi_1 - \varphi_2) = \frac{1}{2\sigma\sqrt{\pi}\sqrt{1 - \mu(\rho)}} \exp\left(-\frac{(\varphi_1 - \varphi_2)^2}{4\sigma^2(1 - \mu(\rho))}\right)$$

(8.15)

Noting that the variance in this result is just the phase structure function, the final result becomes

$$\mathsf{P}(\Delta\psi) = \frac{1}{\sqrt{2\pi\mathcal{D}_\varphi(\rho)}} \exp\left(-\frac{\Delta\psi^2}{2\mathcal{D}_\varphi(\rho)}\right).$$

(8.16)

Of course, one can establish the same result directly by noting that $\varphi_1 - \varphi_2$ is a linear combination of two Gaussian random variables and hence itself a Gaussian random variable.

### 8.2.2 Experimental Confirmations

The distribution of the phase difference for adjacent receivers has been studied in several measurement programs. Millimeter-wave experiments were conducted in Illinois at 173 GHz over a 1.4-km path [6]. The phase difference was taken for inter-receiver separations ranging from 1.43 to 10.0 m. The distribution was normal unless the data was noisy. A relatively noise-free data set is reproduced in Figure 8.3 and good agreement with a Gaussian model is found.

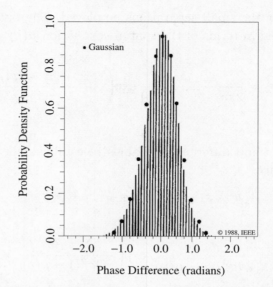

Figure 8.3: The distribution of the phase difference at 173 GHz for receivers 2.86 m apart measured by Hill, Bohlander, Clifford, McMillan, Priestley and Schoenfeld [6].

A normal distribution for the phase difference was measured during 1969 with the Cambridge One-Mile Radio Telescope operating at 5 GHz [7]. Numerous phase-difference-measurement programs were conducted at microwave frequencies and are summarized in Table 5.1. Unfortunately, only one tried to establish the probability density function of the phase difference. This experiment used 9.6-GHz transmissions over a 150-km path between Maui and Hawaii [8]. The distribution of the phase difference between two receivers 100 m apart varied from day to day and hour to hour. It was often Gaussian but sometimes quite different, especially when the fading was greater than 5 dB. This is not surprising because the mode of propagation for this long path was probably a mixture of scintillation and ducting.

An early optical experiment in Colorado measured the phase difference using a helium–neon laser on a 70-m path [9]. The distribution was measured for inter-receiver spacings ranging from 3 mm to 30 cm. The agreement of this data with the Gaussian prediction was only fair. The data was noisy, which had a strong influence on the measured distributions. This experiment was repeated in Japan using a phase-locked interferometer, which provided a great improvement in accuracy [10]. The measured histograms of the phase difference are reproduced in Figure 8.4 and approximate the Gaussian distribution for each distance.

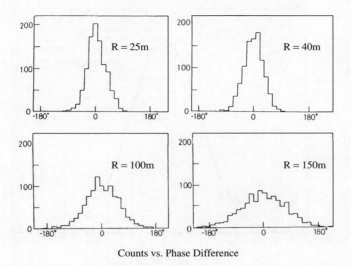

Counts vs. Phase Difference

Figure 8.4: Histograms of fluctuations in phase difference of laser signals between phase-locked receivers separated by 25 mm measured at various distances by Matsumoto and Tsukahara [10].

## 8.3 The Angle-of-arrival Distribution

Independent confirmation of the normal phase distribution comes from angle-of-arrival measurements made with astronomical telescopes. The angular error of a stellar image is related to the phase difference of two arriving waves by (7.1). If the phase front is reasonably flat and the instrument is aimed at the star,

$$\delta\theta = \frac{\lambda}{2\pi\rho}[\,\varphi(x+\rho) - \varphi(x)]$$

where $\rho$ is the ray separation at the telescope primary. Combining this relationship with (8.16) shows that the angle-of-arrival distribution should also be Gaussian:

$$P(\delta\theta) = \frac{1}{q\sqrt{2\pi}} \exp\!\left(-\frac{\delta\theta^2}{2q^2}\right) \tag{8.17}$$

where

$$q^2 = \left(\frac{\lambda}{2\pi\rho}\right)^2 \mathcal{D}_\varphi(\rho) \tag{8.18}$$

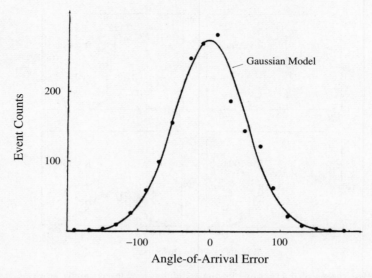

Figure 8.5: The distribution of image displacements about the nominal star trail measured in the focal plane of a stationary telescope by Kolchinskii [11].

Introducing the expression for the phase structure function for parallel rays established in (5.19) gives

$$\mathsf{q}^2 = 8\pi^2 R \int_0^\infty d\kappa\, \kappa \Phi_n(\kappa)\frac{1 - J_0(\kappa\rho)}{\rho^2}. \tag{8.19}$$

The separation between parallel rays is usually small and one can take the limit $\rho \to 0$:

$$\mathsf{q}^2 = 2\pi^2 R \int_0^\infty d\kappa\, \kappa^3 \Phi_n(\kappa)$$

This is precisely the expression for the mean-square angle of arrival computed in (7.8). The identification of the $\delta\theta$ distribution with the Gaussian model is thus complete.

The position of a stellar image in the focal plane of a telescope is proportional to the angle-of-arrival error for the incoming signal. In an early series of experiments, Kolchinskii measured the distribution of image displacements from nominal star trails in photographic plates with a stationary telescope [11][12]. The distribution obtained on 15 November 1950 is reproduced in Figure 8.5 and matches the Gaussian model quite well.

## 8.4 Temporal Distributions

Phase fluctuations measured by a single receiver change stochastically with time because the dielectric variations along the path do. The joint probability distribution (8.15) should describe the phase fluctuations measured at different times if we identify

$$\varphi_1 = \varphi(t) \quad \text{and} \quad \varphi_2 = \varphi(t + \tau).$$

The variances of the two phase shifts are the same if $\tau$ is not too large. The difference distribution now depends on the autocorrelation of phase:

$$\mathsf{P}[\varphi(t) - \varphi(t + \tau)] = \frac{1}{2\sigma\sqrt{\pi}\sqrt{1 - \mu(\tau)}} \exp\left(-\frac{[\varphi(t + \tau) - \varphi(t)]^2}{4\sigma^2[1 - \mu(\tau)]}\right) \quad (8.20)$$

This result can be used to estimate the distribution of frequency variations imposed on a signal by a moving atmosphere. The instantaneous frequency shift is simply the phase derivative:

$$\delta f = \lim_{\tau \to 0}\left(\frac{\varphi(t + \tau) - \varphi(t)}{\tau}\right)$$

The effect of a moving medium is to create a spread of frequency shifts normally distributed about the carrier frequency:

$$\mathsf{P}(\delta f) = \frac{1}{\mathsf{g}\sqrt{2\pi}} \exp\left(-\frac{\delta f^2}{2\mathsf{g}^2}\right) \quad (8.21)$$

The mean-square frequency shift is described in terms of the temporal structure function:

$$\langle \delta f^2 \rangle = \mathsf{g}^2 = \lim_{\tau \to 0}\left(\frac{\mathcal{D}_\varphi(\tau)}{\tau^2}\right) \quad (8.22)$$

If the changing phase fluctuations are caused by a frozen random medium moving past the line of sight, one can use the relationship (6.25) to write the structure function in terms of the time delay:

$$\langle \delta f^2 \rangle = 8\pi^2 R k^2 \int_0^\infty d\kappa\, \kappa \Phi_n(\kappa) \lim_{\tau \to 0}\left(\frac{1 - J_0(\kappa\, v\tau)}{\tau^2}\right)$$

or

$$\langle \delta f^2 \rangle = 2\pi^2 R v^2 k^2 \int_0^\infty d\kappa\, \kappa^3 \Phi_n(\kappa) \quad (8.23)$$

It is significant that the variance of the frequency is proportional to the third moment. When Taylor's hypothesis is valid, there is therefore an important connection between angle-of-arrival errors and variations in frequency:

$$\langle \delta f^2 \rangle = v^2 k^2 \frac{v^2}{2\pi\lambda^2} \langle \delta\theta^2 \rangle \tag{8.24}$$

This means that one can infer the frequency spread from measurements of wind speed and bearing errors. In a similar way, one can use knowledge of frequency spreading and wind speed to anticipate the mean-square angle of arrival.

It has been suggested that frequency shifts caused by a moving interstellar plasma may be responsible for the anomalous red shifts observed for some quasars [13]. The plasmas surrounding these sources are believed to move at speeds approaching one tenth the speed of light. On the other hand, the rms angular error measured at the earth is only $10^{-4} - 10^{-5}$ rad. The apparent Doppler shift due to the motion of plasma is therefore considerably smaller than the red shift due to expansion of the universe.

## 8.5 Problem

*Problem*

Use the probability density function (8.10) which incorporates a linear phase trend to calculate the corresponding cumulative probability and compare the result numerically with the Gaussian reference.

## References

[1] V. P. Lukin, V. V. Pokasov and S. S. Khmelevtsov, "Investigation of the Time Characteristics of Fluctuations of the Phases of Optical Waves Propagating in the Bottom Layer of the Atmosphere," *Izvestiya Vysshikh Uchebnykh Zavedenii, Radiofizika (Soviet Radiophysics)*, **15**, No. 12, 1426–1430 (December 1972).

[2] E. J. Fremouw, R. C. Livingston and D. A. Miller, "On the Statistics of Scintillating Signals," *Journal of Atmospheric and Terrestrial Physics*, **42**, No. 8, 717–731 (August 1980).

[3] P. Fang, M. Bevis, Y. Bock, S. Gutman and D. Wolfe, "GPS Meteorology: Reducing Systematic Errors in Geodetic Estimates for Zenith Delay," *Geophysical Research Letters*, **25**, No. 19, 3583–3586 (1 October 1998).

[4] A. E. Niell, "Global Mapping Functions for the Atmosphere Delay at Radio Wavelengths," *Journal of Geophysical Research*, **101**, No. B 2, 3227–3246 (10 February 1996).

[5] C. J. Naudet, "Estimation of Tropospheric Fluctuations using GPS Data," TDA Progress Report 42–126, Jet Propulsion Laboratory, Pasadena, CA (15 April 1996).

[6] R. J. Hill, R. A. Bohlander, S. F. Clifford, R. W. McMillan, J. T. Priestley and W. P. Schoenfeld, "Turbulence-Induced Millimeter-Wave Scintillation Compared with Micrometeorological Measurements," *IEEE Transactions on Geoscience & Remote Sensing*, **26**, No. 3, 330–342 (May 1988).

[7] R. Hinder, "Fluctuations of Water Vapor Content in the Troposphere as Derived from Interferometric Observations of Celestial Radio Sources," *Journal of Atmospheric and Terrestrial Physics*, **34**, No. 7, 1171–1186 (July 1972).

[8] M. C. Thompson, H. B. Janes, L. E. Wood and D. Smith, "Phase and Amplitude Scintillations at 9.6 GHz on an Elevated Path," *IEEE Transactions on Antennas and Propagation*, **AP-23**, No. 6, 850–854 (November 1975).

[9] S. F. Clifford, G. M. B. Bouricius, G. R. Ochs and M. H. Ackley, "Phase Variations in Atmospheric Optical Propagation," *Journal of the Optical Society of America*, **61**, No. 10, 1279–1284 (October 1971).

[10] H. Matsumoto and K. Tsukahara, "Effects of the Atmospheric Phase Fluctuation on Long-Distance Measurement," *Applied Optics*, **23**, No. 19, 3388–3394 (1 October 1984).

[11] I. G. Kolchinskii, "Optical Instability of the Earth's Atmosphere According to Stellar Observations," originally published in *Naukova Dumka, Kiev* (1967) and translated into English by the Aeronautical Chart and Information Center, St. Louis, Missouri (February 1969). See Figure 54 on page 143 of the Russian version and on page 214 of the English translation.

[12] I. G. Kolchinskii, "Some Results of Observations of the Vibration of Images of Stars at the Main Astronomical Observatory of the Academy of Sciences of the Ukrainian SSR at Goloseyevo," *Astronomicheskii Zhurnal (Soviet Astronomy)*, **1**, No. 4, 624–636 (April 1957).

[13] J. T. Foley and E. Wolf, "Frequency Shifts of Spectral Lines Generated by Scattering from Space–Time Fluctuations," *Physical Review A (General Physics)*, **40**, No. 2, 588–598 (15 July 1989).

# 9

# Field-strength Moments

The electric field strength is the central player in electromagnetic propagation. It is important to estimate its moments averaged over the ensemble of random fluctuations in the medium. We also need to describe its correlations with respect to time, place and frequency. To perform these calculations we need an expression for the electric field strength. In geometrical optics it is related to the field that would be received over a vacuum path as follows:

$$E = E_0(1 + \chi) \exp(i\varphi) \tag{9.1}$$

We found in (3.45) that the logarithmic amplitude variance $\langle \chi^2 \rangle$ is much smaller than unity when geometrical optics is valid. This means that the field is determined primarily by its phase in this approximation [1]:

$$E = E_0 \exp(i\varphi) \tag{9.2}$$

Field strength moments thus depend primarily on the phase distribution. With this *phase-only* expression one can estimate the ensemble average for several important measures of the field using the phase distributions established in the previous chapter.

## 9.1 The Average Field Strength

The average field strength is a benchmark quantity that describes the extent to which the initial wave has been disturbed. It can be expressed as an integral over the phase distribution:

$$\langle E \rangle = \int_{-\infty}^{\infty} d\varphi \, \mathsf{P}(\varphi) E_0 \exp(i\varphi) \tag{9.3}$$

384

Ignoring trends for now, we can use the phase distribution (8.1) to evaluate the ensemble average:

$$\langle E \rangle = E_0 \frac{1}{\sigma\sqrt{2\pi}} \int_{-\infty}^{\infty} d\varphi \exp\left( i\varphi - \frac{\varphi^2}{2\sigma^2} \right) = E_0 \exp\left( -\frac{\sigma^2}{2} \right)$$

The parameter $\sigma^2$ is the phase variance, so

$$\langle E \rangle = E_0 \exp(-\tfrac{1}{2}\langle\varphi^2\rangle). \tag{9.4}$$

It is significant that this result agrees with the predictions of modern theories that include the effects of diffraction and multiple scattering.

For propagation along a path close to the ground one can replace $\langle\varphi^2\rangle$ by the spectrum representation developed in (4.11):

$$\text{Troposphere:} \quad \langle E \rangle = E_0 \exp\left( -2R\pi^2 k^2 \int_0^{\infty} d\kappa \, \kappa \Phi_n(\kappa) \right) \tag{9.5}$$

The average field strength therefore decreases exponentially with distance as the wave passes through the random medium. This attenuation can be understood on physical grounds by using diffraction theory and that model is developed in Chapter 9 of Volume 2.

In the last chapter we found that phase trends broaden and lower the probability density function relative to a normal distribution. We can use the analytical description (8.9) to estimate the average field strength that would be measured in the presence of a linear phase drift:

$$\langle E \rangle = E_0 \frac{1}{\sigma\sqrt{2\pi}} \int_{-\infty}^{\infty} d\psi \exp(i\psi) \frac{1}{2\eta T} \int_{\psi-\eta T}^{\psi+\eta T} d\varphi \exp\left( -\frac{\varphi^2}{2\sigma^2} \right)$$

We shift to a new variable $\varphi = \psi + x$ and change the order of integration:

$$\langle E \rangle = E_0 \frac{1}{2\eta T} \int_{-\eta T}^{\eta T} dx \exp\left( -\frac{x^2}{2\sigma^2} \right) \int_{-\infty}^{\infty} d\psi \exp\left[ \psi\left( i + \frac{x}{\sigma^2} \right) - \frac{\psi^2}{2\sigma^2} \right]$$

The $\psi$ integral is found in Appendix C and the remaining integration on $x$ is easily done:

$$\langle E \rangle = E_0 \frac{\sin(\eta T)}{\eta T} \exp\left( -\frac{1}{2}\langle\varphi^2\rangle \right) \tag{9.6}$$

This approaches (9.4) as the rate of change of phase goes to zero. The measured value should oscillate for large samples but is dominated by the exponential attenuation.

## 9.2  The Mutual Coherence Function

One can also use geometrical optics to calculate the average of the field strength taken at one point times its complex conjugate at another. This is called the *mutual coherence function* and plays a central role in propagation studies [2]. Using the phase-only approximation (9.2) to describe the electric field, we can write this quantity as

$$\langle E_1 E_2^* \rangle = |E_0|^2 \langle \exp[i(\varphi_1 - \varphi_2)] \rangle. \qquad (9.7)$$

This expression depends only on the phase difference $\varphi_1 - \varphi_2$ and phase trends should cancel out if the two receivers are close to each other. Since the individual phase fluctuations are Gaussian random variables, their difference is also. This allows one to write

$$\langle E_1 E_2^* \rangle = |E_0|^2 \exp[-\tfrac{1}{2}\langle (\varphi_1 - \varphi_2)^2 \rangle]$$

or

$$\langle E_1 E_2^* \rangle = |E_0|^2 \exp\left[-\sigma^2(1 - \mu(\rho))\right]. \qquad (9.8)$$

This can also be written in terms of the phase structure function:

$$\langle E_1 E_2^* \rangle = |E_0|^2 \exp[-\tfrac{1}{2}\mathcal{D}_\varphi(\rho)] \qquad (9.9)$$

The same result is predicted by descriptions of the propagation that include the effects of diffraction and multiple scattering, as we shall learn later in these volumes.

The mutual coherence function describes the correlation of light received at different points on the primary mirror of a large telescope. The above results tell one that the light reflected by two points will be coherent only if

$$\mathcal{D}_\varphi(\rho) < 1.$$

Signals reflected by points that are separated by approximately 20 cm therefore should not be coherent. This limitation is observed for large telescopes and substantially reduces their effective apertures. This is the principal reason for placing the Hubble space telescope in earth orbit.

The phase variation that appears in the expression for the field strength also varies with time because the dielectric variations do. The mutual coherence function for time-displaced fields can be written

$$\langle E^*(t)E(t+\tau)\rangle = |E_0|^2 \exp\left[-\sigma^2(1-\mu(\tau))\right] \tag{9.10}$$

where $\mu(\tau)$ is the autocorrelation which is discussed in Chapter 6.

## 9.3 Frequency Coherence

One can use the phase distribution to estimate the maximum information bandwidth which can be communicated over an atmospheric path. This limitation is characterized by the correlation of the field at one frequency and its complex conjugate at another. Using the basic expression for random phase shifts given by (3.24), we see that

$$\langle E(f_1)E^*(f_2)\rangle = |E_0|^2 \left\langle \exp\left(i\pi(k_1-k_2)\int_0^R ds\,\delta n(s,t)\right)\right\rangle.$$

The frequency difference enters only as a factor. The path integration is often a Gaussian random variable, which means that one can express the ensemble average as follows:

$$\langle E(f_1)E^*(f_2)\rangle = \exp\left(-\frac{\pi^2}{2}(k_1-k_2)^2\left\langle \int_0^R ds\int_0^R ds'\,\delta n(s,t)\,\delta n(s',t)\right\rangle\right)$$

The path integrations for various types of propagation were evaluated in Chapter 4 and are proportional to the phase variance. Changing from wavenumber to radio frequency, the frequency coherence can be written in an unusually simple form:

$$\langle E(f_1)E^*(f_2)\rangle = |E_0|^2 \exp\left(-\frac{(f_1-f_2)^2}{2\mathcal{B}^2}\right) \tag{9.11}$$

The *medium bandwidth* $\mathcal{B}$ is defined in terms of the center frequency

$$f_c = \tfrac{1}{2}(f_1+f_2)$$

and the phase variance:

$$\mathcal{B} = \frac{f_c}{\sqrt{\langle\varphi^2\rangle}} \tag{9.12}$$

This important variable defines the frequency separation at which the field strength becomes decorrelated. In combination, these expressions describe how waves of different frequencies respond to random dielectric variations as they travel along the ray path.

The medium bandwidth is independent of frequency in the lower atmosphere and is related to the turbulent spectrum as follows:

$$\text{Troposphere:} \quad \mathcal{B} = c \left( 16\pi^4 R \int_0^\infty d\kappa \, \kappa \Phi_n(\kappa) \right)^{-\frac{1}{2}} \quad (9.13)$$

Like the phase variance, the medium bandwidth is strongly influenced by large eddies. One can bypass the problem of estimating this quantity by using measured values for the single-path phase variance in expression (9.12). The rms phase fluctuation was measured as a few degrees at 1046 MHz over a 16-km path in Colorado [3]. We therefore estimate that

$$\text{Microwave Link:} \quad \mathcal{B} \simeq 10\,000 \text{ MHz}$$

The medium bandwidth is large relative to most carrier frequencies and irregularities in the troposphere should not limit the bandwidth of terrestrial microwave links.

The situation is different for a horizontal optical link because $\langle \varphi^2 \rangle$ is so large. The variance estimated in Section 4.1.5 leads to a medium bandwidth that is considerably smaller than the carrier frequency of optical signals:

$$\text{Optical Link:} \quad \mathcal{B} \simeq 4000 \text{ GHz}$$

This does not pose a problem in practice because present-day optical modulators are limited to much smaller information bandwidths. For satellite-to-ground optical links the general expression for an inclined path is

$$\text{Satellite Link:} \quad \mathcal{B} = c \left( 2\pi \sec \vartheta \int_0^\infty dz \int_0^\infty d\kappa \, \kappa \Phi_n(\kappa, z) \right)^{-\frac{1}{2}}$$

$$(9.14)$$

Using standard models for the turbulent atmosphere, Yura estimates values for $\mathcal{B}$ that are comparable to the optical frequencies themselves [4]. This suggests that very large bandwidths are available for ground-to-satellite links, unless cloud cover prevents transmission from space.

## 9.4 Shortcomings

The remarkable feature of geometrical optics is that, in several cases, it predicts the same field strength moments that modern theories do. We should inquire where it breaks down. Optical experiments usually measure the intensity of a signal and the phase-only approximation for the field strength (9.2) shows that the instantaneous intensity is *constant*:

$$I = |E_0 \exp(i\varphi)|^2 = |E_0|^2 \qquad (9.15)$$

This means that we will be frustrated in attempts to calculate intensity fluctuations with this approximation since they always vanish:

$$\langle \delta I^2 \rangle = \langle (I - \langle I \rangle)^2 \rangle = 0 \qquad (9.16)$$

To obtain meaningful estimates for $\delta I$ one must use the complete expression for the field strength (9.1) and include $\chi$ in the calculations. On the other hand, we found in Chapter 3 that geometrical optics cannot describe logarithmic amplitude fluctuations in most situations of practical interest. This means that we must wait for diffraction theory to provide a satisfactory representation of $\chi$ before we can complete the program begun here.

## 9.5 Problems

### Problem 1

The Poynting vector describes the energy flow per unit area in terms of the electric field strength:

$$\mathbf{S} = \frac{c}{8\pi k} \Im(E^* \, \boldsymbol{\nabla} E)$$

Use the phase-only approximation to show that the ensemble average of this vector is constant, which means that energy is conserved in this approximation. Find the weakness in this demonstration.

### Problem 2

Use (9.2) to evaluate the ensemble average of field components taken at four different points. Show that this can be expressed in terms of $\sigma$ and the phase correlation taken between all possible pairs of points:

$$\langle E_1 E_2^* E_3 E_4^* \rangle = |E_0|^4 \exp[-\sigma^2(2 - \mu_{12} + \mu_{13} - \mu_{14} - \mu_{23} + \mu_{24} - \mu_{34})]$$

When points 1 and 3 coalesce and point 2 approaches point 4, this becomes

$$\langle E_1^2 E_2^{*2} \rangle = |E_0|^4 \exp[-4\sigma^2(1 - \mu_{12})]$$

which is similar to the mutual coherence function. How does this relate to the problem of estimating variations in intensity?

# References

[1] V. I. Tatarskii, *The Effects of the Turbulent Atmosphere on Wave Propagation* (translated from the Russian and issued by the National Technical Information Office, U.S. Department of Commerce, Springfield, VA 22161, 1971), 193.

[2] J. W. Goodman, *Statistical Optics* (John Wiley and Sons, New York, 1985), 170 *et seq.*

[3] J. W. Herbstreit and M. C. Thompson, "Measurements of the Phase of Radio Waves Received over Transmission Paths with Electrical Lengths Varying as a Result of Atmospheric Turbulence," *Proceedings of the IRE*, **43**, No. 10, 1391–1401 (October 1955).

[4] H. J. Yura, private communication, 6 January 1997.

# Appendix A
## Glossary of Symbols

| | |
|---|---|
| $a_{\mathrm{r}}$ | Radius of circular receiver |
| $a$ | Anisotropic scaling parameter in the $x$ direction |
| $A(\mathbf{r})$ | Amplitude of electric field strength |
| $A_n(\mathbf{r})$ | Terms in expansion of amplitude in inverse powers of $k$ |
| $\mathsf{A}$ | Area of circular receiving aperture |
| | Cross sectional area of ray bundle |
| $\mathcal{A}$ | Axial ratio of field-aligned plasma irregularities |
| $\mathbb{A}(\tau)$ | Allan variance, (6.62) |
| $b$ | Anisotropic scaling parameter in the $y$ direction |
| $\mathsf{B}_\varepsilon(\mathbf{r}, \mathbf{r}')$ | Spatial covariance of dielectric fluctuations |
| $\mathsf{B}_\varepsilon(\tau)$ | Temporal covariance of dielectric fluctuations |
| $\mathcal{B}$ | Medium bandwidth, (9.12) |
| $\mathbf{B}$ | Induction field in Maxwell's equations |
| $\mathsf{c}$ | Speed of light |
| $c$ | Anisotropic scaling parameter in the $z$ direction |
| $\mathsf{C}(\tau)$ | Autocorrelation of dielectric variations, (2.19) |
| $\mathsf{C}(\rho)$ | Spatial correlation of dielectric variations, (2.33) |
| $C_{\Delta\varphi}(D)$ | Spatial correlation function of phase difference |
| $C_n^2$ | Strength of turbulent refractive-index irregularities |
| $\mathcal{C}_u$ and $\mathcal{C}_v$ | Combination of ionospheric parameters, (7.78) and (7.79) |
| $d$ | Differential of following quantity |
| $D_r$ | Diameter of circular receiver |
| $D$ | Diffusion constant |
| $D_{12}$ | Separation between two points on adjacent rays |

| | |
|---|---|
| $\mathcal{D}_\varphi(\rho)$ | Phase structure function for receivers spaced a distance $\rho$ apart |
| $\mathcal{D}_n(D)$ | Refractive index structure function for separation $D$ |
| $\mathbf{D}$ | Displacement field in Maxwell's equations |
| $e$ | Electric charge |
| $\mathrm{erf}(x)$ | Error function defined in Appendix B |
| $\mathbf{E}$ | Electric field strength vector at $\mathbf{r}$ |
| $E(\mathbf{r})$ | Principal scalar component of electric field strength vector |
| $E_0(\mathbf{r})$ | Unperturbed field strength |
| $f$ | Electromagnetic frequency in cycles per second |
| $\delta f$ | Frequency fluctuation caused by random medium |
| $f_\mathrm{L}$ | Focal length of a beam wave |
| $f(x)$ | Token function used in Appendix F |
| $F(\kappa R)$ | Finite-path wavenumber-weighting function, (4.10) |
| $F(\omega)$ | Fourier transform of time series of signal phase |
| $_2F_1(a,b,c;z)$ | Hypergeometric function defined in Appendix H |
| $\mathcal{F}(\kappa \ell_0)$ | Spectrum factor that describes the dissipation region |
| $g(x)$ | Token function used in Appendix F |
| $\mathsf{g}$ | Variance of frequency fluctuations, (8.22) |
| $\mathcal{G}(\kappa, \rho, v\tau)$ | Phase-difference weighting function, (6.70) |
| $h$ | Receiver height |
| $\mathsf{h}$ | Line-of-sight height above ground |
| $H$ | Effective height of troposphere |
| | Layer height for ionospheric irregularities |
| $\mathbf{H}$ | Magnetic field vector |
| $\mathcal{H}(\mathfrak{y})$ | Phase-variance weighting function for satellite paths |
| $i$ | $\sqrt{-1}$ |
| $I$ | Intensity of electric field: $I = |E|^2$ |
| $I_0$ | Intensity of unperturbed field: $I_0 = |E_0|^2$ |
| $I_\nu(x)$ | Modified Bessel function of the first kind |
| $\Im(z)$ | Imaginary part of $z$ |
| $j$ | Power-spectrum frequency-dependence exponent, (6.46) |
| $J_\nu(x)$ | Ordinary Bessel function of the first kind |
| $\mathbf{j}(\mathbf{r})$ | Current-density distribution at a transmitter |
| $\mathcal{J}$ | Symbol used for integrals to be evaluated |
| $k$ | Electromagnetic wavenumber: $k = 2\pi/\lambda$ |
| $K(\eta)$ | Weighting function in $\mathcal{D}_\varphi(\rho)$ on satellite paths, (5.48) |
| $K_\nu(x)$ | Modified Bessel function of the second kind |
| $\mathcal{K}(\omega, \omega')$ | Sample-size factor for phase power spectrum, (6.44) |
| $\ell$ | Eddy size |
| $\ell_0$ | Inner scale length |
| $L_0$ | Outer scale length |

| | |
|---|---|
| $L$ | Upper limit for power-spectrum expression, (6.119) |
| $\mathbb{L}_\nu(x)$ | Struve function |
| $\mathcal{L}(\kappa_0 R/a)$ | Finite-path-distance factor for phase variance, (4.20) |
| $m$ | Ratio of phase trend to phase variance, (8.11) |
| $M(a, b; x)$ | First solution of Kummer's differential equation |
| $\mathcal{M}(\kappa)$ | Aperture-averaging factor for satellite paths, (7.35) |
| $M^2$ | Sample-averaged mean value for phase difference, (5.52) |
| $n$ | Refractive index |
| $\delta n$ | Variation in refractive index |
| $\mathbf{n}$ | Vector normal to ray path |
| $N$ | Electron density in ionosphere or interstellar plasma |
| $N_0$ | Ambient electron density |
| $\delta N$ | Fluctuation in electron density |
| $\mathcal{N}$ | Normalization for correlation functions |
| $p$ | Image-centroid displacement, (7.27) |
| $\mathsf{p}$ | Anisotropy parameter for satellite paths, (4.37) |
| $\mathsf{P}(x)$ | Probability density function of $x$ |
| $\mathcal{P}(x)$ | Cumulative probability of $x$ |
| $q$ | Argument of characteristic function, (8.6) |
| | Stretched coordinate to describe anisotropic media |
| $\mathsf{q}$ | Astronomical angle-of-arrival variance, (8.18) |
| $\mathsf{q}(x)$ | Interferometer baseline and sample-length function, (5.64) |
| $Q_\nu$ | Normalization for ionospheric spectrum models |
| $\mathcal{Q}(t, t', \rho)$ | Ensemble average of time-shifted phase differences, (5.58) |
| $\mathsf{Q}$ | Constant in interferometer scaling law, (5.62) |
| $r$ | Radial distance to point in a random medium |
| $r_\mathrm{E}$ | Radius of the earth: $r_\mathrm{E} = 6378$ km |
| $r_i$ | Radial coordinate defining position on receiver surface |
| $r_\mathrm{e}$ | Electron radius: $r_\mathrm{e} = 2.8 \times 10^{-13}$ cm |
| $r_0$ | Fried's coherence radius: $r_0 = 2.099 \rho_0$ |
| $\mathbf{r}$ | Position vector defining a point in a random medium |
| $\mathsf{r}(x)$ | Interferometer baseline and sample-length function, (5.68) |
| $R$ | Length of propagation path |
| $\Delta R$ | Entry plane-to-focus distance for parabolic receiver |
| $\mathbf{R}$ | Vector position of receiver |
| $\mathcal{R}_1(a, b, c, \vartheta)$ | Anisotropy factor for in-plane angular error, (7.63) |
| $\mathcal{R}_2(a, b, c, \vartheta)$ | Anisotropy factor for out-of-plane angular error, (7.63) |
| $\mathsf{R}$ | Constant in interferometer scaling law, (5.66) |
| $\Re(z)$ | Real part of $z$ |
| $s$ | Arc length along ray trajectory |
| $S(\kappa, \tau)$ | Self motion term in turbulence spectrum |

| | |
|---|---|
| $S$ | Signal phase for geometrical optics |
| $S_0$ | Signal phase in absence of a random medium |
| $\mathbf{S}$ | Poynting vector |
| $S(\kappa v T)$ | Sample-length wavenumber-weighting function, (4.25) |
| $\mathcal{S}(\kappa_0 v T)$ | Sample-length factor for phase variance, (4.27) |
| $t$ | Time in seconds or minutes |
| $\mathbf{t}$ | Tangent vector to ray-path trajectory |
| $T$ | Sample length in seconds or minutes |
| $\mathcal{T}(\mathbf{r}, t)$ | Temperature measured at point $\mathbf{r}$ and time $t$ |
| $\delta \mathcal{T}$ | Temperature fluctuation about mean value |
| $\mathbf{T}$ | Decorrelation time for turbulent velocity components |
| $\mathbf{T}$ | Vector defining transmitter position |
| $u$ | Difference coordinate |
| $U(a, b; z)$ | Second solution of Kummer's differential equation |
| $\mathcal{U}(u, \rho, \tau, \vartheta, \mathbf{v})$ | Terms in time-shifted structure function, (6.90) |
| $v$ | Surface wind speed normal to the propagation path |
| | Sum coordinate |
| $\mathbf{v}$ | Wind vector |
| $\mathbf{v}_0$ | Average wind vector |
| $\delta \mathbf{v}$ | Wind-velocity fluctuation |
| $\mathcal{V}(\omega)$ | Combination of terms in Problem 6 of Chapter 6 |
| $w(x)$ | Width of beam wave as a function of distance |
| $w_0$ | Initial width of beam wave |
| $W_\varphi(\omega)$ | Power spectrum of single-path phase fluctuations |
| $W_{\Delta\varphi}(\omega)$ | Power spectrum of phase-difference fluctuations |
| $\mathcal{W}_\nu(a_r)$ | Ionospheric aperture factor for the angle of arrival, (7.76) |
| $x$ | Horizontal coordinate along propagation path |
| $\mathcal{X}(\xi, p)$ | Spectrum factor describing variable wind speed, (6.53) |
| $y$ | Horizontal coordinate normal to propagation path |
| $Y_\nu(x)$ | Second solution of Bessel's equation for a real argument |
| $\mathcal{Y}(\zeta, \rho, \tau, \vartheta, \mathbf{v})$ | Weighting function for covariance of |
| | phase difference, (6.93) |
| $z$ | Vertical coordinate and height above ground |
| $\mathcal{Z}(\zeta, \omega)$ | Weight function for astronomical |
| | power spectrum, (6.118) |
| $\mathcal{Z}'(\zeta, \omega)$ | Combination of terms in Problem 7 of Chapter 6 |
| $\nabla$ | Gradient operator |
| $\nabla_\perp$ | Gradient operator in direction normal to path |
| $\nabla^2$ | Laplacian operator |
| $\nabla_\perp^2$ | Laplacian operator taken normal to path |

| | |
|---|---|
| $\alpha$ | Local inclination of ray path |
| $\alpha_0$ | Inclination of transmitted signal |
| | Elevation angle of received signal |
| $\beta$ | Azimuth orientation of receiver baseline or velocity |
| | Angular width of source |
| $\gamma$ | Angle between magnetic field and line-of-sight vectors |
| $\delta$ | Variation of following quantity |
| $\delta$ | Angle between two nearby stars |
| $\delta(x)$ | Dirac delta function defined in Appendix F |
| $\epsilon$ | Small parameter usually taken to zero |
| $\varepsilon$ | Dielectric constant |
| $\varepsilon_0$ | Initial or average value of dielectric constant |
| $\Delta\varepsilon$ | Variation in the dielectric constant |
| $\varepsilon_n$ | One or two in series of Bessel functions |
| $\zeta$ | Generic integration variable |
| $\eta$ | Slope of phase trend |
| | Parameter used in distance scaling: $\eta = \kappa_0 R / a_\mathrm{r}$, (4.18) |
| | Kolmogorov microscale, (2.60) |
| $\theta$ | Polar angle defining a point in a medium |
| | Angle between rotated ray paths |
| $\theta_i$ | Polar angle defining points on extended source |
| $\vartheta$ | Zenith angle |
| $\delta\boldsymbol{\theta}$ | Vector defining angular error components |
| $\delta\theta_n$ | Angular error in the ray plane and normal to the ray |
| $\delta\theta_\perp$ | Angular error out of the ray plane |
| $\kappa$ | Turbulence wavenumber: $\kappa = 2\pi/\ell$ |
| $\boldsymbol{\kappa}$ | Turbulence wavenumber vector |
| $\kappa_0$ | Outer-scale wavenumber cutoff: $\kappa_0 = 2\pi/L_0$ |
| $\kappa_\mathrm{s}$ | Inner-scale wavenumber cutoff: $\kappa_\mathrm{s} = 2\pi/\ell_0$ |
| $\kappa_\mathrm{m}$ | $\kappa_\mathrm{m} = 0.942\kappa_\mathrm{s}$ |
| $\lambda$ | Wavelength of electromagnetic radiation |
| $\mu$ | Exponent in structure-function scaling: $\mathcal{D}_\varphi(\rho) = A\rho^\mu$ |
| $\mu(\tau)$ | Autocorrelation of phase, (6.22) |
| $\mu(\rho)$ | Space correlation of phase, (5.7) |
| $\mu(\tau, T)$ | Phase autocorrelation for sample length $T$, (6.27) |
| $\mu_\mathrm{m}$ | Magnetic permeability connecting $\mathbf{E}$ and $\mathbf{H}$, (2.7) |
| $\nu$ | Exponent in ionospheric spectrum model |
| | Index of Bessel functions |
| $\xi$ | Parameter leading to (6.53) |
| $\pi$ | Numerical constant: $\pi = 3.141\,59$ |

| | |
|---|---|
| $\boldsymbol{\rho}$ | Vector connecting adjacent receivers |
| $\rho$ | Scalar separation between adjacent receivers |
| $\rho_e$ | Charge density in Maxwell's equations |
| $\rho_0$ | Coherence radius: $\rho_0 = r_0/2.099$ |
| $\sigma^2$ | Phase variance |
| $\sigma_v$ | Dispersion in Gaussian wind-speed distribution |
| $d^2\sigma$ | Surface element on receiver aperture |
| $\tau$ | Time delay used in autocorrelation functions |
| $\upsilon$ | Exponent in zenith-angle scaling law, (7.40) |
| $\phi$ | Azimuth angle defining a point in a random medium |
| $\phi_i$ | Polar coordinate of points on receiving aperture |
| | Polar angle defining point on an extended source |
| $\varphi$ | Phase fluctuation of signal: $\varphi = S - S_0$ |
| $\overline{\varphi}$ | Average phase shift estimated from sample length $T$ |
| $\Delta\varphi$ | Phase difference measured between adjacent receivers |
| $\chi$ | Logarithmic amplitude fluctuation of signal |
| $\psi$ | Polar angle defining wavenumber vector |
| | Measured phase shift in presence of a trend: $\psi = \varphi + \eta t$ |
| $\omega$ | Frequency in radians per second: $\omega = 2\pi f$ |
| | Azimuthal angle defining wavenumber vector |
| $\omega_0$ | Threshold frequency in single-path phase spectrum |
| $\omega_c$ | Corner frequency in phase-difference spectrum |
| $\omega_p$ | Plasma frequency of ionized media |
| $\boldsymbol{\varrho}$ | Displacement for locally frozen random medium, (6.12) |
| $\varrho_n$ | Auxiliary functions used in Problem 7 of Chapter 6 |
| $\Gamma_\nu$ | Combination of gamma functions, (7.76) |
| $\Gamma(x)$ | Gamma function |
| $\Theta$ | Angle between magnetic field and wavenumber vector |
| $\Sigma$ | Summation of following terms |
| $\Sigma(\kappa,\omega)$ | Combination of terms used in (6.74) |
| $\Upsilon(\rho)$ | Difference of vertical and horizontal angular errors, (7.54) |
| $\Psi(r)$ | Eikonal function |
| $\Psi_N(\kappa)$ | Wavenumber spectrum of electron-density variations |
| $\Phi_n(\kappa)$ | Wavenumber spectrum of refractive-index variations |
| $\Phi_\varepsilon(\kappa)$ | Wavenumber spectrum of dielectric variations |
| $\Omega(\kappa)$ | Normalized wavenumber spectrum: $\Phi_n(\kappa) = C_n^2\,\Omega(\kappa)$ |
| $\Lambda(u,t,\rho)$ | Combination of terms, (5.61) |
| $\Lambda(\kappa,\tau,T)$ | Combination of terms, (6.30) |

$$\Phi_\varepsilon(\kappa) = \begin{cases} 4\Phi_n(\kappa) & \text{in the troposphere} \\ r_0^2\lambda^4\Psi_N(\kappa) & \text{in the ionosphere} \end{cases}$$

# Appendix B
## Integrals of Elementary Functions

Numerical values for the following integrals with specific exponents are needed in our work:

1. $\displaystyle\int_0^\infty \frac{\sin x}{x^{\frac{2}{3}}}\,dx = 1.339\,47$

2. $\displaystyle\int_0^\infty \frac{\sin x}{x^{\frac{11}{6}}}\,dx = 6.451\,98$

3. $\displaystyle\int_0^\infty \frac{\sin^2 x}{x^{\frac{11}{6}}}\,dx = 1.540\,19$

4. $\displaystyle\int_0^\infty \frac{\sin^2 x}{x^{\frac{17}{6}}}\,dx = 6.270\,61$

5. $\displaystyle\int_0^\infty \frac{1}{x^{\frac{5}{3}}}\left(1 - \frac{\sin x}{x}\right)dx = 1.205\,52$

6. $\displaystyle\int_0^\infty \frac{1}{x^{\frac{11}{6}}}\left(1 - \frac{\sin x}{x}\right)dx = 0.942\,98$

7. $\displaystyle\int_0^\infty \frac{1 - \cos x}{x^{\frac{5}{3}}}\,dx = 2.009\,20$

8. $\displaystyle\int_0^\infty \frac{1 - \cos x}{x^2}\,dx = \frac{\pi}{2} = 1.570\,80$

9. $\displaystyle\int_0^\infty \frac{1 - \cos x}{x^{\frac{11}{6}}}\,dx = 1.728\,80$

10.    $\displaystyle\int_0^1 x^{\frac{5}{6}}(1-x)^{\frac{5}{6}}\,dx = 0.220\,54$

11.    $\displaystyle\int_0^1 x^{\frac{8}{3}}(1-x)^{\frac{8}{3}}\,dx = 0.011\,88$

12.    $\displaystyle\int_1^{\infty} \frac{1}{x^{\frac{8}{3}}\sqrt{x^2-1}}\,dx = 0.841\,31$

13.    $\displaystyle\int_0^{\infty} \frac{1}{(x^2+a^2)^{\frac{11}{6}}}\,dx = 0.841\,31 a^{-\frac{8}{3}}$

14.    $\displaystyle\int_0^{\infty} \frac{x^2}{(x^2+a^2)^{\frac{11}{6}}}\,dx = 1.261\,96 a^{-\frac{2}{3}}$

15.    $\displaystyle\int_0^{\infty} \frac{dx}{x^{\frac{2}{3}}(a^2+x^2)^{\frac{1}{3}}} = 5.782\,87 a^{-\frac{1}{3}}$

16.    $\displaystyle\int_0^{\infty} \left[(x^2+a^2)^{\frac{1}{3}} - \left(x^2\right)^{\frac{1}{3}}\right] dx = 1.457\,19 a^{\frac{5}{3}}$

In some applications it is important to have general expressions for these integrals that are valid for arbitrary exponents:

1.    $\displaystyle\int_0^{\infty} x^{\mu-1}\sin(ax)\,dx = a^{-\mu}\Gamma(\mu)\sin\!\left(\frac{\pi}{2}\mu\right)$      $0 < \mu < 1, a > 0$

2.    $\displaystyle\int_0^{\infty} x^{\mu-1}\cos(ax)\,dx = a^{-\mu}\Gamma(\mu)\cos\!\left(\frac{\pi}{2}\mu\right)$      $0 < \mu < 1, a > 0$

3.    $\displaystyle\int_0^{\infty} x^{\mu-1}\sin^2(ax)\,dx = -a^{-\mu}\frac{\Gamma(\mu)}{2^{1+\mu}}\cos\!\left(\frac{\pi}{2}\mu\right)$

$$-2 < \mu < 0,\ a > 0$$

4.    $\displaystyle\int_0^{\infty} \frac{1}{x^{\mu}}\left(1 - \frac{\sin(ax)}{ax}\right)dx = a^{\mu-1}\frac{-\pi}{2\Gamma(1+\mu)\cos\!\left(\dfrac{\pi}{2}\mu\right)}$

$$1 < \mu < 3,\quad a > 0$$

5. $\displaystyle\int_0^\infty \frac{1-\cos(ax)}{x^\mu}\,dx = a^{\mu-1}\frac{-\pi}{2\Gamma(\mu)\cos\left(\dfrac{\pi}{2}\mu\right)}$

$$1 < \mu < 3, a > 0$$

6. $\displaystyle\int_0^\infty \frac{dx}{(x^2+a^2)^\mu} = a^{1-2\mu}\frac{\sqrt{\pi}\,\Gamma(\mu-\frac{1}{2})}{2\Gamma(\mu)} \qquad |\arg a| < \frac{\pi}{2}, \quad \mu > \frac{1}{2}$

7. $\displaystyle\int_0^1 [x\,(1-x)]^\mu\,dx = \frac{[\Gamma(\mu+1)]^2}{\Gamma(2\mu+2)}$

8. $\displaystyle\int_0^\infty \frac{\sin(bx)}{(x^2+a^2)^\mu}\,dx = (2a)^{\frac{1}{2}-\mu}(b)^{\mu-\frac{1}{2}}\frac{\sqrt{\pi}\,\Gamma(1-\mu)}{2}$

$$\times[I_{\mu-\frac{1}{2}}(ab) - \mathbf{L}_{\mu-\frac{1}{2}}(ab)]$$

$$b > 0,\ \mu > 0,\ a > 0$$

9. $\displaystyle\int_0^\infty \frac{\cos(bx)}{(x^2+a^2)^\mu}\,dx = \left(\frac{2a}{b}\right)^{\frac{1}{2}-\mu}\frac{\sqrt{\pi}}{\Gamma(\mu)}K_{\frac{1}{2}-\mu}(ab)$

$$b > 0,\ \mu > 0,\ a > 0$$

10. $\displaystyle\int_0^{\frac{\pi}{2}} (\sin x)^{2\mu-1}(\cos x)^{2\nu-1}\,dx = \frac{1}{2}B(\mu,\nu) = \frac{1}{2}\frac{\Gamma(\mu)\Gamma(\nu)}{\Gamma(\mu+\nu)}$

$$\mu > 0,\ \nu > 0$$

11. $\displaystyle\int_0^{\frac{\pi}{2}} \frac{1}{a^2\cos^2\omega + b^2\sin^2\omega}\,d\omega = \frac{\pi}{2ab}$

12. $\displaystyle\int_0^{\frac{\pi}{2}} \frac{\sin^2\omega}{(a^2\cos^2\omega + b^2\sin^2\omega)^2}\,d\omega = \frac{\pi}{4ab^3}$

13. $\displaystyle\int_0^{\frac{\pi}{2}} \frac{\cos^2\omega}{(a^2\cos^2\omega + b^2\sin^2\omega)^2}\,d\omega = \frac{\pi}{4a^3b}$

14. $\displaystyle\int_0^{\frac{\pi}{2}} \frac{1}{(a^2\cos^2\omega + b^2\sin^2\omega)^2}\,d\omega = \frac{\pi}{4ab}\left(\frac{1}{a^2} + \frac{1}{b^2}\right)$

15. $\displaystyle\int_0^{\frac{\pi}{2}} \frac{1}{c^2 + a^2 \cos^2 \omega + b^2 \sin^2 \omega}\, d\omega = \frac{\pi}{2\sqrt{(c^2 + a^2)(c^2 + b^2)}}$

16. $\displaystyle\int_0^{\infty} \cos(\omega x)\frac{\sin\!\left(b\sqrt{x^2 + a^2}\right)}{\sqrt{x^2 + a^2}}\, dx = \begin{cases} \dfrac{\pi}{2} J_0(a\sqrt{b^2 - \omega^2}), \\[2mm] \qquad \text{for } 0 < \omega < b \\[2mm] 0, \qquad \text{for } \omega > b \end{cases}$

The following parametric representation is useful in evaluating the small difference of large quantities:

1. $\displaystyle (u^2 + a^2)^{\frac{1}{3}} - u^{\frac{2}{3}} = \frac{1}{3}\int_0^{a^2} \frac{dx}{(u^2 + x)^{\frac{2}{3}}}$

2. $\displaystyle \int_0^z \left[(u^2 + a^2)^{\frac{1}{3}} - u^{\frac{2}{3}}\right] du = \frac{3}{5}z[(z^2 + \rho^2)^{\frac{1}{3}} - z^{\frac{2}{3}}]$

$$+ \frac{2}{5}a^2 \int_0^z (x^2 + a^2)^{-\frac{2}{3}}\, dx$$

# Appendix C
## Integrals of Gaussian Functions

The *error function* is the simplest example of an integral involving the Gaussian function:

$$\operatorname{erf}(x) = \frac{2}{\sqrt{\pi}} \int_0^x dt \, \exp(-t^2)$$

and approaches unity for large values of $x$. The following expansion can be used for small values:

$$\operatorname{erf}(x) = \frac{2x}{\sqrt{\pi}} \left( 1 - \frac{x^2}{1!\,3} + \frac{x^4}{2!\,5} - \frac{x^6}{3!\,7} + \cdots \right)$$

Tables of numerical values are provided in many texts and references.[1]

The following definite integrals are encountered in our work. They depend primarily on the error function and the Kummer function $M(a, b, z)$, which is discussed in Appendix G:

1. $$\int_0^\infty x^\mu e^{-ax^2} \, dx = \frac{1}{2} \Gamma\left( \frac{1+\mu}{2} \right) \frac{1}{a^{\frac{\mu+1}{2}}} \qquad \mu > -1, \quad a > 0$$

2. $$\int_0^\infty \cos(bx) \, e^{-ax^2} \, dx = \frac{1}{2} \sqrt{\frac{\pi}{a}} \, \exp\left( -\frac{b^2}{4a} \right) \qquad a > 0$$

3. $$\int_0^\infty x \sin(bx) \, e^{-ax^2} \, dx = \frac{b}{4a} \sqrt{\frac{\pi}{a}} \, \exp\left( -\frac{b^2}{4a} \right) \qquad a > 0$$

4. $$\int_0^\infty \frac{\sin(bx)}{x} e^{-ax^2} \, dx = \frac{\pi}{2} \operatorname{erf}\left( \frac{b}{2\sqrt{a}} \right) \qquad a > 0$$

---

[1] For example, see M. Abramowitz and I. A. Stegun, *Handbook of Mathematical Functions* (Dover, New York, 1964), 295–329.

5. $\displaystyle\int_0^\infty \frac{1-\cos(bx)}{x^2}e^{-ax^2} = \frac{\pi}{2}b\,\mathrm{erf}\!\left(\frac{b}{2\sqrt{a}}\right)$

$$-\sqrt{\pi a}\left[1-\exp\!\left(-\frac{b^2}{4a}\right)\right] \qquad a>0$$

6. $\displaystyle\int_0^\infty x^{\mu-1}\sin(bx)\,e^{-ax^2}\,dx = \frac{b}{2\,a^{\frac{\mu+1}{2}}}\Gamma\!\left(\frac{1+\mu}{2}\right)$

$$\times\, M\!\left(\frac{1+\mu}{2},\frac{3}{2},-\frac{b^2}{4a}\right)$$

$$a>0,\ \mu>-1$$

7. $\displaystyle\int_0^\infty x^{\mu-1}\cos(bx)\,e^{-ax^2}\,dx = \frac{1}{2a^{\frac{\mu}{2}}}\Gamma\!\left(\frac{\mu}{2}\right)M\!\left(\frac{\mu}{2},\frac{1}{2},-\frac{b^2}{4a}\right)$

$$a>0,\ b>0,\ \mu>0$$

8. $\displaystyle\int_0^\infty \frac{1}{x^{\frac{5}{3}}}\left(1-\frac{\sin(bx)}{bx}\right)e^{-ax^2}\,dx = \frac{3}{2}\Gamma\!\left(\frac{2}{3}\right)a^{\frac{1}{3}}$

$$\times\left[M\!\left(-\frac{1}{3},\frac{3}{2},-\frac{b^2}{4a}\right)-1\right]$$

9. $\displaystyle\int_{-\infty}^\infty e^{-ax^2+bx}\,dx = \sqrt{\frac{\pi}{a}}\,\exp\!\left(\frac{b^2}{4a}\right)$

10. $\displaystyle\int_{-\infty}^\infty xe^{-ax^2+bx}\,dx = \sqrt{\frac{\pi}{a}}\left(\frac{b}{2a}\right)\exp\!\left(\frac{b^2}{4a}\right)$

# Appendix D
## Bessel Functions

Bessel functions play an important and pervasive role in descriptions of propagation through random media. This occurs because we often need to solve the wave equation in cylindrical coordinates. There are two families of Bessel functions. They differ by having real or imaginary arguments and we discuss them separately.

### D.1 Ordinary Bessel Functions

The ordinary Bessel functions satisfy the following second-order differential equation:

$$\frac{d^2 w}{dz^2} + \frac{1}{z}\frac{dw}{dz} + \left(1 - \frac{\nu^2}{z^2}\right)w = 0$$

which has two independent solutions:

$$w(z) = A J_\nu(z) + B Y_\nu(z)$$

The first solution is regular at the origin and is described by a power- series expansion:

$$J_\nu(z) = \left(\frac{z}{2}\right)^\nu \sum_{n=0}^{\infty} \frac{\left(-\frac{z^2}{4}\right)^n}{n!\,\Gamma(\nu + n + 1)} \qquad |\arg z| < \pi$$

Tables of numerical values for integral and half-integral values of the index $\nu$ are available in standard references.[1] The Bessel functions

---

[1] M. Abramowitz and I. A. Stegun, *Handbook of Mathematical Functions* (Dover, New York, 1964), 355–433 and 435–478.

403

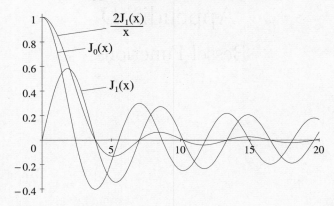

Figure D.1: A graph comparing the Bessel functions $J_0(x)$ and $J_1(x)$, and a common combination that occurs in descriptions of aperture averaging.

corresponding to $\nu = 0$ and $\nu = 1$ occur frequently in our work

$$J_0(x) = 1 - \frac{1}{(1!)^2}\left(\frac{x}{2}\right)^2 + \frac{1}{(2!)^2}\left(\frac{x}{2}\right)^4 - \frac{1}{(3!)^2}\left(\frac{x}{2}\right)^6 \cdots$$

$$J_1(x) = \frac{x}{2} - \frac{1}{2}\left(\frac{x}{2}\right)^3 + \frac{1}{12}\left(\frac{x}{2}\right)^5 \cdots.$$

The following combination occurs in descriptions of aperture averaging:

$$\frac{2J_1(x)}{x} = 1 - \frac{1}{8}x^2 + \frac{1}{192}x^4 \cdots$$

These three functions are plotted together in Figure D.1.

For integral values of the index:

$$J_n(-x) = (-1)^n J_n(x)$$

One can establish the following expressions for the derivatives of the first two Bessel functions from the power series:

$$\frac{d}{dx}J_0(x) = -J_1(x)$$

$$\frac{d}{dx}J_1(x) = J_0(x) - \frac{J_1(x)}{x}$$

The zeroth-order Bessel function has the following important property, which is called the *addition theorem*:

$$J_0\left(\kappa\sqrt{r^2 + \rho^2 - 2r\rho\cos(\alpha - \beta)}\right) = \sum_{0}^{\infty} \varepsilon_n J_n(\kappa r) J_n(\kappa\rho) \cos[n(\alpha - \beta)]$$

where

$$\varepsilon_n = \begin{cases} 1 & n = 0 \\ 2 & n \geq 1. \end{cases}$$

An important property of Bessel functions is their orthogonality, which is expressed in terms of the Dirac delta function described in Appendix F:

$$\int_0^\infty J_\nu(xa) J_\nu(xb) x \, dx = \frac{1}{a} \delta(a - b) \qquad \text{for} \quad \nu \geq -\frac{1}{2}$$

The second solution of Bessel's equation is not regular at the origin and we shall seldom need it in our studies. For non-integral values of the index $\nu$ it is defined by the relationship[2]

$$Y_\nu(z) = \frac{\cos(\nu\pi) \, J_\nu(z) - J_{-\nu}(z)}{\sin(\nu\pi)} \qquad |\arg z| < \pi$$

The following definite integrals lead to ordinary Bessel functions:

1. $\displaystyle\int_0^{2\pi} \exp[i(a \sin x + b \cos x)] \, dx = 2\pi J_0\left(\sqrt{a^2 + b^2}\right)$

2. $\displaystyle\int_1^\infty \frac{\sin(ax)}{\sqrt{x^2 - 1}} \, dx = \frac{\pi}{2} J_0(a)$

3. $\displaystyle\int_0^1 \frac{\cos(ax)}{\sqrt{1 - x^2}} \, dx = \frac{\pi}{2} J_0(a)$

4. $\displaystyle\int_{-\infty}^\infty \frac{\sin(\kappa\sqrt{x^2 + a^2})}{\sqrt{x^2 + a^2}} \, dx = \pi J_0(\kappa a)$

5. $\displaystyle\int_0^{\frac{\pi}{2}} \cos(a \cos \varphi) (\sin \varphi)^{2\nu} \, d\varphi = \frac{\Gamma\left(\frac{1}{2} + \nu\right)\Gamma\left(\frac{1}{2}\right)}{2} \left(\frac{2}{a}\right)^\nu J_\nu(a)$
   $$\nu > -\tfrac{1}{2}$$

We must carry out integrations that involve Bessel functions in many parts of our work. More than 1500 integrals involving Bessel functions of all kinds have been solved analytically.[3] The following indefinite integral is

---

[2] $Y_\nu(z)$ is denoted by $N_\nu(z)$ in some references.

[3] A. P. Prudnikov, Yu. A. Brychkov and O. I. Marichev, *Integrals and Series, Volume 2: Special Functions* (Gordon and Breach, New York and London, 1986), 168–420.
I. S. Gradshteyn and I. M. Ryzhik, *Table of Integrals, Series, and Products* (Academic Press, New York, 1980), 665–788.
A. D. Wheelon, *Tables of Summable Series and Integrals Involving Bessel Functions* (Holden-Day, San Francisco, 1968).

valid for integral indices:

$$\int_0^z x^{n+1} J_n(x)\, dx = z^{n+1} J_{n+1}(x)$$

and the special case

$$\frac{1}{a^2} \int_0^a x J_0(x\kappa)\, dx = \frac{J_1(\kappa a)}{\kappa a}$$

is often used in describing aperture averaging by circular receivers.

Definite integrals are needed frequently and the vast majority of the available analytical results falls in this category. In many cases the index $\nu$ can assume complex values but we need only real values in our applications. The other parameters that occur in these integrals can also be complex but are usually real in our work. We need only a small fraction of the available integrals and they are listed below:

1. $\displaystyle\int_0^\infty \frac{1}{x^{\frac{5}{3}}} J_1(ax)\, dx = 1.863\,88 a^{\frac{2}{3}}$

2. $\displaystyle\int_0^\infty \frac{1}{x^{\frac{2}{3}}} J_1^2(ax)\, dx = 0.660\,28 a^{-\frac{1}{3}}$

3. $\displaystyle\int_0^\infty \frac{1}{x^{\frac{8}{3}}} J_1^2(ax)\, dx = 0.864\,37 a^{\frac{5}{3}}$

4. $\displaystyle\int_0^\infty \frac{1}{x^{\frac{8}{3}}} \left(1 - \frac{4 J_1^2(ax)}{(ax)^2}\right) dx = 1.064\,98 a^{\frac{5}{3}}$

5. $\displaystyle\int_0^\infty \frac{1}{x^{\frac{8}{3}}} \left[1 - J_0^2(ax) - J_1^2(ax)\right] dx = 1.037\,28 a^{\frac{5}{3}}$

6. $\displaystyle\int_0^\infty \frac{1}{x^{\frac{8}{3}}} [1 - J_0(ax)]\, dx = 1.118\,33 a^{\frac{5}{3}}$

7. $\displaystyle\int_0^\infty \frac{1}{x^{\frac{8}{3}}} \left(1 - \frac{2 J_1(ax)}{ax}\right) dx = 0.609\,98 a^{\frac{5}{3}}$

8. $\displaystyle\int_0^\infty \frac{1}{x^\mu} J_\nu(ax)\, dx = \frac{\Gamma\left(\dfrac{\nu - \mu + 1}{2}\right)}{2^\mu \Gamma\left(\dfrac{\nu + \mu + 1}{2}\right)} a^{\mu-1}$

$$-\tfrac{1}{2} < \mu < \nu + 1$$

9. $\displaystyle\int_0^\infty \frac{1}{x^\mu} J_\nu^2(ax)\, dx = \frac{\Gamma(\mu)\Gamma\left(\dfrac{2\nu + 1 - \nu}{2}\right)}{2^\mu\left[\Gamma\left(\dfrac{\mu + 1}{2}\right)\right]^2 \Gamma\left(\dfrac{2\nu + 1 + \mu}{2}\right)} a^{\mu-1}$

$$0 < \mu < 2\nu + 1, \quad a > 0$$

10. $\displaystyle\int_0^\infty \frac{1}{x^\mu}[1 - J_0(ax)]\, dx = \frac{\Gamma\left(\dfrac{3 - \mu}{2}\right)}{2^{\mu-1}(\mu - 1)\Gamma\left(\dfrac{\mu + 1}{2}\right)} a^{\mu-1}$

$$1 < \mu < 3$$

11. $\displaystyle\int_0^\infty J_\nu(ax) x^\mu\, dx = \frac{\Gamma\left(\dfrac{\nu + 1 + \mu}{2}\right)}{2\Gamma\left(\dfrac{\nu + 1 - \mu}{2}\right)}\left(\frac{2}{a}\right)^{\mu+1}$

$$\mu + \nu > -1, \quad a > 0, \quad \mu < \tfrac{1}{2}$$

12. $\displaystyle\int_0^\infty [1 - J_0(ax)]\frac{x}{(x^2 + 1)^{\frac{11}{6}}}\, dx = \frac{3}{5}\left(1 - \frac{2^{\frac{1}{6}}}{\Gamma\left(\frac{5}{6}\right)} a^{\frac{5}{6}} K_{\frac{5}{6}}(a)\right)$

13. $\displaystyle\int_0^\infty J_0(ax) \sin\left(bx^2\right) x\, dx = \frac{1}{2b} \cos\left(\frac{a^2}{4b}\right)$

14. $\displaystyle\int_0^\infty J_0(ax) \cos\left(bx^2\right) x\, dx = \frac{1}{2b} \sin\left(\frac{a^2}{4b}\right)$

15. $\displaystyle\int_0^\infty \frac{x^{\nu+1} J_\nu(ax)}{(x^2 + b^2)^{\mu+1}}\, dx = \frac{a^\mu b^{\nu-\mu}}{2^\mu \Gamma(\mu + 1)} K_{\mu-\nu}(ab)$

$$a, b > 0, \quad -1 < \nu < 2\mu + \tfrac{3}{2}$$

16. $\displaystyle\int_0^\infty J_\nu(ax)\sin(bx)\,dx = \frac{\sin\left[\nu\sin^{-1}(b/a)\right]}{\sqrt{a^2-b^2}}$

$$0 < b < a, \qquad \nu > -2$$

$$= \frac{a^\nu\cos(\nu\pi/2)\left(b+\sqrt{b^2-a^2}\right)^{-\nu}}{\sqrt{b^2-a^2}}$$

$$0 < a < b, \qquad \nu > -2$$

17. $\displaystyle\int_0^\infty J_\nu(ax)\cos(bx)\,dx = \frac{\cos\left[\nu\sin^{-1}(b/a)\right]}{\sqrt{a^2-b^2}} \quad 0 < b < a$

$$= \frac{-a^\nu\sin(\nu\pi/2)\left(b+\sqrt{b^2-a^2}\right)^{-\nu}}{\sqrt{b^2-a^2}}$$

$$\text{for} \quad 0 < a < b, \ \ \nu > -1$$

18. $\displaystyle\int_0^{\frac{\pi}{2}} J_{\mu+\nu}(2a\cos\varphi)\cos[(\mu-\nu)\varphi]\,d\varphi = \frac{\pi}{2}J_\mu(a)J_\nu(a)$

$$\mu+\nu > -1$$

19. $\displaystyle\int_0^{\frac{\pi}{2}} J_{m-n}(2a\cos\varphi)\cos[(m+n)\varphi]\,d\varphi = \frac{\pi}{2}(-1)^n J_m(a)J_n(a)$

$$n \text{ and } m \text{ integers}, \qquad m-n > -1$$

20. $\displaystyle\int_0^{\pi} J_0(2a\cos\varphi)\cos(2n\varphi)\,d\varphi = (-1)^n\frac{\pi}{2}J_n^2(a) \qquad n \text{ an integer}$

21. $\displaystyle\int_0^{\frac{\pi}{2}} J_0(2a\cos\varphi)\,d\varphi = \frac{\pi}{2}J_0^2(a)$

22. $\displaystyle\int_0^{\frac{\pi}{2}} J_0(2a\cos\varphi)\cos(2\varphi)\,d\varphi = -\frac{\pi}{2}J_1^2(a)$

23. $\displaystyle\int_0^{\pi} J_{2n}(2a\sin\varphi)\,d\varphi = \pi J_n^2(a)$

24. $\displaystyle\int_0^{\frac{\pi}{2}} J_\mu(a\sin\varphi)(\sin\varphi)^{\mu+1}(\cos\varphi)^{2\nu+1}\,d\varphi$

$$= \frac{2^\nu\Gamma(\nu+1)}{a^{\nu+1}}J_{\mu+\nu+1}(a) \qquad\qquad \mu>-1,\quad \nu>-1$$

25. $\displaystyle\int_0^\infty J_0(a\sqrt{x^2+y^2})\cos(bx)\,dx \;=\; \begin{cases} \dfrac{\cos\!\left(y\sqrt{a^2-b^2}\right)}{\sqrt{a^2-b^2}} & a>b \\[2ex] 0 & a\le b,\, y>0 \end{cases}$

26. $\displaystyle\int_0^\pi J_0(a\sin\varphi)\exp(ib\cos\varphi)\sin\varphi\,d\varphi = 2\frac{\sin(\sqrt{a^2+b^2})}{\sqrt{a^2+b^2}}$

## Weber's integrals[4]

27. $\displaystyle\int_0^\infty J_0(ax)\exp\!\left(-p^2x^2\right)x\,dx = \frac{1}{2p^2}\exp\!\left(-\frac{a^2}{4p^2}\right) \quad \Re\!\left(p^2\right)>0$

28. $\displaystyle\int_0^\infty J_\nu(ax)\exp\!\left(-p^2x^2\right)x^{\mu-1}\,dx$

$$= \frac{a^\nu\Gamma\!\left(\dfrac{\mu+\nu}{2}\right)}{2^{\nu+1}p^{\mu+\nu}\Gamma(\nu+1)}M\!\left(\frac{\mu+\nu}{2},\nu+1,-\frac{a^2}{4p^2}\right)$$

$$= \frac{a^\nu\Gamma\!\left(\dfrac{\mu+\nu}{2}\right)}{2^{\nu+1}p^{\mu+\nu}\Gamma(\nu+1)}\exp\!\left(-\frac{a^2}{4p^2}\right)M\!\left(\frac{\nu-\mu}{2}+1,\nu+1,\frac{a^2}{4p^2}\right)$$

$$\mu+\nu>0,\quad \Re(p^2)>0$$

---

[4] G. N. Watson, *A Treatise on the Theory of Bessel Functions* (Cambridge University Press, 1952), 393.

## Lamb's integral[5]

29. $$\int_0^\infty J_0(ax) \frac{\exp\left(ik\sqrt{x^2+y^2}\right)}{\sqrt{x^2+y^2}} x\,dx$$

$$= \frac{i}{\sqrt{k^2-a^2}} \exp\left(i|y|\sqrt{k^2-a^2}\right) \qquad a < k$$

$$= \frac{1}{\sqrt{a^2-k^2}} \exp\left(-|y|\sqrt{a^2-k^2}\right) \qquad a > k$$

## Sommerfeld's integral[6]

There is another way to describe the basic relationship suggested by the integral above and it is known as Sommerfeld's integral:

30. $$\int_0^\infty \frac{J_0(xu)}{\sqrt{u^2-k^2}} \exp\left(-|y|\sqrt{x^2-k^2}\right) u\,du = \frac{\exp\left(ik\sqrt{x^2+y^2}\right)}{\sqrt{x^2+y^2}}$$

$$0 \le \arg k < \pi, \quad -\frac{\pi}{2} \le \arg\sqrt{u^2-k^2} < \frac{\pi}{2}, \quad x \text{ and } y \text{ real}$$

It can be established by taking the Fourier Bessel transform of Lamb's integral and using the orthogonality of the Bessel functions indicated earlier.

## Series of Bessel functions

Sometimes infinite series of Bessel functions are needed:

1. $\sum_0^\infty \varepsilon_{2n} J_{2n}(a) \cos(2n\theta) = \cos(a\sin\theta)$

2. $\sum_0^\infty \varepsilon_{2n}(-1)^n J_{2n}(a) \cos(2n\theta) = \cos(a\cos\theta)$

3. $\sum_0^\infty J_{2n+1}(a) \sin[(2n+1)\theta] = \sin(a\sin\theta)$

4. $\sum_0^\infty (-1)^n J_{2n+1}(a) \cos[(2n+1)\theta] = \sin(a\cos\theta)$

[5] H. Lamb, "On the Theory of Waves Propagated Vertically in the Atmosphere," *Proceedings of the London Mathematical Society* (2), **7**, 122–141 (10 December 1908).
H. Bateman, *Partial Differential Equations of Mathematical Physics* (Dover, New York, 1944), 411.
[6] W. Magnus and F. Oberhettinger, *Formulas and Theorems for the Special Functions of Mathematical Physics* (Chelsea, New York, 1949), 34.

## D.2 Modified Bessel Functions

These functions satisfy the modified Bessel differential equation:

$$\frac{d^2w}{dz^2} + \frac{1}{z}\frac{dw}{dz} - \left(1 + \frac{\nu^2}{z^2}\right)w = 0$$

and the two independent solutions are written as

$$w(z) = AI_\nu(z) + BK_\nu(z)$$

The first solution is regular at the origin and can be computed from its power-series expansion:

$$I_\nu(z) = \left(\frac{z}{2}\right)^\nu \sum_0^\infty \frac{1}{n!\,\Gamma(n+\nu+1)}\left(\frac{z^2}{4}\right)^n$$

From this it is apparent that

$$\lim_{z \to 0}\left[I_\nu(z)\right] = \frac{1}{\Gamma(\nu+1)}\left(\frac{z}{2}\right)^\nu$$

provided that $\nu$ is not a negative integer. For real $z$ this solution is related to the ordinary Bessel function by the relationship

$$I_\nu(z) = \exp\left(-i\frac{\nu\pi}{2}\right) J_\nu\left[z\exp\left(i\frac{\nu\pi}{2}\right)\right]$$

as one can verify by changing the independent variable in the power series or in the differential equation.

The second solution is known as the *MacDonald function* and diverges at the origin. It can be computed from the first solution using the following relationship, provided that $\nu$ is not an integer:

$$K_\nu(z) = \frac{\pi}{2}\frac{\left[I_{-\nu}(z) - I_\nu(z)\right]}{\sin(\nu\pi)}$$

We are usually interested in fractional values of the index $\nu$, so this form can be used in conjunction with the power series for $I_\nu(z)$ to calculate $K_\nu(z)$. When the argument is small

$$\lim_{z \to 0}\left[K_\nu(z)\right] = \frac{1}{2}\Gamma(\nu)\left(\frac{z}{2}\right)^{-\nu} \qquad \nu > 0$$

We often need the second term in the small-argument expansion and the following results are useful:

1. $\lim_{x\to 0}[K_0(x)] = -\left[\ln\left(\dfrac{x}{2}\right) + \gamma\right]\left(1 + \dfrac{1}{4}x^2 \cdots\right) + \dfrac{1}{4}x^2$

$$\text{Euler's constant } \gamma = 0.577\ 215\ 7$$

2. $\lim_{x\to 0}\left[K_{\frac{1}{3}}(x)\right] = \dfrac{1}{2}\Gamma\left(\dfrac{1}{3}\right)\left(\dfrac{2}{x}\right)^{\frac{1}{3}}\left[1 - \left(\dfrac{x}{2}\right)^{\frac{2}{3}}\dfrac{\Gamma(\frac{2}{3})}{\Gamma(\frac{4}{3})}\cdots\right]$

3. $\lim_{x\to 0}\left[K_{\frac{5}{6}}(x)\right] = \dfrac{1}{2}\Gamma\left(\dfrac{5}{6}\right)\left(\dfrac{2}{x}\right)^{\frac{5}{6}}\left(1 - x^{\frac{5}{3}}\dfrac{\Gamma(\frac{1}{6})}{\Gamma(\frac{11}{6})}\cdots\right)$

4. $\lim_{x\to 0}\left[K_{\frac{4}{3}}(x)\right] = \dfrac{1}{2}\Gamma\left(\dfrac{4}{3}\right)\left(\dfrac{2}{x}\right)^{\frac{4}{3}}\left(1 - x^2\dfrac{\pi}{2\sqrt{3}\,\Gamma(\frac{2}{3})\Gamma(\frac{4}{3})}\cdots\right)$

5. $\lim_{x\to 0}\left[K_{\frac{11}{6}}(x)\right] = \dfrac{1}{2}\Gamma\left(\dfrac{11}{6}\right)\left(\dfrac{2}{x}\right)^{\frac{11}{6}}\left(1 - x^2\dfrac{\pi}{2\Gamma(\frac{1}{6})\Gamma(\frac{11}{6})}\cdots\right)$

We encounter the modified Bessel functions primarily as definite integrals of elementary functions:

1. $\displaystyle\int_0^\pi \cos(n\varphi)\exp(b\cos\varphi)\,d\varphi = \pi I_n(b)$   $n$ an integer

2. $\displaystyle\int_{-1}^1 \left(1 - x^2\right)^{\nu-\frac{1}{2}}\exp(-bx)\,dx = \sqrt{\pi}\,\Gamma(\nu + \tfrac{1}{2})\left(\dfrac{2}{b}\right)^\nu I_\nu(b)$

$$\nu > -\tfrac{1}{2}$$

3. $\displaystyle\int_0^\infty \exp\left(-ax - \dfrac{b}{x}\right)x^{\mu-1}\,dx = 2\left(\dfrac{b}{a}\right)^{\frac{\mu}{2}}K_\mu(2\sqrt{ab})$

4. $\displaystyle\int_1^\infty \dfrac{\exp(-bx)}{(x^2-1)^{\nu+\frac{1}{2}}}\,dx = \dfrac{1}{\sqrt{\pi}}\Gamma(\tfrac{1}{2}-\nu)\left(\dfrac{z}{2}\right)^{-\nu}K_\nu(b)$

$$\nu > -\tfrac{1}{2}, \quad |\arg b| < \pi/2$$

5. $\displaystyle\int_0^\infty \dfrac{\cos(bx)}{(x^2+a^2)^{\nu+\frac{1}{2}}}\,dx = \dfrac{\sqrt{\pi}}{\Gamma(\nu+\frac{1}{2})}\left(\dfrac{2a}{b}\right)^{-\nu}K_\nu(ab)$

$$\nu > -\tfrac{1}{2}, \quad b > 0, \quad |\arg a| < \pi/2$$

On rare occasions we need integrals involving the modified Bessel functions and the following integrals are helpful:

1. $\displaystyle\int_0^z dx\, x^\nu K_{\nu-1}(x) = -z^\nu K_\nu(z) + 2^{\nu-1}\Gamma(\nu)$ $\qquad R(\nu) > 0$

2. $\displaystyle\int_0^\infty dx\, x^\nu K_\nu(x) = 2^{\nu-1}\sqrt{\pi}\,\Gamma(\nu+\tfrac{1}{2})$

3. $\displaystyle\int_0^\infty dx\, \frac{x^{2\mu+1}}{\left(\sqrt{x^2+z^2}\right)^{\frac{\nu}{2}}} K_\nu\!\left(a\sqrt{x^2+z^2}\right) = \frac{2^\mu\Gamma(\mu+1)}{a^{\mu+1}z^{\nu-\mu-1}}K_{\nu-\mu-1}(az)$

$$a > 0, \quad R(\mu) > -1$$

4. $\displaystyle\int_{-\infty}^\infty dx\, |x|^\nu e^{-ax} K_\nu(b|x|) = \frac{\sqrt{\pi}\,\Gamma\!\left(\nu+\tfrac{1}{2}\right)(2b)^\nu}{(b^2-a^2)^{\nu+\frac{1}{2}}}$

$$a < b, \quad \nu < \tfrac{1}{2}$$

# Appendix E
## Probability Distributions

### E.1 The Gaussian Distribution

The Gaussian distribution is frequently used to describe random processes. The Central Limit theorem tells us that a signal that is the additive result of a large number of independent contributions will be distributed as a Gaussian random variable. This is often the case in our work and the probability density function is written

$$\mathsf{P}(x) = \frac{1}{\sigma\sqrt{2\pi}}\exp\left(-\frac{1}{2\sigma^2}(x-x_0)^2\right)$$

and is symmetrical around the most likely value, $x_0$. The width and height of the distribution are determined by the parameter $\sigma$. The area under this function is unity for all values of $x_0$ and $\sigma$, as one can show by using the integrals in Appendix C:

$$\int_{-\infty}^{\infty} dx\, \mathsf{P}(x) = 1$$

This model is widely used to describe thermal noise in electronic circuits. In our explorations, it provides a good description of phase fluctuations over a wide range of conditions. We will also use it to describe the distribution of wind speeds about a mean value. The mean value of this distribution is independent of the parameter $\sigma$:

$$\langle x \rangle = \int_{-\infty}^{\infty} dx\, x \mathsf{P}(x) = x_0$$

Notice that the average value coincides with the most likely value for a Gaussian distribution.

The mean square value depends on both parameters:

$$\langle x^2 \rangle = \int_{-\infty}^{\infty} dx\, x^2 \mathsf{P}(x) = x_0^2 + \sigma^2$$

414

This means that $\sigma^2$ is the variance of $x$ and defines the spread of the distribution:

$$\langle (x - \langle x \rangle)^2 \rangle = \sigma^2$$

The Gaussian distribution becomes narrow and increases at the reference point $x = x_0$ as $\sigma$ becomes small. In the limit as $\sigma \to 0$ this provides one definition for the delta function, as we shall see in Appendix F.

Using the integrals in Appendix C, one can establish the following expressions for higher-order moments:

$$\langle x^3 \rangle = 3x_0\sigma^2 + x_0^3$$
$$\langle x^4 \rangle = 3\sigma^4 + 6\sigma^2 x_0^2 + x_0^4$$
$$\langle x^5 \rangle = 15x_0\sigma^4 + 10\sigma^2 x_0^3 + x_0^5$$
$$\langle x^6 \rangle = 15\sigma^6 + 45\sigma^4 x_0^2 + 15\sigma^2 x_0^4 + x_0^6$$

The characteristic function plays an important role in analyzing random variables and is defined by the ensemble average:

$$\langle \exp(i\kappa x) \rangle = \int_{-\infty}^{\infty} dx \, \exp(i\kappa x) \, \mathsf{P}(x)$$

We can estimate this function by introducing the probability density function and carrying out the integrations:

$$\langle \exp(i\kappa x) \rangle = \exp\left( i\kappa x_0 - \frac{1}{2}\kappa^2 \sigma^2 \right)$$

With this result we can express the moments of $x$ as derivatives of the characteristic function:

$$\langle x^n \rangle = \int_{-\infty}^{\infty} dx \, \mathsf{P}(x) \left( -i \, \frac{\partial}{\partial \kappa} \right)^n \exp(i\kappa x) \bigg|_{\kappa=0}$$

so that

$$\langle x^n \rangle = \left( -i \, \frac{\partial}{\partial \kappa} \right)^n \exp\left( i\kappa x_0 - \frac{1}{2}\kappa^2 \sigma^2 \right) \bigg|_{\kappa=0}$$

Sometimes we need to average Gaussian random variables that occur in the exponent. We can do so by assigning special values to $\kappa$ in the characteristic function or differentiating with respect to $\kappa$:

$$\langle \exp(\eta x) \rangle = \exp\left( \eta x_0 + \frac{1}{2}\eta^2 \sigma^2 \right)$$
$$\langle x \exp(\eta x) \rangle = (x_0 + \eta \sigma^2) \exp\left( \eta x_0 + \frac{1}{2}\eta^2 \sigma^2 \right)$$

The *probability distribution function* is defined as the integral of the probability density function. It gives the probability that the random variable is somewhere in the range $-\infty < x < \eta$:

$$\mathcal{P}(x \leq \eta) = \int_{-\infty}^{\eta} dx\, \mathsf{P}(x)$$

This is also called the *cumulative probability* and is the quantity that is often measured. When the underlying variable $x$ is Gaussian we can estimate it as follows:

$$
\begin{aligned}
\mathcal{P}(x \leq \eta) &= \frac{1}{\sigma\sqrt{2\pi}} \int_{-\infty}^{\eta} dx \exp\left(-\frac{1}{2\sigma^2}(x - x_0)^2\right) \\
&= \frac{1}{\sqrt{\pi}}\left(\int_{-\infty}^{0} du\, e^{-u^2} + \int_{0}^{\frac{\eta - x_0}{\sigma\sqrt{2}}} du\, e^{-u^2}\right)
\end{aligned}
$$

The second integral is the error function defined in Appendix C, which is widely available in tabulated form:[1]

$$\mathcal{P}(x \leq \eta) = \frac{1}{2}\left[1 + \mathrm{erf}\left(\frac{\eta - x_0}{\sigma\sqrt{2}}\right)\right]$$

The complementary measure is the *exceedance probability* which describes the probability that the level of the signal exceeds a prescribed value. Since the two measures must add up to unity,

$$\mathcal{P}(x \geq \eta) + \mathcal{P}(x \leq \eta) = 1$$

and we can write

$$\mathcal{P}(x \geq \eta) = \frac{1}{2}\left[1 - \mathrm{erf}\left(\frac{\eta - x_0}{\sigma\sqrt{2}}\right)\right].$$

## E.2  The Bivariate Gaussian Distribution

In our work we often need to describe the statistical properties of two quantities that are each distributed as Gaussian random variables. The bivariate probability density function is the key to these descriptions. It is represented quite generally as the exponential of a quadratic function in the excursions of the two variables about their mean values:

$$\mathsf{P}(x, y) = \frac{1}{N} \exp\{-[A(x - x_0)^2 + B(y - y_0)^2 + 2C(x - x_0)(y - y_0)]\}$$

---

[1] For example, see M. Abramowitz and I. A. Stegun, *Handbook of Mathematical Functions* (Dover, New York, 1964), 295–329.

This joint distribution must reduce to the simple Gaussian expression when it is integrated over all values of either $x$ or $y$. This condition allows one to express the constants in terms of their variances and cross correlation:

$$\sigma_x^2 = \langle (x - x_0)^2 \rangle$$
$$\sigma_y^2 = \langle (y - y_0)^2 \rangle$$
$$\mu \sigma_x \sigma_y = \langle (x - x_0)(y - y_0) \rangle$$

In terms of these parameters:

$$N = 2\pi \sigma_x \sigma_y \sqrt{1 - \mu^2}$$
$$A = \left[ 2\sigma_x^2 (1 - \mu^2) \right]^{-1}$$
$$B = \left[ 2\sigma_y^2 (1 - \mu^2) \right]^{-1}$$
$$C = -\mu \left[ \sigma_x \sigma_y (1 - \mu^2) \right]^{-1}$$

In this volume we have identified the variables $x$ and $y$ with phase fluctuations measured at adjacent receivers:

$$x = \varphi(R) \qquad \text{and} \qquad y = \varphi(R + \rho)$$

We have also identified them with phase fluctuations measured at the same receiver but displaced in time:

$$x = \varphi(t) \qquad \text{and} \qquad y = \varphi(t + \tau)$$

The average phase shift vanishes in both cases. One can assume that the variances are the same if the receivers are not too far apart or the time delay is not too great:[2]

$$\sigma_x = \sigma_y = \sigma$$

The bivariate probability density function which corresponds to these assumptions is

$$\mathsf{P}(x, y) = \frac{1}{2\pi \sigma^2 \sqrt{1 - \mu^2}} \exp\left( \frac{-1}{2\sigma^2 (1 - \mu^2)} \left( x^2 + y^2 - 2\mu\, xy \right) \right).$$

We can show that this probability density function is normalized to unity for all values of $\sigma$ and $\mu$ using the integrals in Appendix C:

$$\int_{-\infty}^{\infty} dx \int_{-\infty}^{\infty} dy\, \mathsf{P}(x, y) = 1$$

---

[2] In the next volume we will need to consider the bivariate distribution for fluctuations both of phase and of amplitude. Their variances are very different and the generalization of the expression above to describe that case will be developed in Appendix E of Volume 2.

One can use the same techniques to evaluate the following moments and correlations:

$$\langle x^3 y \rangle = 3\mu\sigma^4$$
$$\langle x\, y^3 \rangle = 3\mu\sigma^4$$
$$\langle x^2 y^2 \rangle = \sigma^4(1 + 2\mu^2)$$

The joint characteristic function for the two-dimensional Gaussian distribution is calculated as the double Fourier integral of the bivariate probability density function:

$$\langle \exp[i(px + qy)]\rangle = \int_{-\infty}^{\infty} dx \int_{-\infty}^{\infty} dy\, \mathsf{P}(x, y) \exp[i(px + qy)]$$

On introducing the expression for $\mathsf{P}(x, y)$ and carrying out the integrations, we obtain

$$\langle \exp[i(px + qy)]\rangle = \exp\left(-\frac{\sigma^2}{2}(p^2 + 2\mu pq + q^2)\right).$$

A useful special case results on setting $p = 1$ and $q = -1$:

$$\langle \exp[i(x - y)]\rangle = \exp[-\sigma^2(1 - \mu)]$$

One can also express the moments of $x$ and $y$ as partial derivatives of the characteristic function. For instance,

$$\langle x^2 y^2 \rangle = \frac{\partial^2}{\partial p^2}\, \frac{\partial^2}{\partial q^2} \langle \exp[i(px + qy)]\rangle \Big|_{p=q=0}$$
$$= \frac{\partial^2}{\partial p^2}\, \frac{\partial^2}{\partial q^2} \exp\left(-\frac{\sigma^2}{2}(p^2 + 2\mu pq + q^2)\right)\Big|_{p=q=0}$$

and this confirms the result given above.

With two independent random variables there are other probability density functions for various combinations of $x$ and $y$. In Section 8.2.1 we showed how the bivariate Gaussian model leads to the following probability density function for their difference:

$$\mathsf{P}(x - y) = \frac{1}{2\sigma\sqrt{\pi}\sqrt{1 - \mu}} \exp\left(-\frac{(x - y)^2}{4\sigma^2(1 - \mu)}\right)$$

The *conditional probability density function* for $x$ given that its companion $y$ has a fixed value is another useful measure:

$$\mathsf{P}(x\,|\,y) = \frac{\mathsf{P}(x, y)}{\mathsf{P}(y)} = \frac{1}{\sigma\sqrt{2\pi}\sqrt{1 - \mu^2}}\exp\left(-\frac{(x - \mu y)^2}{2\sigma^2\,(1 - \mu^2)}\right)$$

In writing this we have used the simple Gaussian distribution for $y$ given previously.

# Appendix F
## Delta Functions

The delta function was first widely used by Dirac in quantum mechanics[1] and now plays an important role in descriptions of propagation through random media. Of all the special functions used in mathematical physics, the delta function is the most unusual. It is not a proper function in the mathematical sense since it is neither continuous nor differentiable. Our first approach to understanding it is to consider its influence when it is paired with a continuous function in the integrand of a definite integral:

$$\int_a^b \delta(x - x_0) f(x) \, dx = f(x_0) \qquad a < x_0 < b$$

We regard this expression as the basic definition of the delta function. Its unique property is that it allows one to select the value of a companion function at the reference point and completely disregard all other values in the range. For this reason it is sometimes called the *sifting integral*. The following normalization condition is an immediate consequence:

$$\int_{-\infty}^{\infty} \delta(x - x_0) \, dx = 1$$

This follows upon setting $f(x) = 1$ and noting that the reference point must occur somewhere in the infinite range of integration.

Another way to think about the delta function is to imagine that it is an infinitely high and infinitely narrow pulse. It would satisfy the basic definition if it had this characteristic. We approximate this unusual

---

[1] P. A. M. Dirac, *The Principle of Quantum Mechanics* (Clarendon Press, Oxford, 1930). The delta function was known to Oliver Heaviside and others much earlier.

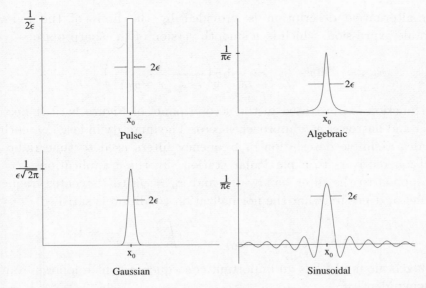

Figure F.1: Graphical representations of four models that approach the Dirac delta function in the limit $\epsilon \to 0$.

behavior by a limiting process applied to the square wave illustrated in Figure F.1:

$$\delta(x - x_0) = \lim_{\epsilon \to 0} \begin{cases} \dfrac{1}{2\epsilon} & x_0 - \epsilon < x < x_0 + \epsilon \\ 0 & \text{otherwise} \end{cases}$$

Introducing this expression into the basic definition and expanding the companion function in a Taylor series about the reference point gives

$$\int_a^b \delta(x - x_0) f(x)\, dx = \lim_{\epsilon \to 0} \left( \frac{1}{2\epsilon} \int_{x_0 - \epsilon}^{x_0 + \epsilon} dx \left[ f(x_0) + (x - x_0) f'(x_0) \right.\right.$$

$$\left.\left. + \tfrac{1}{2}(x - x_0)^2 f''(x_0) \right] \right).$$

The second term is as often negative as it is positive in the integration interval and vanishes. The third term is of order $\epsilon^2$ when it is integrated and disappears as $\epsilon \to 0$. The combined result is the original definition:

$$\int_a^b \delta(x - x_0) f(x)\, dx = f(x_0)$$

This description is useful for describing short-pulse signals. On the other hand, it is neither continuous nor differentiable, both of which are desirable physical properties.

An alternative description is provided by the limit of the following algebraic expression, which is a smooth version of the sharp pulse:

$$\delta(x - x_0) = \frac{1}{\pi} \lim_{\epsilon \to 0} \left( \frac{\epsilon}{\epsilon^2 + (x - x_0)^2} \right)$$

This function is illustrated in the second panel of Figure F.1. It becomes higher and narrower as $\epsilon$ approaches zero. The quantity in large parentheses provides a simple description of frequency filters used to tune radio and television receivers to a particular station. In these applications $\epsilon = \Delta f$ corresponds to the filter bandwidth and $x_0 = f_0$ to its center frequency. The factor $\pi$ ensures that the normalization condition is satisfied:

$$\frac{1}{\pi} \int_{-\infty}^{\infty} \frac{\epsilon \, dx}{\epsilon^2 + (x - x_0)^2} = \frac{2\epsilon}{\pi} \int_0^{\infty} \frac{du}{\epsilon^2 + u^2} = 1$$

The algebraic model has an important consequence, which follows from the following identity:

$$\frac{1}{z - i\epsilon} = \frac{z}{z^2 + \epsilon^2} + i\frac{\epsilon}{z^2 + \epsilon^2}$$

Taking the limit of both sides, we can express the result in terms of the *principal value*[2] of $z$ and the delta function:

$$\lim_{\epsilon \to 0} \left( \frac{1}{z - i\epsilon} \right) = \mathcal{P}\left( \frac{1}{z} \right) + i\pi\delta(z)$$

This means that we can also represent the delta function as

$$\delta(x - x_0) = \Im\left[ \frac{1}{\pi} \lim_{\epsilon \to 0} \left( \frac{1}{x - x_0 - i\epsilon} \right) \right].$$

One encounters this combination in some physical problems in which $\epsilon$ represents the influence of a small attenuation term on the wavenumber and hence on the Green's function.

Another representation of the delta function is to describe it as the limit of a sinusoidal function:

$$\delta(x - x_0) = \frac{1}{\pi} \lim_{\epsilon \to 0} \left( \frac{\sin\left[\epsilon^{-1}(x - x_0)\right]}{(x - x_0)} \right)$$

Like the previous models, it has a very large value at $x_0$ and falls rapidly to zero away from the reference point. This behavior is illustrated in the fourth panel of Figure F.1. The factor $\pi^{-1}$ ensures that its integration over

---

[2] E. T. Whittaker and G. N. Watson, *A Course of Modern Analysis* (Cambridge University Press, Cambridge, 1947), 75.

all values satisfies the normalization condition. This formulation leads to two important integral representations of the delta function. Noting that

$$\int_{-\epsilon^{-1}}^{\epsilon^{-1}} d\kappa \exp[i\kappa(x - x_0)] = 2\frac{\sin\left[\epsilon^{-1}(x - x_0)\right]}{x - x_0}$$

the delta function can be expressed as

$$\delta(x - x_0) = \lim_{\epsilon \to 0} \left( \frac{1}{2\pi} \int_{-\epsilon^{-1}}^{\epsilon^{-1}} d\kappa \exp[i\kappa(x - x_0)] \right)$$

or finally

$$\delta(x - x_0) = \frac{1}{2\pi} \int_{-\infty}^{\infty} d\kappa \exp[i\kappa(x - x_0)].$$

Since the imaginary term integrates to zero, an equivalent version is

$$\delta(x - x_0) = \frac{1}{\pi} \int_0^{\infty} d\kappa \cos[\kappa(x - x_0)].$$

These integral definitions of the delta function provide a convenient way to represent the current distribution of a point source. In such cases we can write the current density as the product of three delta functions that localize the current density to the immediate position of the transmitter:

$$j(\mathbf{r}) = j_0 \delta(x - x_0)\delta(y - y_0)\delta(z - z_0)$$

The appropriate generalization of the infinite integral above allows us to express a point source in terms of a Fourier integral in three dimensions:

$$j(\mathbf{r}) = j_0 \frac{1}{8\pi^3} \int d^3\kappa \exp[i\boldsymbol{\kappa} \cdot (\mathbf{r} - \mathbf{r}_0)]$$

This result is used to generate solutions for a point source (i.e., spherical waves) and to develop explicit expressions for Green's function.

There is a fourth way to describe the delta function. It follows from taking the limit of a Gaussian probability density function as the variance goes to zero:

$$\delta(x - x_0) = \lim_{\epsilon \to 0} \left\{ \frac{1}{\epsilon\sqrt{2\pi}} \exp\left[ -\left( \frac{1}{2\epsilon^2}(x - x_0)^2 \right) \right] \right\}$$

This is also illustrated in Figure F.1, in which one sees that $\epsilon$ represents the width of the distribution. It is normalized to unity for all possible values of $x_0$ and $\epsilon$. This formulation occurs because the Gaussian distribution is quite common and the dispersion of the measured quantity is often small.

We have now found four analytical models that lead to the same conclusion in the appropriate limit. When we encounter any one of them we will replace it by $\delta(x - x_0)$. The resulting calculations are simplified by using the basic definition to collapse one of the remaining integrations.

Several important properties of the delta function can be established with the models advanced here. Sometimes the argument of the delta function is multiplied by a parameter. Appealing to the integral definition and rescaling the transform variable, we see that

$$\delta(\eta x) = \frac{1}{|\eta|}\delta(x)$$

and, in particular,

$$\delta(-x) = \delta(x).$$

By factoring the argument and using the rescaling relation we can also show that

$$\delta(x^2 - a^2) = \frac{1}{2|a|}[\delta(x - a) + \delta(x + a)].$$

In some applications we encounter delta functions whose arguments are more complicated functions of $x$:

$$\delta[g(x)]$$

If $g(x)$ vanishes at a point $x_0$ in the integration range, we can expand it in a Taylor series about that point:

$$g(x) = (x - x_0)g'(x_0) + \tfrac{1}{2}(x - x_0)^2 g''(x_0)$$

When the first derivative does not vanish at the reference point, one can use the rescaling relation to write

$$\delta[g(x)] = \frac{1}{|g'(x_0)|}\delta(x - x_0).$$

The basic definition imagines that the reference point lies within the range of integration but does not coincide with the end points. If it does, only half of the area under the symmetrical delta function is included in the range of integration:

$$\int_a^b \delta(x - a)f(x)\,dx = \frac{1}{2}f(a)$$

$$\int_a^b \delta(x - b)f(x)\,dx = \frac{1}{2}f(b)$$

The derivative of a delta function occurs inside the integration in some applications:

$$\int_a^b f(x)\,\frac{d}{dx}\delta(x-x_0)\,dx \qquad a < x_0 < b$$

We integrate by parts to obtain

$$\int_a^b f(x)\,\frac{d}{dx}\delta(x-x_0)\,dx = f(x)\delta(x-x_0)\big|_a^b - \int_a^b \delta(x-x_0)\,\frac{df(x)}{dx}\,dx.$$

The integrated term vanishes because $x_0$ falls in the range between $a$ and $b$:

$$\int_a^b f(x)\,\frac{d}{dx}\delta(x-x_0)\,dx = -\,\frac{df(x)}{dx}\bigg|_{x=x_0}$$

The delta function occurs in expressing the inherent orthogonality of many special functions. In Appendix D we showed that two Bessel functions with different arguments, integrated over all positive values, give a delta function:

$$\int_0^\infty J_\nu(xa)J_\nu(xb)x\,dx = \frac{1}{a}\delta(a-b) \qquad \text{for} \quad \nu > -\frac{1}{2}$$

This is a special case of a general class of *eigenfunctions* that have the following property:[3]

$$\int_0^\infty \psi(x,a)\psi(x,b)r(x)\,dx = \frac{1}{r(a)}\delta(a-b)$$

where $r(x)$ is a density function that is associated with the family of eigenfunctions $\psi(x,a)$.

[3] P. M. Morse and H. Feshbach, *Methods of Theoretical Physics: Part I* (McGraw-Hill, New York, 1953), 764.

# Appendix G
## Kummer Functions

Kummer functions play a small but important role in our studies of electromagnetic propagation through random media. The second-order differential equation

$$z \, \frac{d^2 w}{dz^2} + (b - z) \, \frac{dw}{dz} - aw = 0$$

has two independent solutions:

$$w(z) = AM(a, b, z) + BU(a, b, z)$$

The variable $z$ can be real or complex and we will use both versions in our work. The first solution is well behaved at the origin but the second is not.

Our applications depend primarily on the first solution, which is called the *Kummer function*. It is also known as the *confluent hypergeometric function* and has two equivalent notations:

$$M(a, b, z) = \; {}_1F_1(a, b, z)$$

These functions reduce to elementary forms for various combinations of the parameters $a$ and $b$.[1] This is seldom the case in our applications and we must use the power-series expansion to compute numerical values:

$$M(a, b, z) = 1 + \frac{a}{b}z + \frac{a(a + 1)}{b(b + 1)} \frac{z^2}{2!} + \cdots \frac{(a)_n}{(b)_n} \frac{z^n}{n!} + \cdots$$

Here

$$a_0 = 1 \quad \text{and} \quad (a)_n = a(a + 1)(a + 2) \cdots (a + n - 1)$$

---

[1] M. Abramowitz and I. A. Stegun, *Handbook of Mathematical Functions* (Dover, New York, 1964), 509–510.

with a similar expression for $(b)_n$. Several useful properties emerge from this series:

$$M(a, b, 0) = 1 \quad \text{for b} \neq -\text{n}$$

$$M(a, b, z) = e^z M(b - a, b, -z)$$

$$\frac{d}{dz} M(a, b, z) = \frac{a}{b} M(a + 1, b + 1, z)$$

$$\frac{d^n}{dz^n} M(a, b, z) = \frac{(a)_n}{(b)_n} M(a + n, b + n, z)$$

The second function is defined in terms of the Kummer function as follows:

$$U(a, b, z) = \frac{\pi}{\sin(b\pi)} \left( \frac{M(a, b, z)}{\Gamma(1 + a - b)\Gamma(b)} - z^{1-b} \frac{M(1 + a - b, 2 - b, z)}{\Gamma(a)\Gamma(2 - b)} \right)$$

The following asymptotic expansions of these functions are sometimes needed:

1. $\lim_{z \to \infty} M(a, b, z) = \dfrac{\Gamma(b)}{\Gamma(a)} e^z z^{a-b} [1 + O(|z|^{-1})] \quad \Re(z) > 0$

2. $\lim_{z \to \infty} M(a, b, z) = \dfrac{\Gamma(b)(-z)^{-a}}{\Gamma(b - a)} [1 + O(|z|^{-1})] \quad \Re(z) < 0$

3. $\lim_{z \to \infty} U(a, b, z) = z^{-a} [1 + O(|z|^{-1})]$

Kummer functions occur in our work primarily as definite integrals that express measured quantities. They also occur as weighting functions in integrals over the turbulence spectrum:

1. $\displaystyle\int_0^1 e^{zx} x^{a-1} (1 - x)^{b-a-1} \, dx = \frac{\Gamma(b - a)\Gamma(a)}{\Gamma(b)} M(a, b, z)$

$$\Re(b) > \Re(a) > 0$$

2. $\displaystyle\int_{-1}^1 e^{-\frac{zx}{2}} (1 - x)^{a-1} (1 + x)^{b-a-1} \, dx$

$$= 2^{b-1} e^{-\frac{z}{2}} \frac{\Gamma(b - a)\Gamma(a)}{\Gamma(b)} M(a, b, z)$$

$$\Re(b) > \Re(a) > 0$$

3. $\displaystyle\int_1^\infty e^{-zx} x^{b-a-1}(x-1)^{a-1}\,dx = e^{-z}\Gamma(a)U(a,b,z)$

$$\Re(a) > 0, \quad \Re(z) > 0$$

4. $\displaystyle\int_0^\infty e^{-zx} x^{a-1}(x+1)^{b-a-1}\,dx = \Gamma(a)U(a,b,z)$

$$\Re(a) > 0, \quad \Re(z) > 0$$

The following results are useful when the Kummer functions must be integrated over the independent variable:

1. $\displaystyle\int_0^\infty x^{b-1} M(a,c,-x)\,dx = \frac{\Gamma(b)\Gamma(c)\Gamma(a-b)}{\Gamma(a)\Gamma(c-b)}$

$$\Re(a) > \Re(b) > 0$$

2. $\displaystyle\int_0^z x^{b-1}(z-x)^{c-b-1} M(a,b,x)\,dx$

$$= z^{c-1}\frac{\Gamma(b)\,\Gamma(c-b)}{\Gamma(c)}M(a,c,z) \qquad \Re(c) > \Re(b) > 0$$

# Appendix H
## Hypergeometric Functions

The hypergeometric functions of Gauss play a small but important role in descriptions of electromagnetic propagation. They are solutions of the following second-order differential equation:

$$z(1-z)\frac{d^2w}{dz^2} + [c - (a+b+1)z]\frac{dw}{dz} - abw = 0$$

The solution which is regular at the origin is denoted by

$$w(z) = {}_2F_1(a,b,c\,;z)$$

When the parameters $a$, $b$ and $c$ take on integral or half-integral values it reduces to elementary functions or one of the special functions of mathematical physics. These cases are seldom encountered in our applications and we must use the following series expansion to compute numerical values:

$$_2F_1(a,b,c\,;z) = 1 + \frac{a\,b}{c}\frac{z}{1!} + \frac{a(a+1)b(b+1)}{c(c+1)}\frac{z^2}{2!} + \cdots$$

or, more compactly,

$$_2F_1(a,b,c\,;z) = \frac{\Gamma(c)}{\Gamma(a)\,\Gamma(b)}\sum_0^\infty \frac{\Gamma(a+n)\Gamma(b+n)}{\Gamma(c+n)}\frac{z^n}{n!} \qquad \text{for} \qquad |z| < 1$$

Values of $z$ outside the unit circle can be established by analytical continuation.

From the series expression one can establish the following special values:

$$_2F_1(a,b,c\,;0) = 1$$

$$_2F_1(a,b,c\,;1) = \frac{\Gamma(c)\Gamma(c-a-b)}{\Gamma(c-a)\Gamma(c-b)}$$

$$\Re(c-a-b) > 0, \qquad c \neq 0 \text{ or } -n$$

The derivative relationship also follows from the power series:

$$\frac{d}{dz}\,_2F_1(a,b,c\,;z) = \frac{ab}{c}\,_2F_1(a+1,b+1,c+1;z)$$

In our work the hypergeometric function is encountered primarily as a way to express certain definite integrals that describe measured quantities. The following example is quite general and occurs in several situations of interest:

1.  $$\int_0^1 x^{b-1}(1-x)^{c-b-1}(1-xz)^{-a}\,dx$$

$$= \frac{\Gamma(b)\Gamma(c-b)}{\Gamma(c)}\,_2F_1(a,b,c\,;z) \qquad \Re(c) > \Re(b) > 0$$

Sometimes we need integrals of the hypergeometric functions:

2.  $$\int_0^z {}_2F_1(a,b,c\,;x)\,dx$$

$$= \frac{c-1}{(a-1)(b-1)}\left[{}_2F_1(a-1,b-1,c-1\,;z) - 1\right]$$

3.  $$\int_0^1 x^{c-1}(1-x)^{\mu-c-1}\,_2F_1(a,b,c\,;xz)\,dx$$

$$= \frac{\Gamma(c)\Gamma(\mu-c)}{\Gamma(\mu)}\,_2F_1(a,b,\mu\,;z)$$

$$|\arg(1-z)| < \pi, \quad z \neq 1, \quad \Re(\mu) > \Re(c) > 0$$

The function $_2F_1(a,b,c\,;x)$ discussed above is the most important member of a *family* of hypergeometric functions. Some integrations we encounter are expressed in terms of a less familiar member of this family:

$$_1F_2(a,b,c\,;x)$$

The reversal of indices means that it has the following series expansion:

$${}_1F_2(a,b,c\,;z) = 1 + \frac{a}{bc}\frac{z}{1!} + \frac{a(a+1)}{b(b+1)c(c+1)}\frac{z^2}{2!} + \cdots$$

This series is valid for $|z| < 1$ and the function can be calculated outside the unit circle by continuation. Other properties of the hypergeometric function are found in the following standard references:

1. M. Abramowitz and I. A. Stegun, *Handbook of Mathematical Functions* (Dover, New York, 1964), 556–565.

2. W. Magnus and F. Oberhettinger, *Formulas and Theorems for the Special Functions of Mathematical Physics* (Chelsea, New York, 1949), 7–11.

3. E. T. Whittaker and G. N. Watson, *A Course of Modern Analysis* (Cambridge University Press, Cambridge, and Macmillan, London, 1947), 280–301.

# Appendix I
## Aperture Averaging

It is often necessary to analyze the influence of aperture averaging on electromagnetic signals that have traveled through random media. The measured quantities can be expressed in terms of the correlation of the desired property averaged over the surface of the receiver aperture. The vast majority of optical and microwave receivers uses circular apertures and this simplifies the analysis considerably. The cylindrical coordinates identified in Figure 4.11 provide the natural description for the surface integrals required and the separation between typical points on the receiving surface. The quantity to be averaged is related to the spatial correlation of the measured quantity by

$$\mathcal{J}(\kappa, a_{\mathrm{r}}) = \frac{1}{\pi^2 a_{\mathrm{r}}{}^4} \int_0^{a_{\mathrm{r}}} r_1 \, dr_1 \int_0^{a_{\mathrm{r}}} r_2 \, dr_2 \int_0^{2\pi} d\phi_1 \int_0^{2\pi} d\phi_2$$
$$\times \, C\left( \kappa \sqrt{r_1^2 + r_2^2 - 2r_1 r_2 \cos(\phi_1 - \phi_2)} \right)$$

where $\kappa$ is an inverse scale length.

It is surprising that this four-fold integration can be reduced to a single integral without specifying the correlation function. There are several ways to do so. The simplest derivation[1] is to represent the spatial correlation as a Fourier Bessel transform.[2]

$$C(z) = \int_0^\infty du \, u J_0(uz) f(u)$$

---

[1] V. I. Tatarskii, *Wave Propagation in a Turbulent Medium* (Dover, New York, 1967), 233.

[2] G. N. Watson, *A Treatise on the Theory of Bessel Functions* (Cambridge University Press, Cambridge, 1952), 576 *et seq.*

On substituting this expression for the correlation function into the aperture average, we have

$$
J(\kappa, a_{\rm r}) = \int_0^\infty du\, u f(u)\, \frac{1}{\pi^2 a_{\rm r}^4} \int_0^{a_{\rm r}} r_1\, dr_1 \int_0^{a_{\rm r}} r_2\, dr_2
$$
$$
\times \int_0^{2\pi} d\phi_1 \int_0^{2\pi} d\phi_2\, J_0\!\left( u\kappa \sqrt{r_1^2 + r_2^2 - 2r_1 r_2 \cos(\phi_1 - \phi_2)} \right).
$$

The four-fold integration was evaluated in Section 4.1.8 in order to describe phase averaging. The result depends only on the product $\kappa a_{\rm r}$ and is given by (4.33):

$$
J(\kappa a_{\rm r}) = \int_0^\infty du\, u f(u) \left( \frac{2 J_1(u\kappa a_{\rm r})}{u\kappa a_{\rm r}} \right)^2
$$

The orthogonality of the Bessel function indicated in Appendix D allows the transform function $f(u)$ to be expressed in terms of $C(z)$ by a similar relation:

$$
\int_0^\infty dz\, z J_0(z\zeta) C(z) = \int_0^\infty du\, u f(u) \int_0^\infty dz\, z J_0(z\zeta) J_0(uz)
$$
$$
= \int_0^\infty du\, u f(u) \frac{\delta(u-\zeta)}{u}
$$
$$
= f(\zeta)
$$

We introduce this expression for $f(u)$ and reverse the order of integration:

$$
J(\kappa a_{\rm r}) = \frac{4}{\kappa^2 a_{\rm r}^2} \int_0^\infty dz\, z C(z) \int_0^\infty \frac{1}{u} J_1^2(u\kappa a_{\rm r}) J_0(uz)\, du
$$

The integral involving three Bessel functions can be done analytically:

$$
\int_0^\infty \frac{1}{x} J_1^2(x) J_0(xw)\, dx = \begin{cases} \dfrac{1}{\pi}\left[ \arccos\!\left(\dfrac{w}{2}\right) - \dfrac{w}{4}\sqrt{4-w^2} \right] & w < 2 \\ 0 & w > 2 \end{cases}
$$

The aperture average is therefore presented as a single integral over the spatial correlation function:

$$
J(\kappa a_{\rm r}) = \frac{4}{\pi} \int_0^2 dw \left[ \arccos\!\left(\frac{w}{2}\right) - \frac{w}{2}\sqrt{1-\frac{w^2}{4}} \right] C(\kappa a_{\rm r} w) w
$$

This result is related to the problem of finding the common area of two overlapping circles.[3] An equivalent solution has two parametric integrations but is often easier to use:[4]

$$\mathcal{J}(\kappa a_{\mathrm{r}}) = \frac{8}{\pi} \int_0^1 ds \, \sqrt{1 - s^2} \int_0^{2s} dw \, C(\kappa a_{\mathrm{r}} w) w$$

[3] D. L. Fried, "Aperture Averaging of Scintillation," *Journal of the Optical Society of America*, **57**, No. 2, 169–175 (February 1967).
[4] E. Levin, R. B. Muchmore and A. D. Wheelon, "Aperture-to-Medium Coupling on Line-of-Sight Paths: Fresnel Scattering", *IRE Transactions on Antennas and Propagation*, **AP-7**, No. 2, 142–146 (April 1959).

# Appendix J
## Vector Relations

1. $\mathbf{a} \cdot \mathbf{b} \times \mathbf{c} = \mathbf{b} \cdot \mathbf{c} \times \mathbf{a}$
   $$= \mathbf{c} \cdot \mathbf{a} \times \mathbf{b}$$

2. $\mathbf{a} \times (\mathbf{b} \times \mathbf{c}) = (\mathbf{a} \cdot \mathbf{c})\mathbf{b} - (\mathbf{a} \cdot \mathbf{b})\mathbf{c}$

3. $(\mathbf{a} \times \mathbf{b}) \cdot (\mathbf{c} \times \mathbf{d}) = \mathbf{a} \cdot \mathbf{b} \times (\mathbf{c} \times \mathbf{d})$
   $$= \mathbf{a} \cdot [(\mathbf{b} \cdot \mathbf{d})\mathbf{c} - (\mathbf{b} \cdot \mathbf{c})\mathbf{d}]$$
   $$= (\mathbf{a} \cdot \mathbf{c})(\mathbf{b} \cdot \mathbf{d}) - (\mathbf{a} \cdot \mathbf{d})(\mathbf{b} \cdot \mathbf{c})$$

4. $(\mathbf{a} \times \mathbf{b}) \times (\mathbf{c} \times \mathbf{d}) = (\mathbf{a} \times \mathbf{b} \cdot \mathbf{d})\mathbf{c} - (\mathbf{a} \times \mathbf{b} \cdot \mathbf{c})\mathbf{d}$

5. $\boldsymbol{\nabla}(\phi + \psi) = \boldsymbol{\nabla}\phi + \boldsymbol{\nabla}\psi$

6. $\boldsymbol{\nabla}(\phi\psi) = \phi\boldsymbol{\nabla}\psi + \psi\boldsymbol{\nabla}\phi$

7. $\boldsymbol{\nabla} \cdot (\mathbf{a} + \mathbf{b}) = \boldsymbol{\nabla} \cdot \mathbf{a} + \boldsymbol{\nabla} \cdot \mathbf{b}$

8. $\boldsymbol{\nabla} \times (\mathbf{a} + \mathbf{b}) = \boldsymbol{\nabla} \times \mathbf{a} + \boldsymbol{\nabla} \times \mathbf{b}$

9. $\boldsymbol{\nabla} \cdot (\phi\mathbf{a}) = \mathbf{a} \cdot \boldsymbol{\nabla}\phi + \phi\boldsymbol{\nabla} \cdot \mathbf{a}$

10. $\boldsymbol{\nabla} \times (\phi\mathbf{a}) = \boldsymbol{\nabla}\phi \times \mathbf{a} + \phi\boldsymbol{\nabla} \times \mathbf{a}$

11. $\boldsymbol{\nabla}(\mathbf{a} \cdot \mathbf{b}) = (\mathbf{a} \cdot \boldsymbol{\nabla})\mathbf{b} + (\mathbf{b} \cdot \boldsymbol{\nabla})\mathbf{a} + \mathbf{a} \times (\boldsymbol{\nabla} \times \mathbf{b}) + \mathbf{b} \times (\boldsymbol{\nabla} \times \mathbf{a})$

12. $\boldsymbol{\nabla} \cdot (\mathbf{a} \times \mathbf{b}) = \mathbf{b} \cdot \boldsymbol{\nabla} \times \mathbf{a} - \mathbf{a} \cdot \boldsymbol{\nabla} \times \mathbf{b}$

13.   $\boldsymbol{\nabla} \times (\mathbf{a} \times \mathbf{b}) = \mathbf{a}\,\boldsymbol{\nabla} \cdot \mathbf{b} - \mathbf{b}\,\boldsymbol{\nabla} \cdot \mathbf{a} + (\mathbf{b} \cdot \boldsymbol{\nabla})\mathbf{a} - (\mathbf{a} \cdot \boldsymbol{\nabla})\mathbf{b}$

14.   $\boldsymbol{\nabla} \times \boldsymbol{\nabla} \times \mathbf{a} = \boldsymbol{\nabla}\boldsymbol{\nabla} \cdot \mathbf{a} - \boldsymbol{\nabla}^2 \mathbf{a}$

15.   $\boldsymbol{\nabla} \times \boldsymbol{\nabla}\phi = 0$

16.   $\boldsymbol{\nabla} \cdot \boldsymbol{\nabla} \times \mathbf{a} = 0$

17.   $(\boldsymbol{\nabla}\psi \cdot \boldsymbol{\nabla})\,\boldsymbol{\nabla}\psi = \frac{1}{2}\,\boldsymbol{\nabla}(\boldsymbol{\nabla}\psi)^2$

# Appendix K
## The Gamma Function

The gamma function occurs frequently in this work. It often results when one is evaluating integral expressions or Laplace transforms. Mellin transforms provide a powerful way to analyze propagation in random media[1] and they are usually expressed in terms of gamma functions. They are defined by the following integral:

$$\Gamma(z) = \int_0^\infty x^{z-1} \exp(-x)\, dx \qquad \Re(z) > 0$$

The gamma function is defined for complex values of the argument[2] but we usually encounter it for real values of $x$. It is finite with well-defined derivatives unless the argument is zero or a negative integer, for which it has simple poles. This is evident from Figure K.1. When $z$ is a positive integer it is related to the *factorial*:

$$\Gamma(n) = (n-1)! = 1 \times 2 \times 3 \cdots (n-1)$$

A brief table of non-integer values is given below, together with some useful mathematical constants. Complete tabulations of this function are available in standard references.[3] The following special values are

[1] R. J. Sasiela, *Electromagnetic Wave Propagation in Turbulence* (Springer-Verlag, Berlin, 1994).
[2] P. M. Morse and H. Feshbach, *Methods of Theoretical Physics: Part I* (McGraw-Hill, New York, 1953), 419–425.
[3] M. Abramowitz and I. A. Stegun, *Handbook of Mathematical Functions* (Dover, New York, 1964), 255–293.

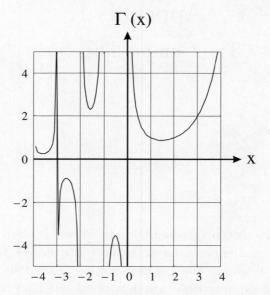

Figure K.1: A plot of the gamma function for positive and negative real values of the argument.

often helpful:

$$\Gamma\left(\tfrac{1}{2}\right) = \sqrt{\pi}$$

$$\Gamma(1) = 1$$

$$\Gamma\left(\tfrac{3}{2}\right) = \tfrac{1}{2}\sqrt{\pi}$$

$$\Gamma(2) = 1$$

The *recurrence* formula can be used to extend the table to other positive and negative values:

$$\Gamma(z + 1) = z\Gamma(z)$$

The *duplication* formula can also be used to extend the tables to larger values:

$$\Gamma(2z) = \frac{1}{\sqrt{\pi}}2^{2z-1}\Gamma(z)\Gamma\left(z + \frac{1}{2}\right)$$

The following *reflection* formula is often used to simplify expressions that involve several gamma functions:

$$\Gamma(z)\Gamma(1 - z) = \frac{\pi}{\sin(\pi z)}$$

For very large values one can use Sterling's approximation:

$$\lim_{z\to\infty}[\Gamma(z)] = \sqrt{2\pi}z^{z-\frac{1}{2}}e^{-z}\left(1 + \frac{1}{12z} + \frac{1}{288z^2} + \cdots\right) \qquad |\arg(z)| < \pi$$

The beta function is closely related to the gamma function:

$$B(a,b) = \frac{\Gamma(a)\Gamma(b)}{\Gamma(a+b)} \qquad \Re(a) > 0,\ \Re(b) > 0$$

This connection allows the following two common definite integrals to be expressed in terms of $\Gamma(a)$ and $\Gamma(b)$:

$$B(a,b) = \int_0^1 x^{a-1}(1-x)^{b-1}\,dx$$

$$B(a,b) = 2\int_0^{\frac{\pi}{2}} (\sin\theta)^{2a-1}(\cos\theta)^{2b-1}\,d\theta$$

Numerical values for constants that are often encountered are provided in Table K.1.

Table K.1: Numerical values for commonly used mathematical constants

| | |
|---|---|
| $\Gamma(\frac{1}{6}) = 5.56632$ | $\Gamma(2) = 1.00000$ |
| $\Gamma(\frac{1}{4}) = 3.62561$ | $\Gamma(\frac{7}{3}) = 1.19064$ |
| $\Gamma(\frac{1}{3}) = 2.67894$ | $\Gamma(\frac{17}{6}) = 1.72454$ |
| $\Gamma(\frac{2}{3}) = 1.35412$ | $\Gamma(\frac{11}{3}) = 4.01220$ |
| $\Gamma(\frac{1}{2}) = 1.77245$ | $\Gamma(\frac{25}{6}) = 7.42606$ |
| $\Gamma(\frac{3}{4}) = 1.22542$ | $\Gamma(-\frac{1}{2}) = -3.54490$ |
| $\Gamma(\frac{5}{6}) = 1.12879$ | $(2)^{\frac{1}{3}} = 1.25992$ |
| $\Gamma(1) = 1.00000$ | $(\pi)^{\frac{1}{3}} = 1.46459$ |
| $\Gamma(\frac{7}{6}) = 0.92772$ | $(2\pi)^{\frac{1}{3}} = 1.84527$ |
| $\Gamma(\frac{4}{3}) = 0.89298$ | $(10)^{\frac{1}{3}} = 2.15443$ |
| $\Gamma(\frac{5}{3}) = 0.90275$ | $(2)^{\frac{1}{6}} = 1.12246$ |
| $\Gamma(\frac{11}{6}) = 0.94066$ | $(2\pi)^2 = 39.47842$ |

# Author Index

# Subject Index

447